The Animal Kingdom

VOLUME 3:
THE CLASS MAMMALIA 3

GEORGES CUVIER
EDITED AND TRANSLATED BY
EDWARD GRIFFITH

CAMBRIDGE
UNIVERSITY PRESS

CAMBRIDGE UNIVERSITY PRESS

Cambridge, New York, Melbourne, Madrid, Cape Town,
Singapore, São Paolo, Delhi, Mexico City

Published in the United States of America by Cambridge University Press, New York

www.cambridge.org
Information on this title: www.cambridge.org/9781108049566

© in this compilation Cambridge University Press 2012

This edition first published 1827
This digitally printed version 2012

ISBN 978-1-108-04956-6 Paperback

CAMBRIDGE LIBRARY COLLECTION

Books of enduring scholarly value

Life Sciences

Until the nineteenth century, the various subjects now known as the life sciences were regarded either as arcane studies which had little impact on ordinary daily life, or as a genteel hobby for the leisured classes. The increasing academic rigour and systematisation brought to the study of botany, zoology and other disciplines, and their adoption in university curricula, are reflected in the books reissued in this series.

The Animal Kingdom

Georges Cuvier (1769–1832), made a peer of France in 1819 in recognition of his work, was perhaps the most important European scientist of his day. His most famous work, Le Règne Animal, was published in French in 1817; Edward Griffith (1790–1858), a solicitor and amateur naturalist, embarked in 1824, with a team of colleagues, on an English version which resulted in this illustrated sixteen-volume edition with additional material, published between 1827 and 1835. Cuvier was the first biologist to compare the anatomy of fossil animals with living species, and he named the now familiar 'mastodon' and 'megatherium'. However, his studies convinced him that the evolutionary theories of Lamarck and St Hilaire were wrong, and his influence on the scientific world was such that the possibility of evolution was widely discounted by many scholars both before and after Darwin. Volume 3 is the third of four books on mammals.

Cambridge University Press has long been a pioneer in the reissuing of out-of-print titles from its own backlist, producing digital reprints of books that are still sought after by scholars and students but could not be reprinted economically using traditional technology. The Cambridge Library Collection extends this activity to a wider range of books which are still of importance to researchers and professionals, either for the source material they contain, or as landmarks in the history of their academic discipline.

Drawing from the world-renowned collections in the Cambridge University Library and other partner libraries, and guided by the advice of experts in each subject area, Cambridge University Press is using state-of-the-art scanning machines in its own Printing House to capture the content of each book selected for inclusion. The files are processed to give a consistently clear, crisp image, and the books finished to the high quality standard for which the Press is recognised around the world. The latest print-on-demand technology ensures that the books will remain available indefinitely, and that orders for single or multiple copies can quickly be supplied.

The Cambridge Library Collection brings back to life books of enduring scholarly value (including out-of-copyright works originally issued by other publishers) across a wide range of disciplines in the humanities and social sciences and in science and technology.

THE

ANIMAL KINGDOM

ARRANGED IN CONFORMITY WITH ITS
ORGANIZATION,

BY THE BARON CUVIER,

MEMBER OF THE INSTITUTE OF FRANCE, &c. &c. &c.

WITH

ADDITIONAL DESCRIPTIONS

OF

ALL THE SPECIES HITHERTO NAMED, AND OF
MANY NOT BEFORE NOTICED,

BY

EDWARD GRIFFITH, F.L.S, A.S., *&c.*

AND OTHERS.

VOLUME THE THIRD.

LONDON:
PRINTED FOR GEO. B. WHITTAKER,
AVE-MARIA-LANE.

MDCCCXXVII.

THE

CLASS MAMMALIA

ARRANGED BY THE

BARON CUVIER,

WITH

SPECIFIC DESCRIPTIONS

BY

EDWARD GRIFFITH, F.L.S., A.S., &c.

MAJOR CHARLES HAMILTON SMITH, F.R.S., L.S., &c.

AND

EDWARD PIDGEON, Esq.

———

VOLUME THE THIRD.

———

LONDON:

PRINTED FOR GEO. B. WHITTAKER,

AVE-MARIA-LANE.

———

MDCCCXXVII.

LONDON:
Printed by WILLIAM CLOWES,
14, Charing Cross.

THIRD ORDER

OF THE

MAMMALIA,

CONTINUED.

SUPPLEMENT ON THE MARSUPIATA.

It is a great misfortune to science that zoological systems are necessarily not merely the creatures of human inventions, but, to a certain extent also, of human fancy and caprice: they have none of the certainty of mathematics, but partake more of that indefinite point of excellence which belongs to painting; the *chef d'œuvres* of a Raphael might be better, and the best zoological arrangement will ever be capable of improvement.

It is painful to observe the defects which are, and ever will be, found in artificial systems of the most acknowledged merit; but it is still more so to contemplate the endless efforts of naturalists to ameliorate and improve them: so long, indeed, as there is no acknowledged standard, it is extremely difficult to fix the arbitrary rovings of fancy; though in natural science, as in political economy, attempts at reformation are dangerous in practice, and uncertain in the result.

Hence the great deference which has long been paid, more especially in this country, to the scientific works of the great Linnæus, may not have been without its beneficial effects; nor are the fashionable, though as we firmly believe necessary and proper, departures from his orthodoxy without their inconvenient consequences.

Such reflections naturally suggest themselves on entering

upon the consideration of the Marsupial animals. It is easy to see that inasmuch as they include groups or genera possessed of teeth as various as in the whole class, and, consequently, of manners and habits equally different, there is an impropriety in treating them merely as one division of the flesh-eating Mammalia, but it is not so easy to discover any other mode of arrangement that will not be without its objections.

If the Marsupiata be treated as a distinct order, (which, if we were to express an opinion, appears to us the least objectionable arrangement,) it might be said that the order includes animals which by their several habits might, with propriety, be referred to all the other orders of Mammalia ; and therefore that the better plan would be to make the Marsupiata a class, and to divide its genera into orders concurrent with those of the Mammalia now established. In short, opinions and plans might be as various as individuals, if every one were to assume the office of zoological legislator.

Allowing, therefore, to our author the credit of seeing and feeling these and similar difficulties which attend his arrangement, the most prudent course appears to be, to guide one's judgment by his, and to conclude that, all things considered, his determination is most likely to be the least objectionable in practical application.

The principal trait in which these animals all agree (however various in other points), is the existence of the abdominal pouch, from which their name is taken, being derived from the Latin word *Marsupium*, which signifies a purse.

Pouched animals were known at first only in America : all the species found on that continent agree so completely in general organization, as well as in this peculiar conformation of the genitals, that Linnæus found in them the elements of a single genus, which he called *Didelphis*, or double-wombed.

Afterwards from the East Indies, and still later from the regions of Australasia, animals arrived equally distinguished by the possession of the abdominal pouch ; these were immediately set down as genuine *Didelphes*, and Gmelin has bestowed on them the titles of *Didelphis Orientalis*, *Didelphis Brunii*, &c. ; and even the *Tarsier* of Daubenton he inscribed among them, under the name of *Didelphis Macrotarsus*.

None, however, of these animals answered to the definition of Linné ; all had less than six incisors above, and less than eight below, *&c.* : nevertheless, Pallas, Camper, and Zimmermann still preserved the appellations of Gmelin, and thus prolonged the abuse.

In the mean time, the discoveries of our countrymen in New Holland added other animals to the list of the Didelphes which carried that external mark of resemblance which we have noticed. Our hardy navigators, familiar with the Opossum of America, set down these as animals of the same kind.

Travellers, by their labours, do doubtless much enrich natural history ; but, in proportion to the multiplication of animals, so is the difficulty and confusion consequent on their classification. Animals were found agreeing, it is true, in the character of the pouch, but varying most essentially in other particulars.

We shall lay before the reader the characters of each of the families of the Marsupiata in the table, and confine ourselves here, for the most part, to some general observations upon them, the substance of which be principally derived from M. Geoffroy St. Hilaire.

The anatomy of the Marsupiata is well worth the minutest consideration : the females, as we have seen, have a pouch under the abdomen, at the bottom of which is the mammary apparatus : within this pouch the young receive their nourishment.

The reproduction of living beings, as is well known, takes place in various modes: attention, however, has never perhaps been given to this subject in the degree which it merits. These very various modes, these unusual combinations, have been observed only in animals of a lower order. They were considered as something inherent in the gradation of organic constitutions, a necessary consequence of the scale of animal life; but so assured did naturalists feel of the uniformity existing in this respect among all animals conformed like Man, that in the case of the Marsupiata, they were inclined to reject as inaccurate all that did not square with received opinions on this point, and agree with known analogies. Such a principle is, doubtless, essential to all just philosophizing, but some reservation must be observed in its application.

Thus from the origin of our acquaintance with the Didelphes, an opinion arose that their young were actually produced in the abdominal pouch beside the mammæ of the mother. It is nearly two centuries since Marcgrave has said, " The pouch is properly the matrix of the Carigueya (*Didelphis Opossum*). I have been unable to find any other; this is a point which I have ascertained by dissection. The semen is produced there, and there the young are formed." Pison confirms the same facts, having, as he observes, dissected many of these carigueyas. Valentyn, occupied in ecclesiastical functions in the East Indies, and who never was aware of the existence of pouched animals in America, makes the same assertion, in his account of the Molucca Islands: " The pouch of the Philanders is a matrix, in which the young are conceived. This pouch is not what is usually supposed. The mammæ are, with regard to the young, what stalks are to their fruits." The young remain attached to the mammæ, until they have attained maturity, and then separate from them as the fruit drops from the stalk.

These notions are also common in Virginia, even among physicians. Beverley says, that the young Opossums exist in the false belly, without ever entering the true, and are developed on the teats of the mother. The Marquess of Chastellux makes a similar remark. Hence Pennant says, "That suspended to the mammæ of the mother, they remain there at first without motion : this lasts until they have acquired some development and strength ; but then they undergo a second birth."

A French gentleman, who accompanied La Fayette to America, was taken prisoner by the Creeks, and afterwards became chief of a savage tribe, has often assured M. Geoffroy, that he brought up a number of Opossums, and always observed that the young were born on the teats within the pouch.

Such a number of testimonies had a considerable effect in Europe. Naturalists procured pouched animals; they had never conceived, or thought of admitting, but one hypothesis : being convinced that anatomical inspection was unfavourable to this, they agreed to reject the facts, and deny their possibility. The most celebrated naturalists and anatomists of the period to which we allude, sought and did not find the direct and interior road from the matrix to the pouch. The Marsupiata were then considered as beings, whose premature birth was compensated for by a sort of incubation in the pouch. "It would be desirable," says Buffon, "to observe living Sarigues, and more particularly that their precocious exclusion from the uterus should be examined : by such observation we might doubtless obtain some insight into the method of preserving the lives of children prematurely born ! The gestation of these beings having been proportionally short, the period of their lactation would be lengthened. " So small are they," says Blumenbach, " when born, that they may well be called abortions." Thus persevering in the system of a second, but premature, birth, naturalists imagined that a

second matrix protected the development of animals born
in a state of such considerable debility.

While this theory was predominant, some observations
appeared from the pen of a French Officer of Artillery, in
1786, in favour of the proscribed notions : they are to be
found in the travels of the Marquess of Chastellux, and we
shall present them in an abridged form to our readers.

Two Opossums, (*Didelphis Virginiana*,) male and female, were domesticated in the house of M. d'Aboville, in
1783 ; these animals copulated, and the effects were attentively observed by that gentleman: in about ten days, the
edge of the orifice of the pouch grew thicker, a phenómenon
which afterwards grew still more perceptible. As the
pouch increased in size, the orifice widened. On the
thirteenth day, the female did not quit her retreat except to
eat, drink, and evacuate : on the fourteenth she did not
stir from it.

M. d'Aboville then determined to seize and examine her:
the pouch, the aperture of which had widened before, was
now nearly closed; a slimy secretion moistened the hairs on its
circumference. On the fifteenth day, a finger was introduced
into the pouch, and a round body about the size of a pea
was plainly felt at the bottom. This examination was made
with difficulty, on account of the impatience of the mother,
who had before this been always very mild and tranquil. On
the seventeenth, she permitted a further examination, and
M. d'Aboville discovered two bodies about the size of a pea.
There was, however, a great number of these young ones.
On the twenty-fifth day, they moved very perceptibly, yielding to the touch : on the fortieth, the pouch was sufficiently
open for them to be plainly distinguished ; and on the
sixtieth, when the mother lay down, they were seen hang-
ing to the teats, some outside the pouch, some inside. The
nipple is about two-eighths of an inch in length ; but it soon
dries up, and at last drops off, after the manner of the umbilical cord.

These observations occasioned Professor Reimanus, of Hamburgh, to address himself on the subject to Dr. Barton in America. Roume de St. Laurent, who had already communicated to Buffon, that the nipples of the female Didelphis appeared at a certain time, in the form of little clear bulbs, in which the embryo was commenced, had also excited the zeal and stimulated the researches of Dr. Barton. This learned physician answered the appeals of his correspondents, and in two letters (one addressed to the Professor, the other to M. Roume) states his facts, observations, and conjectures, touching the generation of the Opossum. His observations are of great importance, and the more so, as while he professes never to depart from what he deemed sound physiological views, the facts he brings forward decidedly militate against the theory he proposes to establish : part of his remarks we shall give in substance to the reader.

The Didelphes put forth not fœtuses, but gelatinous bodies, embryos without eyes or ears. The mouth of these embryos is not cut : sprung from parents about the size of a Cat, they weigh at their first appearance generally about a grain, some a little more, and seven of them together weighed ten grains. Barton detached one of these embryos, weighing about nine grains, without producing any wound. In this he contradicts what Pennant and other English naturalists have asserted. Fifteen days of development in their new *domicile* (an expression of Barton's, intended to give what he thinks the true character of the pouch) are sufficient to bring the little ones to the size of a mouse; they do not quit the teats until they are as large as rats. After this they resume suckling, at pleasure, being then equally sustained by the mother's milk, and whatever substances they can eat at that period. It is necessary to the perfect development of these embryos, that the organs of digestion and respiration should be in perfect harmony Accordingly, we find the nostrils considerably widened from the origin, and they consequently become the first conduits of air to the lungs. The stomach of a young Opossum, which weighed forty-one grains, was found considerably distended and dilated, with a white and milky matter. That of a younger one, on the contrary, contained a transparent and colourless liquid.

The eyes are opened after fifty or fifty-two days. The teats are then quitted, and resumed successively. After sixty days, a young opossum weighs 531 grains. Barton relates a surprising circumstance of a female opossum, having a double gestation of two separate litters, one drawing to its close, and the other just commencing. This mother was then nursing seven young ones as large as rats. Though strong enough to live on solid aliments, they still had recourse to the teats for milk. But on a sudden the pouch closed, for it had become the new *domicile* of seven other young ones, weighing each from one to two grains. Nevertheless, the first litter was not deprived of the cares of the mother, who always manifested a constant attention and affection for them all. Her watchfulness always extended to the family already brought up She continued the cry with which she was accustomed to call them back. She assembled them on her back, and withdrew them, on the appearance of danger, to the tops of trees.

From all these facts, Barton in his first letter concludes that these animals have two kinds of gestation : one which he calls *uterine,* and which he considers to last from twenty-two to twenty-six days ; and the other *marsupial,* which commences from the entrance of the embryo into the pouch. This last, physiologically speaking, is the most important ; for the pouch, he adds, is strictly a second uterus, and the most important of the two.

In the interval of the publication of his two letters, Barton was informed that Sir Everard Home had published a paper on the generation of the Kangaroos, in the Philosophical Transactions for 1795, in which he states this remarkable fact; that the fœtus of pouched animals exhibits no trace of the umbilical cord. Barton proceeded immediately to the authentication of this point, by examining young opossums in the pouch, and found it correct. He supposes, however, that this umbilical cord will be discovered on individuals in the uterine gestation; but his researches not enabling him ever to have seen a fœtus in the uterus, he gives himself up to theoretical conjectures. He proposes to refer the mode of generation peculiar to the Didelphes, to that of reptiles and fishes, which, he believes, are also without the umbilical cord.

Dr. Barton furnishes a testimony in contradiction to the assertion of Camper, that man is the only animal capable of lying on his back. This, he says, often happens to the female opossum, especially when she has young ones. Lying on her back, she can touch, with the extremity of the vagina, every point of the interior sides of the pouch, and consequently the little ones, at the moment of birth, are protruded into the pouch without difficulty.

Fœtuses, without any trace of the umbilical cord, which yet have the nostrils widely opened, and the lungs considerably developed, would seem to countenance the opinion of a different system of organization from other animals. M. de Blainville, in an article on the generation and fœtus of the Didelphis, verifies the facts stated by Barton. He was unable, after the strictest examination, to discover any umbilical vein or artery, nor even the suspensory ligament of the liver. The gland of the ethymus was also wanting, and, generally speaking, none of the arrangements observable in the fœtus of other mammiferous animals, such as circulation and respiration depend on, are found here. From these facts, M. de Blainville agrees with Dr. Barton pretty nearly. " There are," he says, " two sorts of gestation, one *uterine,* another *mammary*, and these two sorts of gestation act differently, one supplying the deficiencies of the other." With Barton, however, the meaning of the word gestation is obvious. He applies the term to the simultaneous existence of the uterus and the pouch, to the notion of those two *domiciles*, in which certain phenomena, imperfectly exhibited in the one, are completed in the other. But with M. de Blainville, the idea of *uterine* and *mammary* gestation extends only to the different action of certain modes of nutrition. In the mammalia, he says, the fœtus, before it can be sustained independently, is capable of drawing its nutriment from the mother in two distinct places, and two different manners; *i. e.*, in the uterus from the blood, by means of the vascular system, and at the teats from the milk by the intestinal canal. And the two modes of nutrition, as to their duration, are in an inverse ratio in the various animals. M. de Blainville applies this generalization to the pouched animals. He imagines that one of these two modes of nutrition might be altogether suppressed. If the uterine be done away, we have the pouched animals; if the mammary, we have mammalia without mammæ; *i.e.*, the *monotremes*. That an animal may be born by means of a mammary nutrition, organized like a young one that has its full term of uterine gestation, is a bold conjecture, and accordingly M. de Blainville does not absolutely insist upon it. But he gives some consistency to this idea, when he admits at the end of his observations, that the fœtus probably passes directly from the uterus into the pouch, observing that the round ligament, the use of which in the ordinary mammalia is not known, may be the means of this passage.

M. Geoffroy, lamenting the vagueness and obscurity existing in the science of the Marsupiata, wrote an article on the subject in 1819, with this query as title, " Are the pouched animals born attached to the teats of the mo ther ?" His object was to call the attention of scientific men to the subject, and more especially of those who possess the means of investigation in those countries which form the habitat of the animals. His observations are highly interesting and important. On the pouch, he remarks, that it is not, in the adult female, a cavity of equal capaciousness at all times. M. d'Aboville observed it to increase in magnitude under the influence of the phenomena of generation, and M. Geoffroy himself has observed its relative dimensions in females of the same species. It is small previous to sexual intercourse, large to excess when the young ones are about to drop from the mammæ, and of a moderate size in the period immediately following. Thus the pouch cannot be considered merely as a *second domicile*, without spring or activity ; it is a true place of incubation, extending by degrees, acquiring more and more volume, as happens to every other *domicile* of the fœtus. Well, therefore, might it be called a second uterus, and the most important of the two.

This pouch, however, is external, and entirely formed by the skin and its fleshy pannicle. Its composition is extremely simple. There are either longitudinal wrinkles on each side, making a very imperfect sort of pouch, quite in a rudimentary state, as in the didelphes of the sub-genus *Micouré*, such as the Marmoses, the Cayopollins, the Brachyuri, &c., or there are ample folds round a central point, which being fixed, obliges them to extend circularly, and be confounded together in a large sort of curtain. The mammary gland, placed in the centre of the region of the abdomen, becomes, from its adherence with the skin and its immutability, the point which commands all the rest. All around it the skin contracts, folds over, and is prolonged in a salient edge, small in front, considerable behind, and of moderate dimensions on the sides.

As in this unusual extension of the dermis, in its foldings,

and in investigating the cause of this extraordinary phenomenon, of this new order of things, lies the entire question
concerning the Marsupiata, it is necessary here to take the
arteries into consideration, which are agents of every organic production.   We know that as well as the nutritious
vessels exist, so must also exist organs which they form and
supply.   As there is but a certain quantity of arterial nutriment to dispose of, if there be more than a due proportion
in one place, there will be less to distribute elsewhere. This
is that law of Nature which preserves a due balance between
the various parts of organization.

In relation to the distribution of the arteries, there are
divers arrangements, some of which give to the Marsupiata
considerable affinities with the birds. The principal modification is, that there is no lower mesentery in the abdominal
aorta.   In the birds, this principal artery is carried behind
the iliacs.   But in the Marsupiata it is entirely wanting.

The consequences of such a combination are, that from
the region of the reins to the rectum, there is no branch of
the abdominal aorta that may not, unless accidentally turned
from that purpose, concur in the business of generation.
In other mammalia, the lower mesentery, deriving in the
midst of the sources of life elements of another description
to return to the intestinal canal, is a cause at least of weak-
ening the generative operations.   With the Marsupiata on
the contrary, (as with the birds where all the branches
springing from the abdominal aorta are similar, and employed without interruption to produce the same result,
having nothing to oppose their peculiar function,) they perform it of course with a proportional facility : and the energy
of their functions is not only increased, but each part yields
to a sort of reaction, the effect of which is to produce greater
activity, and more complete development.

Another arrangement of very considerable influence is the
elevation of that point from which the abdominal aorta origi-

nates. We know that the aorta is always divided at the summit of the crest of the iliac bones. As the pelvis is longer in the Marsupiata, this circumstance places the termination of the aorta higher up. The iliac branches in descending form a more acute angle, and for this reason the blood is more drawn into the principal branch extending even into the crural artery. A third branch, equally great in caliber, is that of the middle sacrum. From this results the powerful and prehensile tail of the didelphes.

In man the primitive iliac is divided into two trunks, which, from the very near equality of their volume, have been considered the same, and called secondary iliacs. These are the external and the internal. The internal iliac becomes the hypogastric, after having furnished a tolerably strong branch, the ileo-lumbar. Its volume is but little diminished by this, so that the hypogastric remains a powerful trunk, of large caliber, and in which is involved a vast mass of nutritious fluids.

It is quite different with the marsupiata. As the primitive iliacs spring more high, it follows that the crural artery in parting from the primitive iliac forms a mother-branch, which has but very small branches on the sides. The first which present themselves, and which spring exactly from the same point, one to the right, and the other to the left, are the ileo-lumbar without, and the hypogastric within. There is a striking analogy between these two arteries, from the distribution of their branches, but more especially from the equality of their volume. Thus the hypogastric, so large in man, that it is one of the two bifurcations of the primitive iliac, and is thus the congener of the crural, is very limited in the marsupiata. Now the uterine and vaginal arteries are derived, as is well known, from the hypogastric.

The uterine and vaginal arteries, which are but small branches of the hypogastric, supply capillary tops to their organs. Diminished as they are in caliber, though suffi-

cient to supply these organs, they cannot turn to their advantage the principal afflux of the blood.   The sexual
organ is deprived of this derivative action which would
supply it with an excess of nutriment, and the blood being
deprived of this outlet, opens for itself another passage.
The crural artery is restrained to the hinge (as it were) of
the thigh upon the trunk ; and it is on the branches which
are found in this place, that the superabundance of nutritious fluid is carried.   Thus, by the determination of this
artery, a new order takes place, and we find the essential
elements of a new family of the animal world.

In the ordinary Mammalia, when the uterine artery
ceases to nourish, the epigastric continues the action ; while
the first acts, the superabundant blood is carried from the
primitive iliac to its interior branch, from that to the
hypogastric, and from the hypogastric to the uterine.   In
the second case, the blood comes into the exterior branch,
and subsequently to the epigastric.   Thus the epigastric
concludes by a lacteal alimentation what the uterine had
begun by a sanguine ; the epigastric being the artery which
supplies the abdominal mammæ.   It is, therefore, by a
kind of mathematical necessity, the uterine being deprived
of its generative functions, that the blood employing the
epigastric at first, must produce in this artery among the
Marsupiata, what the progress of organ would produce
somewhat later.

The action of certain imponderable fluids, from the
external world, and fecundation, produce inflammation in
the sexual organs. The organ which the first of these causes
first puts into play is the ovarium, from which this exci-
tation is propagated contiguously through each part.   This
organ having answered its end, in ordinary cases, the
uterus provides, by means of the uterine artery, for the development of the ovarian product.   Let us admit, in the
case of the Marsupiata, that it is an ovulum traversing a

true oviductus that arrives in the pouch, and is ingrafted, as it were, on the nipple. If the uterine artery be without power, the propagated inflammation must be impossible, and nothing, as far as the uterus is concerned. This action then devolves entirely on the epigastric artery ; but the ovule cannot derive any thing from it, for it contains only a germ imperceptible to our senses. This inflammatory action we alluded to, can be operative only at all the points where the epigastric terminates ; that is, at the mammary gland, and the dermis which surrounds it. The dermis can derive nothing from this, unless it be in a state of development beyond what is requisite for its condition as a mere tegumentary organ. On this principle we can explain the folds of which the pouch is formed, and the augmentation of its volume under the influence of the phenomena of generation.

There has been no question in the works of anatomists concerning the existence of an uterus in the Marsupiata. M. Geoffroy enters into a long discussion on this point, in which it would be inconsistent with the nature and limits of our work to follow him. We shall, therefore, proceed to other peculiarities of the Marsupiata, adding only that M. Geoffroy's opinion on the whole is, that they are born in the state of Medusæ in the second uterus, the pouch.

We shall venture a few more general remarks on their organization :—

The organs of sense do not present any very strong peculiarity, or, at least, any thing very different from those of ordinary Mammalia ; and that cannot, to a certain point, be explained by the circumstances in which these animals live.

Thus the skin is always covered with genuine hairs altogether similar to those of the other Mammalia, more or less long according to the different species and different parts of the body. Sometimes these amount to actual prickles, very stiff and hard, and it is probable, that the muscles of the skin have some modification enabling them to raise these prickles; but this is a point not clearly ascertained.

The tongue, in some species, is very soft; in others slightly papillated.

The eyes are somewhat variable in size, and this, as in the other Mammalia, is determined by the habits to which the animal is destined. They are small, for instance, when the animal burrows in the earth, or lives in the water. In some of the aquatic species, as the Ornithorhyncus for example, the crystalline is very convex. As to the rest, they have the essential characters of Mammalia, as to the disposition of the internal lid, and that of the straight and oblique muscles, &c.

It would seem to be otherwise with the organ of hearing, if what Sir Everard Home advances concerning there being but two small bones in this organ in the Ornithorhyncus be correct. As to the other Didelphes, they have certainly three or four, and the general apparatus of this sense is altogether similar to that of the ordinary Mammalia; similar modifications under similar circumstances, as the absence of a conch when the animal is aquatic or subterraneous.

If we find really no material differences in the organs of sense in the Marsupiata from other animals, this is by no means the case with the organs of loco-motion: there are some very remarkable ones in the skeleton, though not in all its parts. Thus, the general form of the vertebræ in the different portions of the vertebral columns, that of the head, or cranium, have nothing different from the common Mammalia. The classic characters are also to be found in the articulation of their bodies, in the form of the articular and transverse apophyses. The cervical vertebræ are pretty similarly modified as in the monodelphes, as also is the articulation of the cranium. Nevertheless, in the two anomalous species of this group, the holes through which the nerves of the spinal marrow pass are pierced in the middle of the body of each vertebra, pretty nearly as in Birds, and the ribs articulate only with the body of the vertebræ, and not with the transverse apophyses.

The limbs have many characters common to all the animals of this group, and some altogether peculiar.

Thus, in the fore-limbs, are constantly complete clavicles, which are remarkable from the manner in which they are articulated with the anterior piece of the sternum, and in the mode of their junction with the acromion. In the anomalous species, they offer a disposition still more singular by the manner in which they are, as it were, doubled by certain lateral appendages from the first piece of the sternum. The shoulder-blade has then always an acromion apophysis tolerably developed, but no coracoid.

A constant character of the humerus is that its internal condyle is pierced by a hole for the passage of the median nerve ; a disposition not unfrequently to be found in other Mammalia, but which is invariable in the Marsupiata.

Though the fore-paw, or hand, be subject to vary, there are never less than four fingers with the rudiment of a thumb. In general, the humero-cubito-radial articulation is tolerably perfect.

In the hinder limbs three principal and essentially characteristic points are to be remarked. The first, and most remarkable is, the presence of a particular bone, found in no other mammiferous animal even in a rudimental state. This is usually named the *os marsupiale*, or *janitor marsupii*, because it is considered as the bone which opens or closes the pouch. This is, however, by no means universally true ; for many species, and among others the whole group of anomalous Didelphes are destitute of this organ. It is a bone usually flattened, curved a little externally, enlarged and thicker towards its base, and is attached, but not articulated on the anterior edge of the os pubis, pretty near the symphysis, so as almost to approach the opposite side. It is more or less large according to the species. Its development seems to be in no degree proportioned to that of the pouch.

This bone is altogether peculiar, and no trace of it can be found in the other Mammalia or in the Birds. It would seem to have more analogy with bones placed in a similar situation in certain reptiles, such as for example the Crocodiles. Here still is a considerable difference, for these last are inserted into the muscles of the thigh, which seems to render it probable that they are the bones of the ilia, while in the Marsupiata, it is the fibres of the abdominal muscles which are fixed there. It may, perhaps, be safely asserted, that there is nothing precisely analogous to this disposition to be found in the skeleton of any of the vertebrata.

Another, but a less important, though very constant character is, the peculiar length of the symphysis of the pubis; in the formation the ossa ischii come in for a considerable share.

The peroneum has also more development than in any other mammiferous animal. Its head is also constantly articulated with the femur; a disposition to be found in the oviparous vertebrata, but not in the Mammalia.

In the muscles, generally, &c., there is nothing peculiar. Of the pouch we have already spoken. As to the marsupial bones, it is obvious that they can be acted on only by the muscles of the abdomen, without, however, producing any action on the pouch, which is wholly independent of them.

The following is the arrangement of the muscles of the abdomen, as reported by M. de Blainville, from dissection :—

The grand oblique muscles, or external oblique ones, are as usual. The small oblique, or oblique internal, presents this remarkable point, that we may regard as belonging to it a muscular bundle, which, from the whole internal edge of the marsupial bone, is carried towards the white line. Its fibres directed from behind to the front, and from without to within, form with those of the muscle on the opposite side a series of chevrons open behind. The grand straight muscle of the abdomen, which is also very thick, is attached but slightly to the symphysis of the pubis, and in the remainder of its breadth, it also comes from the whole internal and concave edge of the marsupial bone. Finally, the muscle of the thigh, called pectineus, has also its insertion at the external base of this bone, the movements of which, M. de Blainville insists, are totally independent of the pouch, and he therefore considers this bone as altogether misnomered.

The organs of digestion offer nothing to our notice strictly peculiar to these animals. Almost all the modifications of the dentary system, witnessed among monodelphous Mammalia, are found in the Marsupiata. From the Sarigues, which of all Mammalia have the greatest number of teeth, to the Echidnæ, which have hardly any traces of them, the gradations are so nice and delicate, that great difficulty must exist in classing them into genera upon this principle, at least until they are much better known.

The intestinal canal, and the organs of circulation and respiration offer nothing very peculiar to our attention.

The teats, in all the species where they are known, vary little in number, but are always abdominal and placed very far back , the nipples, which terminate them, ranged symmetrically on each side, are always enclosed in the pouch, in such species as have this appendage. There is nothing remarkable in the mammæ of those which are without this organ, or have only a lateral fold. In the anomalous species, it is generally asserted that there are none, and the assertion may, perhaps, be true.

The male sex, in the marsupial animals, presents some singularities worthy attention, though in the disposition of its general structure it agrees with the other Mammalia. We can only venture, however, to hint at them, as they are not of a nature fit for any detailed description, and can be instructive or amusing only to professional readers. There is no trace of a pouch in the male Marsupiata, nor probably of mammæ.

M. de Blainville is inclined to disbelieve the existence of the placenta. He also thinks that the fœtus in the anomalous species arrives to more perfection in the uterus than among the other tribes.

As to the mode in which the young are placed in the external pouch, or rather attached to the nipple, M. de Blainville leaves it in doubt. A communication between the external uterus and this pouch has been asserted to exist, but never demonstrated. Some have imagined that the mother placed the young there herself with her hands and feet; but this is not very likely. Another opinion (which we noticed before) was, that the pouch extended to the orifice of the vagina; but the muscles do not seem disposed for such an arrangement, and some species have no pouch. Certainly nothing is accurately known on this subject, though we think that the conjecture of M. Geoffroy, before cited, is the most probable.

We shall conclude this article with a few more general observations relative to the nomenclature, divisions, and habits of these curious animals, and such other peculiarities as may have been unnoticed.

These animals (as we have observed) derive their appellation of *Marsupiata,* or, as some call them, *Marsupiales,* from the character of the pouch. It may, however, be well questioned, whether as a generic or classic term, it be unobjectionable. There are many species in which this character of the pouch does not exist, while, on the contrary, there are none without the double matrix which would render the Linnæan appellation of Didelphis more universally suitable to all the species.

Be that as it may, the Marsupiata are unquestionably the most singular of all known quadrupeds. With the exception of the peculiarities of their generation, there is scarcely any character in common among them. The organs of locomotion and digestion vary considerably, and that in a manner so nicely graduated, that all the shades between the Carnassiers properly so called, and the genuine Rodentia, are discoverable among the animals in question by the character of the teeth. Their extremities are equally modified from those which are designed to dig the earth, to those adapted for climbing with the utmost facility the loftiest trees.

The feet among some, as the Phascolomys, are calculated for digging in the ground. In this case, there are five toes armed with powerful nails on the fore feet, and four only on the hinder, with a small tubercle instead of thumb. With others (as the Kanguroos, Potorocs, and Perameles) the hinder feet are conformed for the execution of rapid leaps; and then they have but four toes, the second of which is very strong, longer than the others, and furnished with a nail almost as thick as a hoof. The two internal ones are small and connected. The metatarsus is very long, as well as the limb to which it belongs. The fore-paws are very short, and terminated by five toes furnished with tolerably long talons. In the Phalangers, which are eminent climbers, the posterior thumb is considerably separated, and without a claw; the two toes which immediately fol-

low it, are connected by the skin as far as the last phalanx
The toes of the fore-feet differ little from those of the common Carnassiers, while in the Koala these same toes are
divided into two groups for the act of seizing; the thumb
and index being on one side, and the three others on the
opposite. The four hinder toes are connected two by two,
and very distinct from the thumb. In the Dasyuri, which
run upon the ground like the Martens, the fore-feet have
five toes, and the hinder four, all separated and armed with
curved claws, while the hinder thumb is but a simple
tubercle. Finally, the Didelphes which climb trees have
toes like the Dasyuri, except that the posterior thumb is
distinct, and without a nail like that of the Phalangers.
The Chironectes which swim differ from the Didelphis,
only in having the hinder feet palmate.

There is no tail in the Phascolomys. In the Koalas it is
a simple tubercle, but considerably long in all the other
genera. In the Didelphes, the Chironectes, and the true
Phalangers, it is naked, scaly, and prehensile. In the
Kanguroos and Potoroos, it is strong, triangular, and conic,
and concurs to locomotion with the long hinder limbs.
The Isoodonta and the Perameles have it of the same form,
but much less robust. Finally, the Dasyuri, and particularly the flying Phalangers, have it much elongated,
and more or less tufted.

In the Petauristæ alone, we find the skin of the sides
extended between the fore and hind legs, serving as a
parachute after the manner of the Galeopitheci and Polatouches.

The Crab-eating Didelphis, the Kanguroo, the Perameles,
the Isoodon, the Potoroo, the Phalangers, the Petauristæ,
and the Phascolomys alone, have the ventral pouch which
has given a denomination to the entire tribe. In the rest
the mammæ are visible without, and some have on each
side the fold of skin which forms the pouch, but scarcely

visible. The number of the mammæ vary, and is especially considerable among the Didelphes.

The physiognomy of these animals is in relation to their natural habits and modes of living.

The Didelphes and Dasyuri have a conic head, elevated ears, mouth deeply cut, and the aspect of carnivora. The Perameles rather resemble Rats, the Long-legged Kanguroos, Hares, and the Phascolomys the Marmot.

Some, such as the Didelphis and Dasyuri, are Carnassiers, living on eggs, small birds, and corrupted flesh, and sometimes crustacea and insects. Others, as the Kanguroo and Phascolomys, are sustained purely on vegetables. The Phalangers are probably both frugivorous and insectivorous.

They are all remarkable for the imperfect development in which the young are born. Even in the species without pouches, and with prehensile tails, the young hang under the belly of the mother for a certain time ; then they mount on her back, and twist their tails round hers to fix themselves. The young of the Koala which has no tail, fixes itself on the parent's back, and fastens there with its hands, in the manner represented in our figure of this species. The number of the young is variable. In the Didelphes, from ten to twelve, and in the Kanguroo, usually but one.

The Marsupiata are generally solitary. Some remain constantly on the trees, the Didelphes, the Phalangers, and Koalas. Others ferret continually in the rocks on the seashore, as the Dasyuri. Others remain constantly at the bottom of their burrows (the Phascolomys.) The Kanguroos, feeble animals, and without means of defence, live in troops. They alone serve for the purposes of nourishment to man, whom they avoid only by means of that activity with which they execute such rapid and extended leaps. Their skins are the only clothing worn by the natives of those countries which they inhabit.

A very remarkable fact is, that the Marsupiata, have,

as yet, been observed only in South America, New Holland, and some islands of the Indian Archipelago. The Didelphes, properly so called, or the Sarigues, and the Chironectes, are proper to the first of these countries. All the others, except the Phalangers, with naked and scaly tails, are peculiar to the second; and those last mentioned Phalangers, and a species of the Kanguroo, are alone to be met with in the Indian Archipelago. It is remarkable that all the Mammalia known in New Holland, to the present day, with the exception of the Dog, and the Hydromys with white belly and that with yellow, belong to the Marsupiata.

To this continent also belong the Ornithorhynci and Echidnæ, which have also the marsupial bones in both sexes, but whose organs of generation are peculiarly conformed, and in which no mammæ have yet been observed. These animals have so great an analogy with the Marsupiata, that M. De Blainville puts them in the same subclass. But M. Geoffroy has separated them from the other Mammalia, to form an order which he calls Monotremes.

The first of the marsupiata, on which we shall hazard a few particular observations, is the *Didelphis Virginiana*, or Virginian Opossum.

When in the study of animals, we consider nothing but their organic structure, we often fail to ascertain a sufficient cause for their peculiar modes of action, or for the part assigned them to perform in the admirable economy of nature. The organization of the Wolf is the same with that of the Dog; the Marten and the Weasel are not to be distinguished from each other without difficulty; yet their destination is far from similar. The lot of one species is cast in the midst of the thickest and the wildest forests, while the wants and instincts of the other are essentially connected with the habitations of men.

THE VIRGINIAN OPOSSUM.

DIDELPHIS VIRGINIANA.

London Published by G. B. Whittaker Dec.ʳ 1834.

But there is another order of pnenomena, which determines in a great measure the local application of the organic system, and which a consideration of that system cannot enable us previously to announce, though we can have little doubt that it originated there.

This order of phenomena consists in those innate moral qualities, those invincible dispositions, which direct the actions of animals in one channel rather than another, and on which their preservation does not less depend, than on their most prominent physical peculiarities. If the different species of the genus canis, for instance, were stripped of all peculiar moral bias, and placed exclusively under the influence of external circumstance, they would probably cease to be different species and become virtually but one. From that moment, the equilibrium of nature would be broken, and till the re-establishment of order, their destruction would be inevitable.

These intellectual qualities are often the only characteristics of species, and it would appear that they form the principal ingredient in the small portion of influence, which the sarigues are destined to exercise upon earth. The opossum, which our author has placed first in this subdivision, is an animal by no means eminent for intelligence. It digs a burrow or den, near thickets not too far removed from the habitations of men, and sleeps there the live-long day. Seeing but badly while the sun is above the horizon, it is in the night that it proceeds in search of food, and of the female during the season of its amours. It mounts trees, penetrates into farm-yards, attacks the small birds and poultry, sucks their blood, devours their eggs, and then returns to conceal itself at the bottom of its retreat. It frequently contents itself with reptiles and insects, and fruits occasionally form a portion of its food. Though its mode of life is very analogous to that of the Foxes and Weasels, it is considerably less sanguinary and cruel. The Opossums are also much worse provided with the means of defence, than these animals. They run badly, and though their mouth is extremely large, and well furnished with teeth, yet it is deficient in strength, and they are wanting in that intelli gence, which might render it an efficient weapon against their enemies. They attempt to bite the stick which strikes them, but not the arm which guides it; very different in this respect from most other mammalia, which, by a very remarkable act of intelligence, distinguish the person from the instrument which he uses, and invariably attack the former. Their chief resource of defence, seems to consist in the disagreeable odour which they exhale when they find themselves in danger. M. D'Azara, who speaks of it from experience, declares it to be really insupportable.

All their desires seem feeble, even that of reproduction. We have already cited the interesting observations of M. D'Aboville on this subject, and our opinion that it involves a problem, of which naturalists have not yet obtained the solution.

A young male Opossum in the French menagerie, would suffer any one to touch him without the least resistance, and continually endeavoured to avoid the light. When his wish was opposed, he would open his large mouth, and hold it in that gaping position without doing any thing more. This organ, so powerful a weapon with most of the carnassiers, seems in the Opossum to be but a simple in- strument of mastication.

The body of this animal was generally of a greyish yel- low, except that some hairs, entirely black, with others en- tirely white, gave a blackish tint to the dorsal line, and a narrow band descending from the neck on the fore-legs. These last, and the hinder were almost entirely covered with black hairs, and the tail furnished with scales, formed of the epidermis, had only some short and scanty hairs springing between these scales. The hands, ears, and extremity of the muzzle were naked. In general, the fur of this animal was not thick, still it had some woolly hairs; few certainly, but well characterized.

As the Opossum is met with in very various latitudes, from the equator to a considerable degree of elevation, it might serve as a tolerably sure guide in our observations re lative to the influence of temperature on the development of these hairs. The skin of the soles of the feet, was of a violet-black, and the toes as well as the nails were flesh-coloured. The external conch of the ear was also black, except at the base and extremity, where a small flesh-coloured spot was discernible. This would seem a pretty constant character with the Opossums, since they have derived from it, the epithet of bi-coloured ears. The extremities of the nose and lips are also flesh-coloured, and the eye is totally black.

It might be deduced from the preceding observations, that the organs of sense and motion in the Opossum are endowed with no great degree of strength or activity. In fact, the organs of this animal are in perfect harmony with its character. The eyes small, and without external lids, project so much that they seem like the extremity of an ellipsis, and their pupil, like that of cats is long, vertically. The nostrils at the extremity of long muzzle, open on the sides of a protuberant, naked, and glandulous surface. The sense of smelling is the most delicate in the Opossum, and from which the animal derives most advantage. The tongue is covered with very rough papillæ. The ears have the faculty of closing. They bend back, from front to back, by means of three longitudinal folds, and then are finally lowered by certain transverse folds, much more numerous than the former, and which cut them at right angles. The motion of both is doubtless determined by some peculiar muscular apparatus. The sense of touch seems to reside principally in the toes, which are covered with a very delicate skin, and furnished underneath with very delicate tubercles, the forms and relations of which are too complicated to admit of description. These toes are five in number on each foot, provided with rather feeble claws, the thumb of the hind foot excepted, which has none. They are strangely separated from each other in walking. The thumb of the hind foot is opposable to the other toes; thus forming a genuine hand, from which these animals have sometimes been called *Pedimana*. There are some weak mustaches on the upper lip, above the eye, and on the cheeks. The tail may be considered as an appendage of the organs of motion. It is prehensile and very strong, but capable of involution only on the under side. The sound of the Opossum's voice resembles the hissing of a Cat in anger. It is about a foot and a half in length, and the mean height about eight inches. The animal, in the French menagerie above alluded to, was fed on raw meat, and bread steeped in milk. It lapped in drinking. Sometimes it would catch with open mouth, the drops of water which fell from the top of its cage, and it seemed to take no small pleasure in this amusement.

The Opossum has been for a long while distinguished from the other species of the genus Didelphis. Buffon has spoken of it in his supplements, under the names of the Sarigue of the Illinois, and the long-haired Sarigue, imagining at the same time, that these were two distinct species. Linnæus having taken all his Didelphes from Seba, could not, with such imperfect materials, exercise that critical discrimination with which he was so eminently gifted. Pennant was the first who clearly separated this Didelphis from the others, under the name of the *Virginian Opossum*, but his figure of it is not good. As to Schreber, all his Didelphes are bad, not indeed to be recognised as intended for the animals, except his *Marsupialis*, the colours of which bear some resemblance to those of the Opossum.

Shaw's figure is from a stuffed specimen in the Leverian Museum, and the tail by some mistake was carried upwards.

Almost all travellers, struck with the singular organization of this animal, and above all, with the peculiar mode in which it brings up its young, have entered into some details respecting its character and structure. None, however, have been more exact on this subject, than M. D'Azara, who was the first to designate the Opossum by the appellation of *Micouré*.

" It is so common," says he, " that they are frequently seen dead about the villages, and even in the streets of Montevideo. The female has the whole length of the belly cleft or slit, and appearing like a person's waistcoat buttoned only at the top and bottom. This cavity the animal has the power of firmly closing. Within it are thirteen teats, extremely small, one in the centre, and the rest ranged round it."

When non-adult, the cleft is scarcely visible; the pouch therefore, it appears, is not developed until the animal has occasion for its use.

D'Azara speaks of one which had thirteen young, similar to the mother, and about half her size. They had ceased to suck, and indeed the pouch, which they no longer attempted to enter, could not have contained them; but the mother carried them about fixed to her tail, legs, and body; in this situation it was with difficulty she could move, and impossible for her to find sustenance for all of them.

In those genera, where the species are strongly distinguished from each other by colours, form, or organic modification, a slight inaccuracy in figuring them, a trivial deflection from the truth, can be of little consequence, and can lead to no material error. This, however, is not universally true. The species of the Sarigues are inter-distinguished by shades of difference so very easy to confound, that the character may be lost by the slightest negligence. The utmost exactitude is indispensably requisite, and the frequent want of it in figures may be clearly observed by those who have an opportunity of comparing the living species. This is true, in a most especial manner of those singular animals that constitute the sub-genus Didelphis.

M. F. Cuvier seems to think, from a consideration of the figures given by Seba, Schreber, and Buffon, that the first time the species of the *Crab-eating Sarigue (Did. Marsupialis et cancrivora)* has been represented from nature, was in his new great Lithographic work on the Mammalia. This accurate naturalist adds, that he had simultaneously, under his immediate inspection, a young female Opossum of the species *Didelphis Virginiana*, a second female Sarigue, to which, but with some hesitation, he gives the name of the young Opossum, and the Crab-eating Sarigue *(D. Marsupialis et cancrivora)*. The colour of the first is whiter, and its size is larger than those of the others ; the second has a good deal of black in its fur, and the third is of a yellowish tint. These two last are nearly of the same size, and have a slenderer muzzle, and straighter forehead than the first. In this the face is large, and there is a very sensible depression at the lower part of the forehead. The young Opossum has a muzzle still slenderer than that of the Crab-eater, and in common with the Didelphis of Virginia, it has a white spot at the extremity of the ears, and white toes. All the mustaches of the Crab-eater are black ; the young Opossum has no black in its mustaches, except above the eyes and nose ; all the rest are white. The Didelphis of Virginia has all its mustaches white.

The head of the Sarigue, which is the subject of this article, *(Did. Marsupialis et cancrivora,)* is yellowish white, while the ears, eyes, and mustaches are black, strongly contrasting with the pale ground of the head. The neck, the

THE CRAB-EATING OPOSSUM.

DIDELPHIS CANCRIVORA

London. Published by G.B.Whittaker, June 1825.

back, and the sides of the body have a yellowish ground, sprinkled with black. These colours are produced by hairs rather short in comparison with the others, which for the most part are of a dirty white, and by very long hairs the upper part of which is black. These latter hairs are more numerous along the spine of the back than elsewhere, and when the animal is irritated they bristle up in the form of a dorsal mane. The limbs are entirely black as far as the unguical phalanges, which are white. The claws are white, the muzzle is flesh-coloured, and the under-lip is edged with black in its posterior half. The ear is without the white spot we have noticed in the Virginian congener. The third part of the tail at the base is black, the rest white.

This species was first established by Linnæus, after the great Philander of Seba, under the name *Marsupialis*. Gmelin, afterwards by mistake, made a second species of the Crab-eater of Buffon, which he named *cancrivora*. At Cayenne, according to M. Barrère, they give it the denomination of Pian, which is only a modification of the French word *Puant*. The term Crab-eater is objectionable as it is already applied to two other animals.

The next species in our author is the *Four-eyed Sarigue*, of Cayenne, (*Did. Opossum* of Linnæus.) The length of its entire body is about one foot two inches, and the tail a little longer. The head is pointed, the forehead and top of the head are on the same line. The ears are large, round, and thin. The fur is of a reddish brown, mixed with gray on the upper part of the animal, from the end of the muzzle to the scaly part of the tail, as well as on the external surface of the thigh, the leg, the arm, and part of the fore-arm.

The head is of a more reddish brown than the other parts. There are hairs of a dirty white at the basis of the ears. There is a spot of the same colour in front of each ear, and above the eye, from which circumstance the name of the animal is derived. The rest of the body and limbs are of a dirty white, with the exception of the belly, which presents a few reddish tints. The tail is covered with hairs similar to those of the back for about two inches and a half, while its extremity is scaly, and partly brown and partly white. There are six or seven teats in the pouch of the female, which sex, by the way, are of a more reddish colour than the males.

This species is one of those most anciently known, and its habits are precisely similar to those of the foregoing. Its habitat is Cayenne, and doubtless many other of the warmer climates of South America.

The *Cayopollin*, or Mexican Opossum, is about eight inches long, and the tail is about a foot. The muzzle is inclining to be thick, and the ears are rather large. The eyes are slightly bordered with blackish. It is marked in the frontal ridge with a longitudinal line of brown, grayish on the edges. All the upper and external parts are of a fawn colour and gray intermingled, the summit of the hairs being of the former, and the rest of them of the latter colour. The fawn colour predominates on the occiput and the neck. The rest of the animal is of a very pale and almost whitish yellow. The ears are naked at their internal face. The tail is covered with hairs for something more than an inch from its origin, and the rest with scales, intermingled with some brush hairs. Part of the tail is variegated with brown and yellow, the point being of this last colour.

The *Marmose*, or more properly *Marmot* (*Did. Muneira*), the Marine Opossum of Pennant and Shaw, is pretty similar in its general forms to the Cayopollin. It is between seven and eight inches long, and the tail about the same. The muzzle is more pointed than in the Cayopollin, the ears more rounded, and the head more convex. The eyes are situated in the middle of a blackish band,

THE TOUAN *OF BUFFON?*

LIDELPHIS BRACHYURA. GM?

London *Published by* G.B. Whittaker, June 1825

which is larger in front and over the upper lid, than be-hind and on the under. The top of the head, occiput, and upper and external parts of the body generally are of a colour composed of ashen and fawn. The rest is whit-ish, slightly tinctured with fawn. The tail is covered with hair for a very little way, and the rest of it with scales, considerably smaller than those of the Cayopollin. The fawn-coloured tint of the fur varies as to depth in the different individuals of the species. There are fourteen nipples in the females, placed between the folds of the skin of the groins.

M. Desmarest seems to think that the *fourth or long-tailed Micouré* of D'Azara is to be referred to this species. The tail however of this last is longer than that of the Marmotte, and the fur of a mineral gray.

The Marmose is found in the woods, and lives there like other animals of the same genus, in search of small prey. The little ones, in number generally from ten to fourteen, are attached to the nipples after birth, and re-main suspended there, like grapes, until they are in some degree developed. Then they mount on the back of the mother, who thus carries them about, having her tail involved with theirs.

The long-tailed Micouré of Paraguay, above-mentioned, remains in the hollow trunks of trees, in rose-trees, bushes, hedges, &c., where it attaches itself by the tail. Don Joseph de Casal, a friend of D'Azara, says, that size and age make no apparent difference in the individuals of the species ; and that the females have the abdominal pouch. If this be the case, it will be necessary to distinguish this animal as forming a particular species.

The Marmose inhabits Cayenne and Surinam.

The *Touan,* or short-tailed Opossum, *(D. Tricolor vel brachyura,)* is the last Sarigue admitted by the Baron in his " Regné Animal." Its length is something better than five inches, and the tail a little more than an inch. Its ears are of moderate size, naked, and of a rounded form. The tail is very short in comparison of the other species of this genus; hairy at the base, and naked and scaly for the rest of its extent. Upper part of the body, back of the head, and hairs on the basis of the tail, blackish. Cheeks, shoulders, flanks, throat, external side of the thighs and paws of a lively red. The breast and under part of the body of a pure white.

M. D'Azara, during his stay at Paris, recognised in the individuals of this species which form part of the collection of the Museum of Natural History, his own *fifth Micouré*, or that with a short tail, which agreed exactly with the animal now described; except that the belly instead of being quite white, was of a whitish yellow. The female of this animal, according to the same naturalist, has fourteen teats between the folds of skin of the groins, which disappear when she ceases to suckle. The male, when irritated, emits an unpleasant odour. The animals of this species live in holes underground.

The Touan is found at Cayenne in the woods. The young are attached to the nipples, as among the other Didelphes. The short-tailed Micouré is a native of Paraguay.

There are other species of Didelphes enumerated by naturalists, which will be found in the table.

The genus or sub-genus CHIRONECTES was established by Illiger on one species, which naturalists have placed sometimes among the Otters, and sometimes among the Didelphes. It is thus characterized: upper incisors, ten; lower, eight. Canines tolerably long. Pointed muzzle; eyes placed on the sides of the head; ears naked and rounded; tail prehensile and scaly; marsupial bones; plantigrade and pentadactylous feet: the hinder ones alone

THE YAPOCK OR PALMATED OPOSSUM.

CHIRONECTES PALMATA.

C.Hamilton Smith Esc.[r] del.[t]

having the toes united by a membrane, and the thumbs without nails; the nails of the other toes sharp and curved.

The name given to these animals is composed of two Greek words χείρ *manus*, and νηκτης *natator*, swimming with hands, a name sufficiently designative of their habits.

They are found in South America only. The only species is the *Chironectes Yapock*, or *little Otter of Guinea*. Buff. Suppl. *Lutra memina*, Bodd; *Didelphis palmata*, Geoff. The length of this animal, measured from the end of the nose to the extremity of the body, is about a foot, or a little more. The head is about three inches long. The tail a little more than nine, and the average height of the animal little more than three. There are however Yapocks of a smaller size, as that, for instance, described by Buffon, which was but eight inches long, and the tail six. The head is pointed, the muzzle rather slender, and the ears large and naked. The tail is hairy on the upper part, and more especially at the base, naked and prehensile underneath. The fur on the upper part of the animal is marked by large patches of a blackish brown, the intervals of which are filled by a yellowish gray. These dark spots are symmetrical. One of them covers the muzzle and extends on each side behind the ears, where a second spot begins which spreads over the hinder part of the neck, and is united by a dorsal line to the third, which is situated nearly over the shoulders. On the back are two of these spots, equally connected by the brown colour of the central part. The last spot is placed on the crupper, and extends over the basis of the tail, and on the external part of the thighs. Behind each eye is a white spot, and all the under part of the body is white. These marks however vary as to their disposition in different individuals, nor did the specimen whence our figure was taken exactly accord in this respect. The moustaches are about an inch long, as also are the long hairs above the eye, and those on the tarsi.

D 8

The fur is soft, woolly close to the body, and harsher towards the points.

The habits of this animal, which has been found on the banks of the Yapoch, a river of Guyana, are not known; nor is it yet ascertained if the female have an inguinal pouch, like the majority of the Didelphes, or only a longitudinal fold of the skin of the belly like the Marmose.

The several species appropriated to the genus or subgenus DASYURUS, though they differ very materially in external appearance, are as assimilated in impulses and habits as are their influential physical characters. The cheek-teeth at once cutting and pointed, display the general nature of their character; fitted for carnivorous regimen, their habits must necessarily be ferocious and predatory, proportioned to their size and strength.

Their general organization is very like that of the Opossums, like them also the Dasyuri live on fish and insects; but being destitute of the thumb to the hind feet, and also of any prehensile power in the tail, they have not, like the Opossums, the means of pursuing their prey in the trees. During day they remain concealed in cavities of rocks, or in hollow trees, and sally forth under the protection of darkness in search of the Ornythorynchi, Echidnæ, and insects. They also seek, with avidity, the half-corrupted bodies of Seals and cetaceous animals on the sea-shore, and are not unfrequently known to intrude into the dwellings of the natives of New Holland and Van Diemen's Land in search of food.

The *Dog-Faced* or *Zebrine Dasyurus*, is a singular looking animal, by no means pretty. The fur is short and soft; the ground colour of which is a dirty yellowish-brown, lighter on the belly, and of a deep gray about the back. From the insertion of the tail, along the spine, to about one half its length, there are equi-distant transverse stripes, longest on the thighs, and becoming gradually

THE ZEBRA OR DOG-FACED DASYURUS.

DIDELPHIS CYNOCEPHALA. HARRIS.

London. Published by G.B.Whitaker June. 1825.

shorter as they approach the middle of the spine. These transverse stripes are of a brownish-black, and are assimilated to the stripes of the Quagga, though much more regular, and only of partial extent. The tail, not so long as the body, is slightly compressed on the sides, which, together with the under part, are denuded, leaving only the superior side of the tail covered with the short smooth fur which envelopes the rest of the animal. The head thick, and muzzle of moderate length, and truncated, give the animal much of a canine appearance.

But little is known of the habits of this singular animal. Its compressed tail seems to indicate that it is a swimmer, and it is known to be an inhabitant of the rocks on the sea shore of Van Diemen's Land, and to feed on flesh.

The *Ursine Dasyurus* or *Ursine Opossum*, of Harris, is furnished with long rough black hair, irregularly marked with one or two whitish spots, which, in some individuals, are found on the shoulders; and in others, toward the bottom of the back. Their eyes are small, and the mouth large. The tail, which is rather short, is denuded on the under side, and is subprehensile.

Like the Dog-faced Dasyurus, this species inhabits the sea shore; it also sits on the haunches, and employs the anterior extremities to convey food to the mouth.

Each of the two species of this division already noticed takes its name, or, at least, an epithet, from other animals, with which, in certain particulars of external character, they bear a slight resemblance: the first, both to the Dog and the Quagga, or Zebra; and the second, to the Bear; the remaining Dasyuri, and of the number of species there is at present some doubt, seem all allied in appearance to the vermiformed or viverrine animals. Hence, the Spotted Dasyurus is named by Phillips, in his New Holland, the Spotted Marten.

The muzzle of this species is elongated and conical, and

the fur rather short and harsh. It is of a deep shining black, with the body closely, but irregularly, covered with pure white patches, differing in shape, size, and disposition. The tail, also, is said to be spotted, in like manner ; but in the specimen, whence our drawing was taken by the accurate pencil of Lewin in New Holland, no such spots appear ; there the belly is palish white.

This species also is said by Peron and Lesueur to be an inhabitant of the sea shore, and to feed on the bodies of Seals, &c., left by the side ; and, accordingly, the figure of the animal in the splendid work of these travellers and naturalists, is represented in the act of devouring the flesh of a Seal; but it appears, by a note of the late Mr. Lewin, on the drawing whence our plate is engraved, that it is an inhabitant of Bathurst Plains; and is known among the European residents in New Holland, by the ill-applied name of the Cat of Bathurst Plains.

The *Tapoo Tafa*, of White, which is uniformly brown, but lighter underneath, is treated by Mr. Hunter and by Shaw, as a mere variety of the preceding. M. Geoffroy, on the other hand, makes six different species, which will be found noticed in the Table. We have yet much to learn on the subject of New Holland Zoology ; and as we can add little or nothing with any certainty as to the habits and instincts of these animals, we shall merely refer to the Table for the differences of physical character which have been presumed to constitute the several species.

The PERAMELES, at first view, are most like the Opossums ; but their head and muzzle are very much longer. To judge of the habits, however, by their physical peculiarities, an unerring criterion indeed, it is obvious that they differ from the Opossums materially in their modes of life ; every thing indicates that they do not live in trees, but, on the contrary, seek their sustenance on the surface of the ground, or under it, like the Badgers; their muzzle

SPOTTED OPOSSUM. WHITE.

DASYURUS MACRURUS.

Lewin delt

THE LONG-NOSED POUCHED BADGER.

PERAMELES NASUTA. Geoff.

is elongated; their hair rough; and their feet terminated by powerful long straight nails; indeed, there seems to be no animal better provided by nature for excavating than these. The thumb and little finger of the fore feet are merely tubercles. The hind feet have much analogy with those of the Kanguroo: the fourth finger is the longest and thickest; the second and third are united, enveloped under one common integument; but they are distinguishable by the nails, which are free. These two fingers are shorter and thinner than the last, and the fifth. They differ from the Kanguroo, nevertheless, by having a thumb, though it is extremely short: this thumb, as in all the Marsupiata, is without a nail.

The tail is in general serviceable to the Marsupiata, either by its strength and stiffness, or by its prehensible quality; but in the Perameles, it is too short and weak to be apparently of much use.

Their elongated muzzle gives them an excessively stupid appearance, but though we have little information as to their habits, it is probable, from their construction, that they are quick and lively animals.

Our author admits but of one *Perameles*, the *Nasutus*, or that with pointed muzzle. The length of the body is about one foot three or four inches. The tail about five inches or more. The ears are strait, oblong, and covered with hair. The hair is but indifferently furnished, more copious and stiff, however, on the withers The general colour of the animal is a clear brown, not unlike the Surmulot. The under part of the body is white, claws yellowish. The tail is of a more decided brown than the body, bordering on the marron above, and on the chestnut underneath. It is an inhabitant of New Holland, and its habits are unknown.

Dr. Shaw notices another species, which he calls *Perameles obesula*. The chief difference between which

which and the last, is the presence of two additional incisors in the lower jaw, and a molar more above, and one less below on each side. The ears also are larger and more rounded, and the fur in general borders on a reddish yellow. The habitat is the same.

This is also called *Isoodon*, from the evenness of the teeth, and is the *Porculine Opossum* of Shaw, so named from a slight similarity to a pig in miniature.

We shall now speak of the PHALANGERS, with which the second grand subdivision of the Marsupiales commences. Their dentary formula will stand thus:—Incisors $\frac{6}{2}$, canines $\frac{1}{0}$, or, $\frac{0}{0}$; false molars or canine $\frac{2 \cdot 2}{3 \cdot 3}$, or, $\frac{2 \cdot 2}{2 \cdot 2}$; molars $\frac{4 \cdot 4}{4 \cdot 4}$, or, $\frac{6 \cdot 6}{5 \cdot 5} = 38$ or 40.

The two intermediate upper incisors are separated at their base, converge towards their point, and are longer and larger than the others. The exterior ones are smaller. The lower incisors are long, inclining, and nearly horizontal, corresponding to the upper by their external edge. The upper canines are sometimes long, conical, curved, and placed immediately after the incisors. Sometimes their place is filled by two small conical teeth very distinct from the incisors, the anterior of which is the largest. Lower canines there are none, their place being supplied by two or three small even teeth, cylindrical, obtuse, and hardly projecting beyond the gum. The molars are sometimes five on each side of the two jaws, the anterior one of which is very strong, conic and obtuse, and the four others almost equal and squared. Sometimes the upper ones are six in number, two of which are false and compressed, and four true, with tubercles. Of the lower ones, five in number, one is false and denticulated, while four are true and tuberculous.

The head of these animals is elongated. Frontal ridge slightly arched. Mouth not very deeply cut. The ears are

THE VULPINE PHALANGER.

DIDELPHIS VULPINA. Shaw

London Published by G. & W. B. Whittaker. Sept.ʳ 1824.

moderate in size and rounded. The feet are pentadactylous, not united to the body by the skin of the flanks. The
fore-feet have the toes separated, and armed with strong
and crooked, but not retractile claws. There is a large
thumb on the hinder foot without a nail, turned backwards,
very distinct from the other toes, of which the two internal
ones, equal in size, and much shorter than the fourth and
fifth, are united by the skin as far as the base of the
claws.

It was in consequence of this union that these Mammalia
received the name of Phalangers, from Buffon and Daubenton. It was a remarkable character at the epoch in which
those writers flourished, and they named from it the only
species then known to exhibit it. Since that period, however, it has been found in many other genera.

The tail is sometimes naked, sometimes covered with
hairs, more or less prehensile, and almost always as long
as the body. There is an abdominal pouch in the female,
tolerably ample.

These animals live almost continually in trees, where
they subsist on fruits and insects. They are slow in their
movements, and emit an unpleasant odour, which proceeds
from a liquor secreted in a gland observable near the
anus

The Phalangers are found in the Moluccas, New Holland,
and Van Diemen's Land.

The Vulpine Phalanger, (*Didelphis Vulpina*,) is the first
species admitted by Cuvier. It is about the size of a large
cat. The general proportions of its body are elegant and
delicate, more so than those of the other Phalangers. The
upper part and sides of the body, as well as the basis of the
tail, are grayish brown, approaching to fawn-colour on the
shoulders. The head is of a grayish fawn, deeper than
that of the belly. The ears are naked within, and covered
with gray and fawn-coloured hairs without. The external

side of the limbs are rather of a more obscure colour than the back. The tail is covered with hair in its entire extent, with the exception of a narrow band placed underneath, which commences about the middle, and continues to the point. The skin which covers this band is slightly granulated. The hairs of the tail are long, and of a very fine black, except at the base, where they are of the same colour as the back.

The females are similar to the adult male. Young males are of a gray colour, similar to that of the small gray squirrel, on the upper parts of the body and external face of the limbs. The ends of the feet are reddish. The muzzle is of a clear colour, bordering on soiled white. Under part of the body is of a yellowish dirty white. The tail is gray at its base, like the back, and grows progressively deeper towards the extremity, which is quite black. The upper part of the tail is marked with a narrow, black, longitudinal line, very distinct, extending to the point, and beginning about an inch and a half from the origin.

The entire number of teeth in this species is eight and thirty. The upper canine is not much pointed, and rather resembles a false molar than a true canine. It is followed after an interval by another small conical tooth, which itself is distant from the molars. These last are six in number, four true and two false. In the lower jaw, on the bar which separates the incisors from the molars, are two small teeth. There are five molars, one of which is false.

According to Mr. Rollin, surgeon of our naval establishment at Port Jackson, this Phalanger lives in burrows, subsists on small prey, and chases the birds like the Didelphes.

Its habitat is the eastern coast, in the environs of Port Jackson.

The Phalanger of Cook, (Phalangista Cookii,) is about one foot two or three inches long. The tail is nearly equal in length to the whole body. The upper part of the body of a reddish grey. The under part white under the chin and on the upper lip. Throat marked with a brownish spot. The ears are covered externally with greyish-red hairs. The cheeks are marked with a small white spot, scarcely visible behind the eye. The tail is reddish at the base, then brown, and the extremity covered with white hairs.

An individual in the collection of the French Museum, smaller and younger probably than the above description, will agree with it; the upper part of the pelt being brown ; also the upper part of the tail, the extremity of which is, however, white. The circle of the eye, the fore paws, and the sides, are tinted with red. The ears are rounded ; red within, and white at the base

Habitat.—Van Diemen's Land.

The second tribe of the Phalangers, of which some naturalists make a genus, is the PETAURISTÆ, or FLYING PHA-LANGERS. In these, the upper incisors are arranged something in the form of a horse-shoe ; are a little compressed, and placed vertically. The two intermediate ones are the longest, and are pointed. They are also separated from each other at their base, and convergent towards the point. The two following are large with flat coronals ; and the last on each side is smaller than the second. The lower incisors are strong, inclining with the external edge trenchant, and resting on the incisors above. The upper canines are long, conic, and curved. The lower canines are sometimes wanting, sometimes replaced by two very small obtuse teeth, cylindrical, and scarcely projecting. There is a bar between the canines or incisors, and molars in the two jaws. Of the upper molars there are four true, with coronals, furnished with triquetted points, and obtuse ; and two or three are false. Of the lower, four are true, and two false.

The head is moderately elongated; the ears and eyes

large, and the nose pointed. The feet are rather short and pentadactylous ; the hinder have a large thumb without a nail, opposable ; and the two first toes much shorter than the others, and united by one common skin.   Claws arched, compressed, and very strong.

The most peculiar character of the Petauristæ is an arrangement common to them and the Galeopitheci. This is an extension of the skin of the sides, whereby the anterior and posterior extremities are united, and a kind of parachute rather than a wing is formed. There is a spacious ventral pouch in the females.

The tail is very long, not prehensile ; is furnished with hair; is sometimes round, sometimes flat. The habits of these animals are pretty similar to those of the Phalangers we have last noticed. They jump with tolerable activity from branch to branch, and are enabled to sustain themselves for a little time in the air by the assistance of their parachute.   They are nocturnal animals, and are found in New Holland and the Island of Norfolk.   The first species on the Baron's list is the *Dwarf Flying Phalanger, (Did. Pygmæus)* of Shaw.   The length of its entire body is scarcely more than two inches. Its form is more thickset than those of the other Petauristæ. The upper part of the back and head are of an uniform mouse-coloured gray, slightly sprinkled with reddish.   The eyes are surrounded with a clear brown.   The upper lip, belly, under part of the membrane of the sides, are of a pure white.   On each side of the tail are reddish-gray hairs, ranged with the most perfect symmetry. The membrane of the sides terminates at the elbow, and has its edges considerably dilated. The habitat is New Holland.

We shall next notice the *Great Flying Phalanger,* (Did. Petaurus, Shaw.)   The entire length of the body is one foot eight inches, and the tail about two inches shorter. The head is small ; the ears large and covered with hair,

THE GREATER FLYING PHALANGER.

DIDELPHIS PETAURUS. *Shaw.*

London. Published by G & W.B. Whittaker: Sep.r 1824.

and oval in their shape. The tail is round at the base, and very villous and beautiful ; more tufted at the extremity, where the hairs are also flatter. The upper part of the body is of a grayish-brown, pretty uniform throughout; rather more obscure on the external face of the arms and legs, and more gray on the anterior part of the lateral membrane. The head is of a grayish-brown, tolerably deep. On the frontal ridge are hairs of a golden fawn-colour, mixed with others. Chin brown. The neck, throat, breast underneath, and a line on the anterior limbs internally are white. The feet are of a brown approaching to black. The toes of the hinder feet are considerably furnished with hair, especially underneath. The tail is of a fawn-coloured brown near its origin, then passes to brown, and becomes very dark towards the extremity.

A white or whitish variety of this Phalanger has been observed. Sometimes completely white, with the exception of several places on the feet and lateral membrane, where some ashen-gray hairs are found mixed with others; and the end of the tail, which is brown. Sometimes they are found of a dirty whitish-yellow, with the back of a very clear ashen-gray.

Habitat—the environs of Port Jackson and Botany Bay, in New Holland.

The Long-tailed Flying Phalanger, (Did. Macroura,) is about the size of a surmulot-rat. The pelt is of an obscure gray, or brownish above and whitish underneath. The head and neck are both whitish. A band of a more brownish hue extends from the top of the head to the nose. The ears are rather large, slightly rounded and whitish. The extremity of the fore-paws white. The last half of the tail of a deep black, which diminishes a little towards the base, which is of the same brown as the body.

This Phalanger is also a native of New Holland.

The POTOROO, or KANGUROO-RATS, is the last division of the Marsupial family which preserves any of the characters of the order Carnassier. The upper middle incisors are longer than the others, and pointed ; the lower incline forwards. The upper canines are large, flattened laterally and pointed. The four hinder molars to the right and left in the two jaws have blunt tubercles. The anterior one is long, trenchant, and denticulated

The head is long and pointed, the ears are long, and the upper lip divided. The fore-paws are very short, with five toes, armed with crooked claws. The hinder are very long, and thin, terminated by four toes, two of which are very small ; internal and joined together as far as the first phalanx. The third is extremely strong, and furnished with a very thick nail. The fourth is external, of moderate thickness, forming a medium between the two first and third.

The tail is long, and tolerably strong.

There is a complete abdominal pouch in the females, which contains two teats.

The hair is soft and woolly.

The stomach is tolerably complicated, divided into two pouches, and provided with many turgescences. The intestines are short, and the cæcum of moderate size and rounded.

There is but one species, the *Kanguroo-Rat* of our author, *Macropus Minor* of Shaw. It is about the size of a rabbit six months old. The pelt is woolly ; the upper lip furnished with moustaches. The tail of moderate size, scaly, and covered with a few scanty hairs.

The habits of this animal are but little known. To judge from its dentary system and digestive organs, it would seem less herbivorous than the Kanguroos, with which, however, the other details of its organization exhibit the closest relation. The disproportion between its fore and hinder limbs sufficiently indicates the facility which it must possess in jumping.

The last subdivision of the Marsupiata is partially edentatous, being destitute of canines.

The Gigantic Kanguroo (of Cook.)—Buffon, whose only errors were those of genius, clearly perceived that every continuent, in its animal productions, presented the appearance of an especial creation ; but he gave an universality to this proposition, of which it is not altogether susceptible. It is nevertheless true, even at the present day, within certain limits. A great number of the Asiatic animals are not found in Africa, and *vice versâ.* The Lemurs seem to exist only in Madagascar. America is peopled with a host of Mammalia, exclusively peculiar to itself, and there are many more in Europe not to be found in the other quarters of the globe. The discovery of Australasia has given an additional support to this opinion of Buffon. The species of animals there discovered, have not only no affinity with those of the other continents, but, in fact, belong for the most part to genera altogether different. Such are those Mammalia which the natives of New Holland call Kanguroo, and which offer to the observation of the naturalist, organic peculiarities perceivable in no other animal, with the exception of one single species. It is in this tribe that for the first time we view the singular phenomenon of an animal using its tail as a third hind leg in standing upright and in walking. The species we are now upon has received the name of Gigantic, because when named, it was supposed to be the largest of all that are known. It was discovered by Cook in his third voyage.

These are the dimensions of a Kanguroo, represented by M. F. Cuvier: the measures are French :—from the origin of the tail to the extremity of the neck, two feet one inch. From the hinder part of the head to the end of the snout five inches. The leg one foot, two inches, eight lines. The foot ten inches ; the arm eight inches ; the hand two inches, four lines; and the tail two feet. There are some of larger proportions.

As the anterior extremities bear no proportion to the posterior, it is plain that the animal cannot properly make use of them for walking. It sets itself upon them when browzing, and when it has no necessity for moving fast ; but, when it wishes to proceed quicker, and especially to run, it employs its hind feet and tail. It thus makes very considerable leaps, and escapes easily from its enemies, which it possesses no means of resisting. Its hands are feeble in the extreme, and its mouth cannot be of the least assistance as a weapon of defence. It is a quiet animal, endowed, apparently, with a very small portion of intelligence ; on this point, however, too little is known to enable us to speak in detail.

The hind feet of this species, at the first glance, appear to have but three toes; there are, however, in reality, four. The two toes that face inwards are remarkably small and short, and are so joined under the skin as to exhibit nothing externally but their very small claws. The two other toes, of which the internal is very long, and the external very short, are armed with thick and strong nails, with which the animal at times attempts to strike. These are the only defensive arms it seems to possess. The hands have five fingers, furnished with very long nails, proper for digging.

The organs of mastication are not less remarkable than those of motion. The lower incisors are two long and large teeth, pointed and trenchant on their edges, and are directed forwards, almost upon a line with the jaws. There are six upper incisors, the two middle ones being larger than the others and pointed. There are no canines. There are five molars on each side of the two jaws. The eyes are

TUFTED-TAILED OR MOUNTAIN KANGUROO.

THE WOOLLY KANGAROO.

MACROPUS LANIGERUS.

Lewin del.!
New Holland.

H.Kearsley fc

London Published by G.B.Whittaker March 1827.

simple, and the pupil round. They resemble most of the other Didelphes in the mode of parturition and the abdominal pouch. The colour is almost universally a dull brown.

We are told that the flesh of these animals is good for eating; and, as they are found to re-produce in Europe, they might easily be naturalized among us. We are also told that they live in troops, conducted by the old males.

There are other species of Kanguroos known. The *Mac. elegans* is beautifully figured in Peron's Voyage to Australasia. Like the Dog-faced Dasyurus, this species is also marked with transverse stripes, going off from the spine for about the lower half of the back. Others will be found in the Table.

We shall here insert a figure of a new species of Kanguroo, discovered in Bathurst Plains in New Holland. The drawing in our possession is by the late Mr. Lewin, who accompanied the expedition to Bathurst Plains in the year 1821, when the animal was first seen and shot.

His note upon the drawing is very brief, and the dimensions are not stated, but it appears by what is said, to be considerably larger than the Macropus Major, and more deserving therefore of the epithet attached to that species. The character of the fur is very different from that of the species last mentioned, being entirely woolly and very thick. The drawing includes two figures, one of a brownish yellow colour rather lighter than of the Mac. major, and the other slate, inclining to puce colour, but whether these wide differences in tint indicate varieties, sexual differences, or distinct species, does not appear; in size, and the woolly character of the hair, they both accord ; and, in reference tothe latter peculiarity, we have conditionally named it the Woolly Kangaroo, *Macropus lanigerus*.

There are several new and interesting specimens of the Marsupiata in the Museum of the Linnæan Society; the most important of which are referred to in the Table.

In the subdivision of the KOALAS, we find the two upper intermediate incisors much longer than the others. The lower incisors are like those of the Kanguroos. There are four small intermediate teeth between the incisors and upper molars. The molars have four tubercles.

The ears are large and pointed, with the conch directed forwards

The feet are pentadactylous. The toes of the anterior extremities are divided into two groups; the thumb and index on one side, and the three others on the opposite. The thumb of the posterior extremities is very large, separated without a claw. The two following toes are smaller, and united as far as the claws. This description we have taken from M. de Blainville. It differs from that of the Baron in giving a thumb to the hinder extremities.

There is but one species—the *Koala* of our author, which the French naturalists term *Phascolarctos fuscus.* The hair is long, tufted, and thick, chocolate-brown, according to M. de Blainville, but ashen according to M. Cuvier. Its size is that of a middling Dog. In its air and gait it is like a small Bear. It climbs trees with facility, and digs burrows. The female carries the young a long time on the back, in the manner represented in our plate, from a drawing made in New Holland by Mr. Lewin. It inhabits the banks of the river Vapaum, in New Holland.

The Wumbat is a species as yet isolated in nature, which forms the type of a genus, and constitutes that genus in itself alone. Its relations with other Mammalia are so remote, that it is a matter of uncertainty to what order it should be referred. It is related to the Didelphes by the organs of generation; the young are born prematurely

THE COOLA OR KOALA

London. Publish'd by G & W. B. Whittaker Feby 1824.

and enveloped in a pouch which is attached to the mammæ. But this family is so far from being governed by the laws that regulate the animal kingdom in general, that its leading character does not necessarily suppose the presence of any other. It approaches the Rodentia in the number and form of the teeth, but their relative situation forbids us to rank it with this order, as they are much less placed for gnawing and cutting than for grinding the food. Neither do these animals resemble the Rodentia in the articulation of the lower jaw. Again, if they approach them in the general form of the body, they are quite remote from them in the proportions of the limbs and mode of locomotion, which have some analogy to those of the Ursine genus. It being impossible to characterize them precisely, systematic writers have placed them between the two orders to which they might form a common connecting link, either as Didelphes, or Rodentia. Unfortunately these orders are composed of genera so imperfectly linked with each other, that their approximation is just as arbitrary as their existence.

The Wumbat, in its full growth, is about the same size as the Racoon. The tail is a very trifling rudiment.

These animals have two very large incisors, flattened above and opposed coronal to coronal, like tuberculous molars, and not face to face like the incisors of the Ro dentia. They have ten molars in each jaw divided by a transverse furrow. The first is the smallest.

They are plantigrade animals, have five toes on the fore feet, and four on the hind, with the rudiment of a thumb, armed with long and powerful nails for digging. The eye is simple, black, and so small, that the form of the pupil is not to be distinguished. The ear is also very simple, small pointed, and concealed in the hair. The fur is rough and rather thick. The colour is generally of a grayish brown,

E 2

rather paler under the neck and round the ears than on any other part.

There were two males of this species in the French Menagerie, brought by Captain Baudin from the south of New Holland. They did not live any length of time ; they were tamed ; but they appeared rather to be habituated to the presence of men in general, than to distinguish or know them as individuals. All their motions were excessively slow. They seemed to be but little attracted by what passed around them. They suffered themselves to be carried off without resistance, and then, when set upon the ground, they moved no faster than before ; not even blows appeared to excite in them either fear or anger, nor is it probable that any animal exists more completely passive. In their natural state they dig and live in burrows, but nothing further is known of their habits in this state. Peron says that the flesh of the Wumbat is tender and delicate, and that this animal is to be met with as familiar as a dog in the cabins of the English fishermen. Animals of so apathetic a nature, and so easily tamed, would certainly reproduce in a state of domestication. Their climate differs so little from that of Europe, that they might easily be imported here and become naturalized among us.

This animal is the *Phascolomys* of M. Geoffroy, and the *Ursine Opossum* of Shaw.

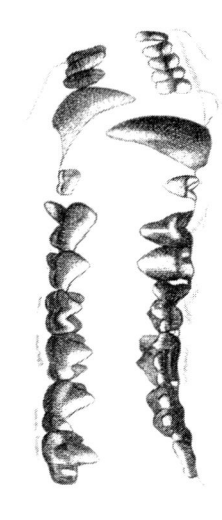

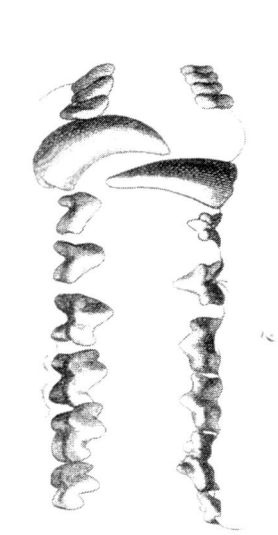

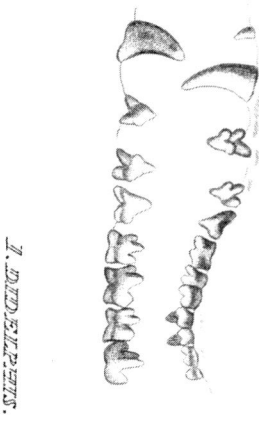

TEETH OF MARSUPIALA.. Pl. I.

1. DIDELPHIS.

2. DASYURUS.

3. PERAMELES.

4. PHILANGISTA.

TEETH OF MARSUPIATA. *Pl. 2.*

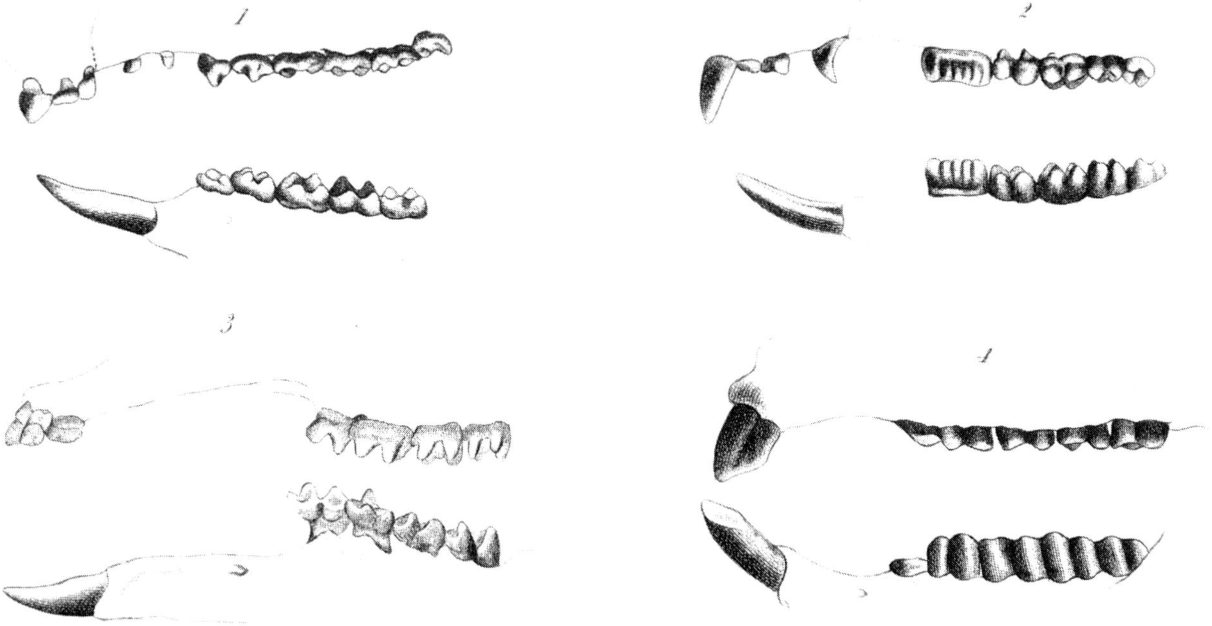

1. *PETAURUS.*

2. *HYPSIPRIMNUS.*

3. *MACROPUS.*

4. *PHASCOLOMYS.*

London Published by C.B Whittaker Jun 1825

Having now gone through this very extensive sub-division of the Animal Kingdom, the Order Carnassier, our readers will permit us to indulge, in conclusion, in a few general observations.

The first remark, which the most superficial glance at the subject suggests to the mind, is, what an immense number and wonderful variety of animals are comprehended in this order ! Families, tribes, genera, sub-genera, species and races, multitudes of which differ from each other in many and strong characteristics, and agree apparently but in a single attribute, that of a greater or less propensity to animal nutriment. Nay, we find included among the *flesh-eaters*, those which are no flesh-eaters, while some whose diet is partly animal, are excluded from this order. Thus, among the Cheiroptera, some are frugivorous, and some of the marsupial tribes are even totally herbivorous; and again, among the Quadrumana, some are insectivorous, and many may be presumed from their dentary and digestive systems, capable of subsisting on flesh. This last presumption derives additional probability from the physical resemblance of several of their tribes to the human species.

The wonderful external difference between many members of this enormous family is indeed, at first view, marvellously striking to the superficial observer. How widely separated, for example, the Bat and the Lion; the Mole and the Seal, the Kanguroo and the Hyæna. Were the more obvious attributes of strength, size, and ferocity, to be regarded as the basis of classification, who would hesitate to unite together,

> " The rugged Russian bear,
> The arm'd rhinoceros, or the Hyrcan tiger,"

preference to joining either of these animals with the

Petaurus or the Koala. Of such arrangement, might not
the uninstructed observer of nature exclaim

" Serpentes avibus geminentur, tigribus agni."

A closer observation, however, and a knowledge of the
true scientific principles of arrangement, will show that a
more accurate classification of those very numerous animals would not have proved a very easy task. All possess
some traces of the three kinds of teeth, down even to the
ultimate divisions of the Marsupiales, which though devoid
of canines, still preserve something of the Carnassial character in the articulation of the lower jaw. They form
a necessary and beautiful link between the order in which
they are classed and the rodentia.

There are certain peculiarities of the cranium mentioned
by our author in which these animals agree, as also in the
power of rotation in the fore arm, and the want of opposa-
bility in the thumb of the anterior extremities. The intestines are likewise less voluminous, and the alimentary canal
less elongated than in the succeeding tribes ; an admirable
provision of nature, to obviate the ill effects of that rapid
decomposition to which animal substances are more peculiarly liable.

On the whole, it seems very doubtful, whether for general
purposes any better arrangement than that of the Baron
can be suggested. Though he has classed together in one
order such various animals, yet, in the sub-divisions of that
order, their variations of character are completely preserved
and clearly distinguished ; and it may well be questioned,
if any of these sub-divisions separately possess such leading
characters, and still more if the animals which are comprehended in them be of such importance as to justify their
classification into distinct orders.

It may be said, for instance, that the Cheiroptera are all

distinguished by one leading character, which exercises the most absolute domination over their destiny and habits of existence; that the single fact of their being *Cheiropterous,* is enough to justify their allocation in a separate order. But we ask in reply, are those animals of sufficient importance in the general economy of animated creation, to entitle them to such a separation ? Certainly not. A mere single physical peculiarity, however important to the creature, is an insufficient basis for one of the grander divisions of the Animal Kingdom. It is necessary, that the animals which constitute an order should act a distinguished and very influential part in the theatre of nature. Why else was MAN formed into a distinct order ? Were physicalities alone in question, the arrangement of Linnæus is, perhaps, as good as that of Cuvier.

If the Cheiroptera are unentitled to be placed in a distinct order, on the score of their comparative unimportance in the animal kingdom, still less can the Insectivora be considered deserving of such a rank ; neither in number nor in power are they sufficiently important for such a purpose, and added to this, they are not distinguished by any characters of striking and prominent peculiarity.

The Carnivora undoubtedly form a very numerous and powerful division of animals, act a conspicuous and important part in creation, and their relative influence on other living beings, even on Man himself, is very considerable. They are likewise prominently characterized in themselves, and the perfection of their physical organization, and, in many instances, their intellectual superiority, or their instinctive acuteness, entitle them to very high consideration. These peculiarities, however, striking as they are in an insulated view, do yet, on a closer examination and comparison with other tribes, seem insufficient for the purpose in question. Regarded in themselves, the

Carnivora strike us as a very peculiar race of animals ; but when we consider them in relation to the tribes which immediately precede and succeed them, we find none of those bars which should separate and mark distinct orders in the living world. Habits, instincts, organization, and other characters, proceed from one family to another with such gentle gradations, with such slender shades of difference, that we are furnished with no basis of sufficient breadth foi a high division of the animal kingdom. When we speak ol the Carnivora, we are apt in general to think only of Lions and Tigers, of Wolves, of Bears and Hyænas ; of creatures of indomitable strength and surpassing ferocity. But it should be remembered, how many members there are in this mighty family, whose share in those distinguished attributes is comparatively limited ; how many whose fierceness is counteracted by deficiency of strength or of sagacity, or whose carnivorous propensities are tempered by mildness of disposition, or regulated by superiority of intellect. All the Carnivora are not those " mighty hunters," which, by their ferocity and destructive instinct, can create a desert around them. Many of them are small, and feeble, and timid, and by no means so superior to the species which precede them as to be considered as belonging to a higher or distinct order.

Besides, were the Carnivora to be made a separate order, the same thing must of necessity be done for the Cheiroptera and Insectivora, and we have already seen that they are not entitled to this distinction.

The preceding arguments will equally apply to the Seals and Morses. Should it be contended that their aquatic life is a reason for separating them into a distinct order, it is answered that they are not the only amphibious animals. If amphibious habits be admitted as a principle of distinction for an order, we are reduced to this dilemma. We must either exclude from the order amphibia animals that are amphibious, or arrange animals together that differ in the most important and essential points.

With the Marsupial tribes, indeed, the case is somewhat different. They differ materially, not only from the foregoing families, but from all the rest of the animal world. The pouch in most, the Marsupial bones, probably in all *, and the double matrix, are high leading characters. But yet these animals exhibit such very various modifications, that, as the Baron observes, if we are to separate them, it should be into a distinct class, rather than an order. They furnish in themselves analogous modifications to all or most of the other orders of Mammalia. But before any such arrangement can take place, or any other arrangement indeed, that shall pretend to permanence and precision, it will be necessary that our knowledge of these animals be considerably enlarged. At present, our acquaintance with them is far too imperfect for this purpose.

Let travellers and naturalists, then, proceed in the grand business of discovery and investigation, and let us meanwhile content ourselves with a classification, which, under existing circumstances, is the least objectionable that can be given. After all, as we have had already frequent occasion to remark, we must recollect that classification is the work of man, not of nature; that general systems are entirely arbitrary, and that however we may pique ourselves on our ingenuity in their formation, they involve a confession of the weakness and imperfection of the human faculties. Classification is intended to assist our apprehension and our memory, which would otherwise be puzzled and overwhelmed by the infinite variety and stupendous extent of the visible creation. The little eye of human intellect

---

* It does not appear that the Dog of New Holland, the Dingo, has ever been examined with reference to the Marsupial bones more or less developed.

must gradually take in parts to be enabled to form any adequate conception of the mighty whole. There is but one Being whom " minuteness cannot perplex, nor magnitude encumber ;" but one ineffable intellect that can at once take in the infinity of the universe, and every minutest atom in the vast creation. That intellect which comprehendeth all things, because it containeth all things. But the intellect of man must proceed in succession, as it is incapable of receiving the whole, and in generalization and abstraction, as it cannot retain all the parts. Systems are valuable, only in so far as they further these intellectual operations, as they assist comprehension and retention. On this principle then, the undue multiplication of divisions and sub-divisions is highly objectionable. It requires almost as great an effort of memory to retain the divisions, as it would to retain the things divided.

But a fondness of system-making and reforming is the grand vice of superior intellects. It is a fault arising out of that power which distinguishes the human above every other animal, and points out its divine affinity. Man, in his way, would be a creator, like him in whose image he exists—nay, he is a creator, and a wonderful one. But as all human works and systems must be imperfect, and as it is the tendency of superior minds to aim at perfection, so when the work of innovation once begins, it is apt to be carried to an extreme. Thus, when men threw off their allegiance to the hierarchy of Rome, the division of sects became endless, and each sect pronounced its own way to be the exclusive path to salvation. They agreed only in abusing the standard they had deserted and reviling one another ; and many conscientious men, puzzled as to choice of faiths, and tired of vacillation, were glad to return and repose themselves on the bosom of a supposed infallible mother. It has been with philosophy and science as with religion. The chair of Aristotle, like that of St. Peter, long thundered forth its dictates to the earned world. The illustrious Linnè was long the Pope of naturalists, and if one man ever deserved such an honour he was that man. But when the infallibility of these mighty spirits was once questioned and shaken, theories after theo-ries arose in quick succession, and fell with the like ra-pidity. Many shone for a time as guiding stars in the intellectual hemisphere, that now, shorn of their beams, are esteemed to have been but delusive meteors.

" Suns sunk on suns, and systems systems crush'd."

In zoological arrangement, however, the safest way is the golden mean. System should certainly be as near to nature as possible, but in endeavouring to approximate it too closely, we introduce complication and perplexity. The grand object, the facilitation of knowledge, is defeated ; the end is altogether frustrated by too fastidious an atten-tion to the means.

The number of species and of sub-divisions enumerated in the Table appended to this order, while they illustrate some of the observations previously made in the present essay, may appear at the same time to militate against opinions therein expressed. We may be said to deprecate the undue multiplications of technicalities in the text, and to make use of all that have been hitherto employed in the Table.

But we beg to remind the reader, that it is by no means intended to adopt the species or divisions, because they are inserted. Far be it from us, unconditionally to criticise the zoological labours of others, or judicially to sanction these on the one hand, or reject those on the other. The object intended is merely to present the reader with a re-pertory for all that has been written on the subject, or at least for the original descriptions of species designated under their several different names. Occasionally, and for the sake of brevity, an observation, subjoined to the description of a species, will be found to supply the place of a sub-divison of such species as contemplated by its describer.

The compilation of the Table involves matter of discretion, principally in the application of the synonyma; though this branch of the subject has been greatly facilitated in the Mammalia by the previous labours of M. Desmarest. In this we can only hope for a reasonable portion of success. Accuracy cannot be expected, when descriptions alone are to be consulted, and not the thing described.

END OF PART VI.

FOURTH ORDER

MAMMALIA.

THE RODENTIA.

WE have just observed in the Phalangers canine teeth so very small, that they can scarcely be considered as such. Accordingly the nutriment of these animals is for the most part vegetable. Their intestines are long, and they have an ample cæcum ; and the Kanguroos, which have no canines at all, subsist entirely on an herbivorous diet.

The Phascolomys might with propriety commence the series of animals of which we are about to speak, and which have a system of mastication still less perfect.

Two large incisors in each jaw, separated from the molars by a wide space, cannot easily seize a living prey, nor tear flesh. They cannot even cut the food, but they serve to file it down, to reduce it by continual labour into fine molecules, in a word to *gnaw* it. From this comes the name of *Rodentia,* given to this order from *rodo,* to gnaw, in French, *rongeurs,* from the verb *ronger,* which bears the same signification. Thus it is that they attack with success the hardest substances, and frequently feed on wood and the bark of trees. The better to fulfil this object, these incisors have no enamel except in front, so that their posterior edge wears out sooner than the anterior. Their prismatic form causes them to grow from the root in proportion as they wear from the edge, and this disposition to grow is so powerful, that if one of them is lost or broken, the opposite one having nothing any longer to comminute, becomes most prodigiously developed. The lower jaw is articulated by a longitudinal condyle, so that it has no horizontal motion, except from rear to front, and *vice versâ,* as is suitable for the action of gnawing. The molars, likewise, have flat coronals, the enamelled eminences of which are always transversal, to be in opposition to the horizontal motion of the jaw, and better answer the purposes of trituration.

Those genera, in which these eminences are simple lines, and where the coronal is perfectly plane, are more exclusively frugivorous. Those which have the eminences of their teeth divided into smooth tubercles, are omnivorous. Finally, the small number of those which have points, attack other animals more readily, and approach a little to the Carnassiers.

The form of body in the Rodentia is generally such, as that the hinder part of the body and limbs exceed the front, so that they may be said to leap

rather than walk. In some sub-genera, this conformation is even as disproportionate as in the Kanguroos.

The intestines of the Rodentia are very long,
their stomach is simple, or very little divided, and
the cæcum is often extremely voluminous, even
more so than the stomach; nevertheless, the sub-
genus *Myoxus* is without this intestine.

In this entire order, the brain is almost entirely
smooth, and without circumvolutions. The orbits
are not separated from the *fossæ temporales*, which
have not much depth. The eyes are altogether directed laterally ; the zygomatic arches being slight
and curved below, indicate the weakness of the
jaws, The fore-arms are scarcely capable of turning, and their two bones are frequently united. In
a word, the inferiority of these animals is manifest
in most of the details of their organization.

However, the more numerous genera which possess stronger clavicles, have a considerable degree
of dexterity, and make use of their fore-feet to carry
their aliments to the mouth. Of them we shall
form our first division.

The most remarkable genus of this division is
that of

The CASTORS *(Castor, L.),*

Which are distinguished from all the other Rodentia by their tail, which is horizontally flattened,
of a form nearly oval, and covered with scales.

They have five toes on every foot ; the hinder ones are united by membranes, and there is a double and oblique claw on that one which follows the thumb ; their cheek-teeth, four in number, and with flat coronals, are so formed as to show one sloping edge at their internal extremity, and three at their external in the upper row, and exactly the reverse in the lower.

The Castors are tolerably large animals, and their life is altogether aquatic. They use equally their feet and tail in swimming. As they live principally on the bark of trees and other hard substances, their incisors are very powerful, and shoot up vigorously from the root in proportion as they become worn in front. They also make use of these teeth in cutting trees of every kind.

Large glandulous pouches which border on the prepuce, produce a pommade of a powerful odour, used in medicine under the name of *castoreum*. In the two sexes, the organs of generation border on the extremity of the rectum, so that there is but a single external aperture.

The Castor of Canada (Castor Fiber), Buff. VII. xxxvi.

Is larger than the Badger. Of all quadrupeds it exhibits most industry in the fabrication of its dwelling. These animals associate together for this work in the most solitary parts of North America.

The Beavers generally choose the deepest waters, which are not likely to be frozen to the bottom, and when they can, they prefer running waters, so that the wood which they cut above may be carried downwards by the course of the stream to where they please. They keep the waters at an equal height by dams made of various branches of trees, mixed with stones and clay, which they strengthen every year, and which concludes at last by germinating, and forming a complete hedge. Each hut serves for two or three families, and has two floors; the upper is dry for the animals; the lower, under water, for the provisions of bark, &c. This last alone is open, and the entrance is under the water, without any communication with the land. These huts are made of branches twined together and cemented with mud. The Beavers have, besides, several burrows along the bank where they take refuge when their huts are attacked. It is only during the winter they make use of these habitations. In summer they are dispersed, and live solitarily.

The Beaver is easily tamed, and accustomed to live on animal substances. The Castor, or Beaver of Canada, is of a uniform reddish-brown. Its fur, as is well known, is in great request for hatting. Some Beavers are of a flaxen colour, some black, and some even white.

We have been unable to ascertain, after the most scrupulous comparisons, if the Castors or Beavers which burrow along the Rhone, the Danube, and the Weser, are different in species from those of North America, or if they are prevented from building by the vicinity of man.

The RATS, *(Mus, L.)*

Linnæus and Pallas seem to have united in a single group, under this appellation, all the Rodentia provided with clavicles, which have not been distinguished by any very obvious external mark.   From this it results, that it is next to impossible to assign them any common character, unless we have recourse to that of the lower pointed incisors, suggested by the first of these naturalists.   Still, it will be necessary, for the sake of precision, to separate, as we do, the *Rat-Mole,* and the *Helamys* or *Pedetes.*   The other Rats may then be properly subdivided by the grinders, into many sub-genera, which may all be reunited into three small groups.

1. Those which have prismatic molars, or with flat coronals, and crossed to the full extent of their height by plates of enamel, a structure which we again discover in Hares and some other animals, and is observable even in the Elephant.   We shall apply to them the generic name of

CAMPAGNOLS, *Cuv. (Arvicola.)*

All of these which are known have three cheek-teeth in each jaw, and each of them are formed of five or six, and sometimes of eight triangular prisms, placed alternately on two lines.

The first subdivision comprehends,

The ONDATRAS *(Fiber, Cuv.),*

Or Campagnols, with palmate feet, and long

compressed and scaly tail; of which, but one species is correctly known.

The Ondatra, or Musk-Rat of Canada, (Castor Zibeticus, Lin. Mus Zibeticus, Gm.) Buff. X. 1.

Of the size of a Rabbit, reddish-gray. They construct in winter, on the ice, a hut of clay, where they inhabit in great numbers, proceeding through a hole, to seek at the bottom the roots *acorus*, on which they subsist. When the ice closes their holes, they are reduced to feed upon each other. This custom of building has induced some writers to class the Ondatras with the Beavers.

The second sub-division is that of the

COMMON CAMPAGNOLS, *(Arvicola, Lacep. Hypudæus, Illig.)*

Which have the tail covered with hair, and nearly of the length of the body.

The Water-Rat (Mus Amphibius), Buff. VII. 11.

Somewhat larger than the common Rat, of a deep grayish-brown; tail as long as the body. It inhabits the banks of streams, and digs in marshy grounds to look for roots. It is but an indifferent swimmer and diver *.

---

* The *Mus Terrestris*, Lin. and the *Schermauss* of Hermann, are only Water-Rats.

The Campagnol, or Small Field-Rat (Mus Arvalis, Lin.),
Buff. VII. XLVII.

As large as a Mouse, of an ashen-red; the tail somewhat less than the body. It lives in holes which it digs in the fields, and where it heaps up grain for the winter. These animals sometimes multiply excessively, and occasion great mischief.

The Meadow Campagnol˙ (Mus Œconomus, Pall.), Glires,
XIV. A., Schreb. Cuv.

A little deeper colour, and tail a little shorter. It inhabits a little chamber formed like an oven, dug under the turf, from whence many narrow and branching canals conduct it in various directions; other canals communicate with a second cavity, where it amasses provision. It inhabits the whole of Siberia. It is also believed to have been found in Switzerland and the South of France †.

The third sub-division will be that of

The LEMMINGS, *Cuv. (Georychus, Illig.* ‡*)*

Which have the tail and ears very short, and the fore-toes peculiarly adapted for digging.

The two first species have five claws very distinct

---

† Here might properly come the *M. Saxatilis, Alliarius, Rutilus, Gregalis,* and *Socialis.* (Pall. Glir.) But the *M. Lagurus* and *Torquatus* are rather Lemmings.

‡ Γεωρυχος, digging the earth.

on the fore-feet, like the Rat-Moles, and Leaping-Hares.

The Lemming (Mus Lemmus, Lin.), Pall. Glir. XII. A., B.,
Schreb. cxcxv.

A northern species, of the size of a Rat, with fur diversified with yellow and black. Very celebrated for the migrations which they make from time to time, without any fixed period, and in innumerable troops. They are said to march in a right line, neither river, mountain, nor any other obstacle impeding their passage ; while they carry devastation into all the lands through which they pass. Their usual habitat seems to be the shores of the Icy Sea.

The Zocor *(Mus Aspalax, Gm.), Pall. Glir. X. Schreb.* ccv.

Reddish-gray ; the three middle claws of the fore-feet long, crooked, compressed, and trenchant, for cutting the soil and roots. The limbs short; scarcely any tail; and the eyes excessively small. It is a native of Siberia, where it always lives underground like the Moles and Rat-Moles, and subsists principally on liliaceous bulbs of various kinds.

The third species, like all the other animals, comprised under the grand genus of Rats, has but the rudiment of a thumb on the fore-foot. This is

The Lemming of Hudson's Bay (Mus Hudsonius, Gm.),
Schreb. cxcvi.

Of a clear pearly ash-colour, without tail or

external ears. The two middle toes on the
fore-feet of the male have the appearance of
possessing double nails, because the skin at the
end of the toe is callous, and projects under the
nail. This is a conformation nowhere to be
met with but in this animal. It is as large as a
Rat, and lives underground in North America.
2d. Come the Rats, whose cheek-teeth are divided
from their base into roots, but the flat coronals of
which present transverse and projecting lines. They
are also particularly frugivorous. We reckon two
sub-genera of them.

The Echimys (Echimys Geoffr.), Loncheres, Illig.

Have four cheek-teeth throughout, each of which
in the lower range present four transverse laminæ
joined two by two. In the upper there are but three,
of which two are thus united :—They are American
animals, which with a form pretty similar to that of
our Rats, often have hairs flattened, wide, stiff, and
terminating in a point, forming in fact genuine flat
spines, after the manner of a sword-blade.

*The Echimys with golden tail, Lerot with golden tail, of
Buffon, (Hystrix, Chrysuros, Schreb.) Buff. Supp. VII.*
LXXII.

Nearly as large as a Rabbit, of a marron-brown,
with white belly, a crest of elongated hairs, and
a white longitudinal band on the head. The
tail is long and black, but its lower moiety is
yellow. A native of Guiana.

The Red Echimys (Prickly Rat of D'Azara), Voy. pl. xiii.

Of the size of a Rat; reddish-gray; tail less than the body. Found in Cayenne and Paraguay. It digs long subterranean passages.

The DORMICE *(Myoxus, Gm.)*

Have also four cheek-teeth, everywhere divided by transverse bands; but the hair is soft; and the tail covered, and even tufted. They live in trees, subsist on fruits, and pass the winter in our climates in a lethargic sleep. In the numerous order of the Rodentia, this is the only genus that is destitute of the cæcum. In France we possess three species.

The Dormouse (Le Loir), Mus Glis, Lin. Buff. VIII. xxiv.

As large as a Rat, a gray-brown ash-colour above; whitish underneath; a deeper brown round the eye; strong mustachios; tail well furnished with hair for its entire length, and nearly disposed like that of a Squirrel. A native of the forests of the South of Europe. This is probably the Rat which the ancients fattened and esteemed a delicacy.

The Lerot (M. Nitela, Gm.), Buff. VII. xxv.

Somewhat less than the preceding; gray-brown above; whitish underneath; a black circle round the eye, which predominates and enlarges towards the shoulder. The tail tufted only at the end, black, with its extremity white.

The Muscardin (Mus Avellanarius), Buff. VII. xxvi.

Of the size of a Mouse ; a cinnamon red above;
whitish underneath ; the hairs of the tail somewhat disposed like the barb of a quill.

3d. The Rats, whose cheek-teeth, more or less
tuberculous, do not so clearly exhibit the transverse
furrows. They are more omnivorous than the others.
Their sub-genera are more numerous.

The Hydromys *(Geoff. An. du Mus tom. VI. page 86 and
following),*

Are distinguished above all the other Rats by their
hind-feet, palmate two-thirds ; their molars have
also a peculiar character, inasmuch as their coronals, obliquely quadrangular, are hollowed in the
middle like a spoon. They are aquatic.

Individuals of this tribe have been sent from
Guiana, some with a white belly, others with a fawn-
coloured, but all deep-brown on the upper part ; the
tail long, black at the base, and white in the posterior
moiety. They are sometimes double the size of the
Surmulot, *(Hydromys leucogaster et H. Chrysogaster,
Geoff.)*

It is also thought that we may refer to this genus,
an animal of North America, whose skin is abundantly used by hatters in Europe, the characters of
which, however, have not yet been examined by
anatomists. This is

The QUOUIYA *of d'Azara (Mus Coypus), Molin et Gm.*

It lives in burrows, on the banks of rivers, in a considerable part of South America. It approaches the Cabiai in size, and resembles it in the colour of the hair, but is distinguished by the fineness of this hair, and above all by the down which constitutes its basis; by its long tail, the number of its toes, *&c.*

The RATS, *properly so called (Mus, Cuv.),*

Have, throughout, three molars, with blunt tubercles, the anterior of which is the largest. Their tail is long and scaly. These species are very hurtful from their fecundity, and the voracity with which they gnaw and devour substances of all kinds. There are three kinds very common in houses, *viz :*

The Mouse (Mus Musculus, Lin.), Buff. VII. xxxix.

Universally known in all times and countries.

The Common Rat (Mus Rattus, Lin.), Buff. VII. xxxvi.

Of which the ancients have not spoken, and which appears to have penetrated into Europe during the middle ages. It is more than double the Mouse in all its dimensions. The skin is blackish.

The Surmulot, or Brown Rat (Mus decumanus, Pall.),
Buff. VIII. xxvii.

Which arrived in Europe only in the eighteenth century, and is at the present day more common than the Rat in Paris, and some other large cities. It is larger than the Rat by one quarter, and differs still more from it by its reddish-brown hair and tail longer in proportion.

These two large species appear to be aboriginal in the East. Like the Mouse, they appear to have been transported into every region by ships.

Oriental Tartary and China have a Rat equal in size to the Surmulot, with a tail a little shorter, stronger jaws; of a flaxen tint. This is the *M. Caraco* of Pallas, Glir. XXIII. Schreb. clxxvii.

There is also another species in the Indies, about one-fourth stronger than the Surmulot; reddish-brown, a little paler in the head, (The *Rat Perchal* of Buffon, Supp. VII. lxix.)

Those species of the size of the Mouse have been less observed.

The Mouse of Cairo (Mus Cahirinus, Geoff. Disc. de l'Eg.
Mammif.),

Has spines instead of hairs on the back. Aristotle has already remarked it.

We know in France but of one species that lives remote from houses. This is the *Mulot* (Field Mouse) *Mus Sylvaticus,* Buff. VIII. xli., which is but little larger than the common Mouse, and is only distinguished by the red colour of the skin. It devastates the woods and fields*.

The HAMSTERS, *(Cricetus, Cuv.)*

Have the same kind of teeth as the Rats, but their tail is short and covered, and the two sides of the mouth are hollowed, as in certain Monkeys, into sacs or cheek-pouches, which serve to transport the grains which they gather, into their subterranean abode.

The Common Hamster, Marmot of Germany, (M. Cricetus, L.) Buff. XIII. xiv.

Is larger than the Rat; reddish-gray above; black in the flanks and underneath, with three white spots on each side; its fore-feet are white; and there is also a white spot under the throat, and another under the chest. There are some individuals entirely black. This animal, so agreeably varied in colour, is one of the most hurtful which exist, in consequence of the quantity of grains which it gathers, and with which it fills its holes, which are at times nearly

---

* To this division probably appertain, *M. Agrarius, M. Minutus, M. Soricinus, M. Vagus, M. Betulinus, M. Pumilio, M. Striatus, M. Barbarus.* We cannot as yet well classify either *M. Pilorides,* or any of the Rats pointed out, rather than described by Molina, for they are not sufficiently well-known.

Here also should come the enormous species, *M. Giganteus,* Lin. Trans. VII. xviii.

seven feet in depth. It is common in all the sandy regions which extend from the North of Germany as far as Siberia.

This last country produces many small species of the Hamster, discovered and published by M. Pallas. Such as M. *Accedula, Arenarius, Phœus, Songarus,* Furunculus, Pall. Glir. et Schreb.

One of the most extraordinary species, if it were completely authenticated, would be the *Mus Bursarius* of Shaw, aboriginal in Canada, ash-coloured, whose pouches, when they are full, would jut out from the two sides of the mouth, and surpass the head in size. It is said to have five claws in front, of which the middle three are very long, and fit for digging; there are four behind; the tail is short, and its size approaches that of the Surmulot.

The JERBOA *(Dipus, Gm.)*

Have the same teeth as the Rats; a long tail tufted at the end; a large head; large projecting eyes; and, above all, the posterior extremities of a disproportioned size in comparison of those in front: this circumstance caused them to be named two-legged Rats by the ancients. In fact, they proceed in general by long leaps on their hinder feet; the fore-feet have five toes; in the hinder, the metatarsus of the three middle toes, is formed only of a single bone, like that which is called the tarsus in birds. There are moreover in certain species small lateral toes* They live in burrows, and fall into a profound lethargy during the winter.

The Gerboa, M. Sagitta, Buff. Supp. VI. xxxix. *et* xi.

Has only three toes; is of the size of a Rat; a clear fawn-colour above, white below; the tufts of the tail black, the end white. Inhabits from Barbary to the North of the Caspian.

The Alactaga, (M. Jaculus.)

Has two small lateral toes; the ears longer than the preceding, but pretty nearly the same colour. M. Pallas has observed three sizes, from that of a Rabbit to that of a Rat. They are perhaps so many species. They are all found from Syria as far as the Eastern Ocean, and the North of India.

We are obliged to separate from the Rats, and to establish altogether as genera, the three following:

The RAT-MOLES, *(Spalax, Guld.)*

Have the same cheek-teeth as the Rats, the Ham-

---

* The *Mus Longipes* of Lin. or *Meridianus* of Pallas, seems to form a new subgenus. The *Tamaracinus* will probably be united to it, if it be not a Dormouse. We have seen neither. It is also probable that we must refer to it, the *Gerbillus* of Olivier. The *M. Canadensis* of Pennant and Shaw, and the *Dipus indicus*, Lin. Trans. VIII. vii. These are the Gerbillus of Desmarest, and Meriones of Illiger.

sters, and the Gerboas, but their incisors are too
large to be covered by the lips. The extremities
of the lower incisors are formed corner-wise ; *i. e.*
trenchant, rectilinear, transverse, and not in a point.
All their feet have five short toes, and five flat and
slender nails; their tail is very short, or rather nothing, as well as their external ear ; they live under
ground, and dig there like Moles, but with much
less powerful instruments, raising the earth like
them, but subsisting only on roots; their eyes are
also extremely small.

The Zanni Slepez, or Blind Rat-Mole, (M. Typhlus, L.)
Pall. Glir. VII. Schreb. CCVI.

Have not even a visible eye externally, but
when the skin is raised, we find a very small
black point, which is organized like an eye,
without serving to the use of vision, since the
skin passes over without opening or even growing thinner, and without having less hairs upon
it than in any other point. This singular animal
has moreover an altogether deformed appearance, from its thick head, angular on the sides,
its short legs, and no tail. It is about the size
of our Rat; of an ash-colour, bordering upon
red. It inhabits all the east of Europe, and
the neighbouring parts of Asia, as far as Persia.
It may be, according to M. Olivier, that the
ancients derived the notion of the Mole being
quite blind from this animal.

The RAT-MOLES *of the Cape,* (*Orycteré*, *F. Cuvier*, *Bath-yergus*, *Illig.*)

With the form, the feet, and the truncated incisors of the preceding species, have four cheek-teeth throughout, and the lower ones profoundly indented on the external side. Their eye is, although small, yet visible, and they have a short tail. Two species are known.

The Maritime Rat-Mole, (*Mus Maritimus, L.*) *Buff. Sup. VI.* xxxviii.

Of a whitish-gray, almost the size of a rabbit, and

The Small Rat-Mole, of the Cape, (*M. Capensis,*) *Buff. Supp. XI.* xxxvi.

Brown; a spot around the eye; one around the ear; and one on the vertex and end of a white muzzle; about as large as a Guinea-Pig. Both these are common in the neighbourhood of the Cape of Good Hope, and they burrow to such an extent in the earth, that it becomes dangerous to ride in these parts *.

* M. Illiger separates the *M. Capensis* from the BATHYERGUS, or M. Maritimus, to place it with the *M. Hudsonius,* and the *Aspalax,* or his own GEORYCHUS; but the conformation of the *Mus Capensis* is absolutely the same as that of the *M. Maritimus,* which we have ourselves authenticated.

H 2

The HELAMYS, *F. Cuv.* (*Pedetes, Illig.*)

Which have been hitherto classed with the Gerboas, resemble them in fact, in the large head, large eyes, long tail, and more especially by the smallness of the fore extremities and the length of the hinder, although the disproportion is much less than in the true Gerboas. The peculiar character of the Helamys, are four cheek-teeth throughout, composed each of two laminæ; five toes on the fore feet, armed with very long and pointed nails, and four upon their large hinder feet, all distinct, even in the bones of the metatarsus, and terminated by large nails, almost resembling hoofs. The number of toes is the inverse of that most general among the rats. Their lower incisors are truncated, and not pointed like those of the true Gerboas, and all other animals comprised under the Rat genus, the Rat-Moles alone excepted.

There is but one species known, belonging to the Cape of Good Hope, as large as a Rabbit, of a clear fawn-colour, tufted tail, black at the end, (*Mus Cafer*, Pall., *Dipus Cafer*, Gm.) *Buff. Supp. VI.* XLI.

Gmelin has already separated from the genus of Rats,

The MARMOTTES, (ARCTOMYS, *Gm.*)

Which with the pointed lower incisors of the other animals comprehended in this great genus, have five cheek-teeth on each side above, and four below; bristling with points; accordingly some species are easily induced to feed on flesh, and will eat insects as well as vegetables. They are short-legged animals, with a tail covered, and rather short; a large and flatted head. They pass the winter in a state of lethargy, in deep holes, the entrance of which they close up with hay. They are gregarious, and easily tamed. These species are known in the Old Continent.

The Marmot of the Alps, (M. Alpinus, L.) Buff. VIII. XXVIII.

As large as a Rabbit, with short tail, and yellowish gray fur, with ash-coloured tints towards the head. It lives in high mountains, immediately under perpetual snows.

The Marmot of Poland, or Bobac, (M. Bobac, Lin.) Pall. Glir. V. Schreb. ccix.

Of the size of the preceding; yellowish-gray, with red tints towards the head. Inhabits mountains slightly elevated, and small hills, from Poland as far as Kamtschatka, and often burrows in the hardest soils.

The Souslik, or Zikel, (M. Citillus, L.) Buff. Supp. III. XXXI.

A pretty little animal, of a grayish-brown, waved or spotted with white; is found from Bohemia as far as Siberia. It has a particular taste for flesh, and does not even spare its own species. There are also some species in America.

The Squirrels, *(Sciurus, L.)*

Which have always been regarded as a separate genus, are recognised by the lower incisors being extremely compressed, and by their long tail, furnished with long and scattered hairs, directed sideways like the barbs of quills. They have four toes in front, and five behind, sometimes the fore-thumb is marked by a tubercle. They have four tuberculous cheek-teeth throughout, and one very small one in front above, which falls very soon. They are light and lively animals, living in trees, nestling there, and subsisting on fruits. The head is large, and the eyes projecting and lively.

There are many species of them in the two continents.

The Common Squirrel, (Sciurus Vulgaris,) Buff. VII. XXXII.

Of a lively red, the ears terminated by a cluster of hairs. Those of the North become of a beautiful bluish ash-colour in winter, and then produce the fur called *minever.* There are also brown and black varieties.

The American species have not the brush of hair at the ears.

The Gray Squirrel of Carolina, (*Sciurus Cinereus, Lin*)
Petit-Gris. de Buff. X. xxv.

Larger than any ; ash-coloured ; white belly.

The Masked Squirrel of the same Country, (*Sci. Capistratus.*) *Basc. Sc. cinereus. Schreb.* ccxiii.

Ash colour, black head ; muzzle, ears, and belly, white. Both these species vary by a greater
quantity of black or white, and become at times
altogether black.

Most of the species of the Old Continent are
also destitute of these brushes. One of the handsomest is,

The Large Squirrel of India, (*Sc. Maximus et Macrourus,
Gm.*) *Buff. Supp. VII.* lxxii.

Almost as large as a Cat; black above; the
flanks and top of the head of a beautiful lively
marron. The head, all the under part of the
body, and inside of the limbs, pale yellow ; a
marron band behind the cheek. It lives on
palm-trees, and is particularly fond of the
milky juice of the cocoa nut.

There are also, in the warm climates, some Squirrels remarkable for the longitudinal bands by which
their fur is varied. Such are

The Barbaresque (*Sc. Getulus, L.*) *Buff. X.* xxvii.

The bands of which extend as far as the tail.

The Palmist, (Sc. Palmarum, L.) Buff. X. xxvi.

It is probable that we should distinguish from the Squirrels certain species which have cheek-pouches, like the Hamsters, and which pass their lives in subterranean holes. (*Tamias*, Illig.) Such is

The Ground Squirrel, (Sc. Striatus, L.) Buff. X. xxviii.

Which is found in the north of Asia and America throughout, especially in the pine forests. Its tail is less furnished with hair than that of the European Squirrel; its ears smooth, and its skin brown, with five black streaks and two white ones.

The Squirrel of Hudson's Bay, (Sc. Hudsonius,) Schreb. ccxiv.

With brown red skin, with a single black streak on each side. Resembles the last closely.

It is probable that the *Guerlinguets,* South American species, should be separated. They are distinguished by a long tail, nearly round, with an enormous and pendant scrotum. Buff. Sup. VII. lxv. lxvi. Their molars, however, are the same as those of the Squirrels and Polatouches: so are those of the Tamias.

Naturalists have already separated the

POLATOUCHES, (*Pteromys, Cuv.*)

To which the skin of their sides, extending between

the fore and hind limbs, imparts the faculty of sustaining themselves in the air for some moments, and of making very great leaps. Their feet have long osseous appendages, which sustain a part of the lateral membrane.

There is one species in Poland, Russia, and Siberia.

Flying Squirrel, (Sciurus volans.) Schreb. ccxxiii.

Ash-coloured above ; white underneath ; large as a Rat ; the tail about half the length of the body. It lives solitarily in the forests.
There is one of North America.

Sc. Voluccella, Buff. X. xxi.

Reddish-gray above ; white underneath ; less than the preceding, with a tail a quarter less than the body. It lives in troops in the temperate prairies of North America. One in the Archipelago of the Indies, nearly as large as a Cat. The male of a beautiful lively marron above, red underneath ; the female brown above, whitish underneath. This is *Sc. Petaurista Taguan.* Buff. Supp. III. xxi. et VII. LXVII.

But this same Archipelago produces also a small species.

Sc. Sagitta.

Deep brown above; white underneath ; which

is distinguished from all the other small species, because its membrane forms, as well as that of the Taguan, a projecting angle extremely acute behind the wrist.

Lastly, M. Geoffroy has also with justice separated from this genus

The AYE-AYE, *Geoffr.* (*Cheiromys, Cuv.*)

In which the lower incisors being still more compressed, and above all more extended from front to rear than in the Squirrels, resemble the sock of a plough; their feet have all five toes, four of which in front are excessively elongated, and in this number the medius is much more slender than the others. In the hind feet the thumb is opposable to the other toes; so that in this respect, among the Rodentia, they are what the Sarigues are among the Carnassiers.

But one species of the Aye-Aye is known, and was discovered at Madagascar by Sonnerat.

Sciurus Madagascariensis, Gm. Buff. Sup. VII. LXVIII.

As large as a Hare; of a brown, mixed with yellow, with a long and thick tail, furnished with thick black bristles; the ears perfectly naked. It is a nocturnal animal, whose movements are painful, and which burrows under ground. It makes use of its slender toe to carry the aliments to its mouth.

The second division of Rodentia comprehends the genera which have only the rudiments of clavicles. The most easy to distinguish is that of

The PORCUPINES, (*Hystrix, Lin.*)

Which are known at the first glance by the rough and sharp spines with which they are armed, like the Hedgehogs among the Carnassiers. The Porcupines have four cheek-teeth in each jaw, on both sides, which are cylindrical, and marked on the crowns with four or five deep impressions. Their tongue is covered with spiny scales; they have four toes before, and five behind, armed with thick nails. They live in burrows, and have many of the habits of Rabbits. Their grunting voice, together with their thick and truncated muzzle, have caused them to be compared to Pigs.

The Common or Maned Porcupine, (Hist. Cristata, L.)
Buff. XII., LI., LII.

Is larger than a Hare, the spines are very long and very strong on the back ; there is a mane of long hair on the head and neck; the tail is short, terminated by two open tubes carried on pedestals, and which make a noise when the animal shakes them. Of Italy, Greece, Barbary, and even the East Indies.

*The Prehensile-Tail Porcupine, (Hist. Prehensilis, L.) Cuendu Marg. Hoitzslaquatzin, Herm.**

Has a long and prehensile tail, without spines on the posterior half, and the spines short everywhere. Of the warm parts of America, where it often suspends itself in trees.

The Pencillated-Tail Porcupine, (Hist. Fasciculata, L.)

With a long tail, terminated by a bundle of spines flatted like strips of parchment; the spines of the body flatted like a sword-blade. Of India, beyond the Ganges†.

The Hairy Porcupine, (Hist. Dorsata, L.) Urson, Buff. XII. iv.

Has a moderate tail; the spines in a great measure hidden under the fur. Of North America.

The HARES, *(Lepus, Lin.)*

Have also a very distinctive character in their upper incisors, which are double, that is to say, each of them has a smaller one behind it. Their molars,

---

* This word implies, in the Mexican language, Spiny Opossum, because it has the prehensile tail of the Sarigue. It is the *Coendou à longue queue*, Buff. Sup. VII. pl. LXXVIII.

† This is the Malacca Porcupine, Buff. Sup. VII. LXXVII. The *Hystrix macroura*, Seb. I. pl. LII. and Schreb. CLXX., is very much like it, only the strips of the tail are represented as if formed of many convexities, resembling so many grains of rice.

five everywhere, are formed each of two vertical
laminæ stuck together, and in the upper jaw is found
a sixth, simple and very small.   They have five toes
before, and four behind ; an enormous cæcum five
or six times as large as the stomach, and furnished
within with a spiral lamina which runs along its
whole length.   The interior of their mouth, and the
under part of the feet have hair like the rest of the
body.

HARES, properly speaking, (Lepus, Cuv.)

Have long ears; a short tail ; the hind feet longest;
imperfect clavicles ; the space under the orbits
reticulated in the skeleton.

The species are numerous, and so similar to one
another as to make it difficult to characterize them.

The Common Hare, (Lepus Timidus, L.) Buff. VII.
XXXVIII.

Of a yellowish gray ; the ears one sixth longer
than the head, ashy behind, black at the point ;
tail as long as the thighs, white, with a black
line above.

Every body knows this animal, whose dark
flesh is good as food, and whose fur is useful.   It
lives isolated, does not burrow, sleeps on the
flat ground.   When hunted it describes a large
circle in running, and has never yet been domesticated.

The Variable Hare, (Lepus Variabilis, Pall.) Schreb.
CCXXXV. *B.*

Is a little larger than the common Hare, with the ears and tail a little shorter, white at all seasons ; the rest of the fur gray in summer, and white in winter. This animal, which is found in the north, and on the high mountains of the south of Europe, has the manners of the common Hare, but its flesh is insipid.

The Rabbit, (Lepus Cuniculus, L.) Buff. VI. L.

Less than the Hare ; the ears a little shorter than the head, and the tail less than the thigh ; fur yellowish gray, red about the neck, throat and belly whitish ; ears gray, without any black ; brown on the tail.

This animal, originally from Spain, is now spread throughout Europe. Lives in troops in burrows, to which it flies when pursued. Its flesh is white and pleasant, but differs considerably from that of the Hare. When domesticated, the Rabbit breeds infinitely, and is considerably varied as to colour and fur.

Foreign countries furnish many species, which are not distinguishable from our Rabbit but by close examination ; of these are

The Siberian Rabbit, (Lepus Tolai, Gm.) Schreb. CCXXXIV.

Which may be considered intermediate between the Hare and Rabbit by its proportions, and sometimes surpasses the former in size. Without digging burrows, it takes refuge in clefts of rocks or other cavities.

The American Rabbit, (Lepus Americanus et Brasiliensis, Gm.) Lepus nanus, Schreb. ccxxxiii. *B.*

About the size and colour of the European species, with reddish feet, without black either on the ears or tail. Nestles in the trunks of trees, and often climbs up in their clefts as far as the branches. Its flesh is insipid and soft.

Others have as strong a resemblance to our Hare, as

The African Hare, (Lepus Capensis, Gm.) Geoff. Quad. d'Egypt.

With the ears a fifth longer than the head; nearly of the size and colour of a Hare; the feet reddish, and a little longer. It appears to be found from one extremity of Africa to the other; at least, that of Egypt does not differ from that of the Cape.

The Lagomys, *Cuv. (Lagomys*.)*

Have the ears moderate; the legs but little differing from each other; the hole under the orbits simple; clavicles, nearly perfect, and without a tail; they often utter a very sharp voice. Hi-

---

* *Lagomys,* Rat-Hares.

therto, they have been found only in Siberia, and
it is Pallas who first made them known. (Glir.
pa. I. et sequa.)

The Dwarf Lagomys, (Lepus Pusillus.) Pall. Glir. I. Schreb.
CCXXXVII.

Gray-brown, as large as the Water-Rat. Lives
in small burrows, in fertile countries, on fruit
and buds.

The Gray Lagomys, (Lepus Alpinus,) Pall. Glir. II. Schreb.
CCXXXVIII.

Very pale gray, with yellowish feet, a little
larger than the preceding ; nestles in the holes
of stones, clefts of rocks, &c. where it collects
hay for winter.

The Lagomys Pica, (Lepus Alpinus,) Pal. Glir. II. Schreb.
CCXXXVIII.

As big as the Guinea-Pig ; yellowish-red ; inhabits the most elevated summits of mountains, where it passes the summer, selecting
and drying herbs for winter provision. Its heaps
of hay, sometimes six or seven feet high, are
valuable resources for the horses of the sable-
hunters.

After the two genera of Porcupines and Hares,
Linnæus and Pallas united the rest under the name
of Cavia, but it is impossible to find any other

common and positive character among them than that of their imperfect clavicles, although the species are not wanting in analogy as to habits and common manners. They are all of the New Continent.

The Cabiais, (Hydrochœrus, *Erxleben.)*

Have four toes before and three behind, armed with large nails and united by membranes; four cheek-teeth, of which the posterior are the longest, composed of numerous simple and parallel laminæ; the anterior of laminæ forked toward the external edge in the upper, and toward the internal in the lower teeth.

But one species is known.

The Capybara of Marg. Capiygoua of d'Azara. Cavia Capibara of Lin. Cabiai of Buff. XII. xlix.

As big as a Siam Pig, with a very thick muzzle; short legs; thick hair, yellowish-brown; without a tail. Inhabits in troops the rivers of Guiana and the Amazons. It is good game, and the largest of the Rodentia. The Castor only approaches it in size.

The Cobayes, *commonly called Guinea-Pigs,* Anœma, *F Cuv.* (Cavia, *Illig.)*

Represent the Cabiais in miniature; but their fingers are separated, and their molars have only each a simple lamina, and a fork on the outside in the upper jaw, and on the inside in the lower.

We know but of one species, Buff. VIII. i. now
much bred in Europe, where they are brought up
in houses, because their odour is thought to drive
away Rats.   They vary in colour, like all domesticated animals.   It seems probable that they proceed
from an animal of America, called *Aperea*, of the
same size and form, but with an uniform reddish-
gray fur.   They are found in the woods of Brazil
and Paraguay.

The AGOUTIS, *Cuv.* (Chloromys, *F. Cuv.* Dasyprocta, *Illig.*)

Have four toes before, three behind; four cheek-
teeth on each side in each jaw, nearly equal, with
flat irregularly ridged crowns, of a circular contour,
indented in the internal edge in the upper jaw, and
on the external of the lower.   They resemble in disposition, and in the nature of the flesh, our Hares
and Rabbits, which they may be said to represent,
in the Antilles and hot parts of America.

The Common Agouti, (Cavia Acuti, L.) Buff. VIII. L.

With the tail reduced to a simple tubercle ; fur
brown-yellow on the crupper in the male; as big
as a Hare.

The Acouchi, (Cavia Acuchi, Gm.) Buff. Sup. III. xxxvi.

The tail with six or seven vertebræ ; fur brown
above and yellow underneath, as big as a
Rabbit.

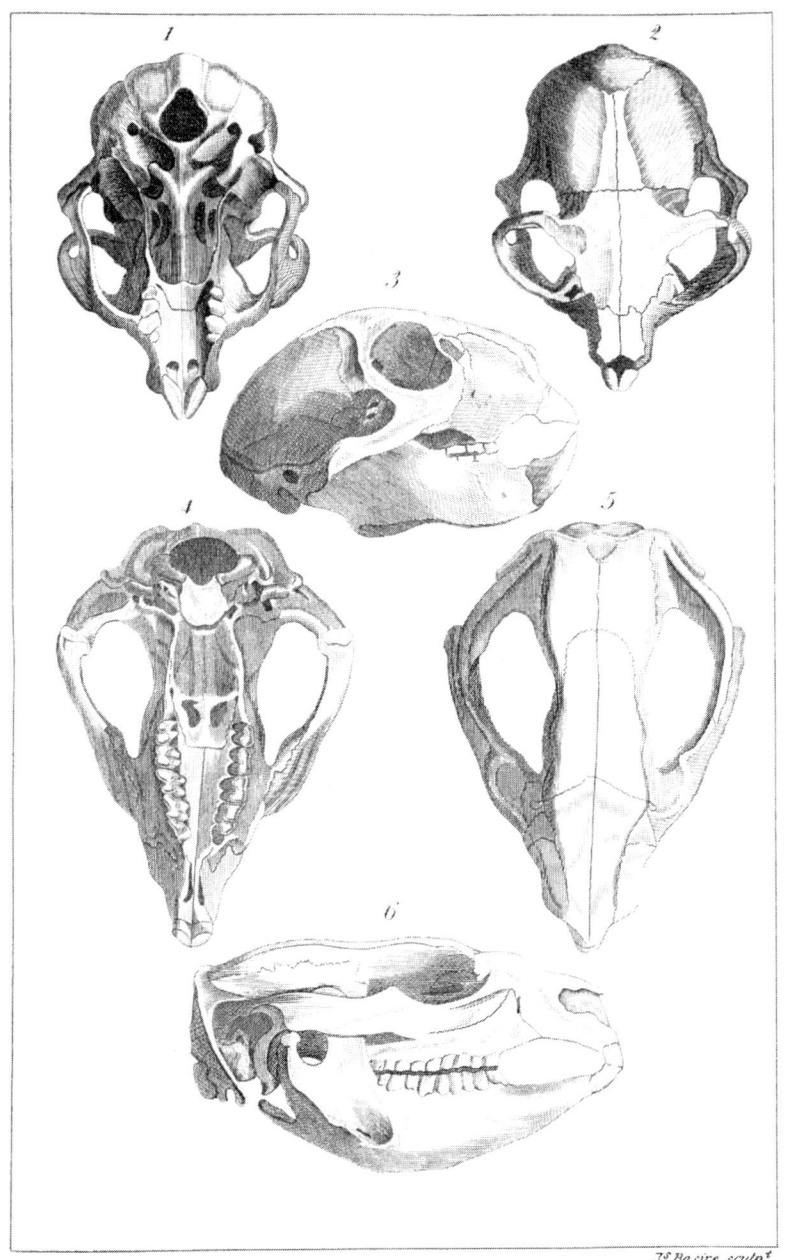

1

2

3

4

5

6

Js Basire sculpt

1.2.3. THE AYE AYE.   4.5.6. THE WOMBAT.

London Published by G.B.Whittaker. Sep. 1825.

The PACAS, (Cœlogenus, *F. Cuv.**)

Have, with the teeth similar to those of the Agouti, one very small toe more than they on the internal side of the fore-feet, and one on each side equally small ; on the hind-feet five toes on each foot. Moreover they have a cavity formed in the jaw, which deepens under a border, formed by a very large and salient zygomatic arch. It is said that their flesh is good food.

There is a species or variety yellow, and another brown, both spotted with white. (*Cavia Paca,* Lin.) Buff. X. XLIII. Sup. III. xxxv.

---

* *Anœma,* without force, *Chloromys,* Yellow-Rat, *Dasyprocta,* hairy buttocks, *Cœlogenus,* hollow-jawed, *Hydrochoerus,* Water-Hog.

I 2

SUPPLEMENT TO THE RODENTIA.

THE leading physical character which distinguishes the order Rodentia, and from which indeed it receives its name, is that of dentition, fitted for the operation of gnawing. The order has been called Rosores, by Stor, and Prensiculantia, by Illiger, and it corresponds with the Glires of Linnæus, so named from one of the genera included in it.

The Rodentia have two principal incisive teeth in each jaw, which are very large, long, and generally bent, with the anterior face in some species flat, in others, subcylindrical. These teeth being exposed by the habit of the animal to frequent and almost constant friction against substances often of considerable hardness, yield in time to this reiterated operation, and wear away ; but the Creator, ever willing to relax in those general laws subject to which his works are constituted, when the necessities of a particular race of creatures may require an exception, has provided that these teeth, as they are exposed to wear, shall be also capable of renovation.

The incisors when they spring from their alveoli are pointed, and grow from the lower part of their posterior side in proportion as they wear away above ; their anterior face is covered with a thick hard enamel, and as the detrition is always oblique, the teeth preserve, in the upper jaw at least, a constant sharp edge of enamel.

To supply matter for this continued detrition, the embedded portion of each incisor, which is not an indurated root, but a mere gelatinous pulp, is much enlarged, and extends in a curve above the upper and beneath the lower cheek-teeth, behind those of the cheek-teeth.   In the upper

Teeth of Rodentia.

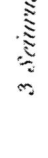

1 *Arctomys.*　　3 *Sciurus.*　　5 *Myoxus.*

2 *Castor.*　　4 *Arvicola.*　　6 *Mus.*

London. Published by G.B.Whittaker March 1.1827.

jaw they are in the maxillary, and not in the intermaxillary bone. This provision for renovation is confined to the incisive teeth.

In some species, as the hares and the pikas, the two large upper incisors have two other smaller, and placed behind them; the use of which to the animal is not apparent.

The lower incisors are sometimes moveable independently of each other, by the nonconsolidation of the symphysis of the lower jaw.

The molars undergo, as to their detrition, the same laws as in other animals; those that feed on vegetables alone, have their sutures deeply divided by small ridges, so that the upper surface of the teeth is flat, and traversed by lines or ridges of enamel, very various in their directions in the different genera. The species which feed on softer substances are less fitted for trituration by having a smoother surface.

In the Herbivorous Rodentia these teeth increase for a very long time, from their base, in proportion to their wear on their crowns; they do not become divided into roots till very late, and in some species it is not certain whether this process ever takes place; in others, again, these teeth cease to grow when the animal attains maturity, and the roots are formed very early.

The cheek-teeth vary in number, and still more in the shapes or directions of the enamel ridges in the different genera or divisions of the order; these will be noticed in their places as generic characters; to which we have added copies of the figures of these teeth given by the Baron in his Ossemens Fossiles, which will display their peculiarities much more effectually than verbal description.

We know very little of the laws and the periods of the succession of teeth of the Rodentia. Like other viviparous quadrupeds, many of them are known to have deciduous, or milk-teeth, and permanent teeth ; but whether the latter have had predecessors in all cases is not ascertained, owing to the very early period, if at all, at which the succession takes place.

The Baron has watched this operation of nature in the hare; the upper posterior incisors only were observed to change; the cheek-teeth continuing a long time, with their permanent successors, during which time the animal appeared to have six incisors above, instead of four, the proper number.

Of the Molars it is certain that there are three out of six in the upper jaw, and two out of five in the lower, which change ; the three posterior cheek-teeth, on both sides in each jaw, come but once, and in this respect the hares are like the horses, and the ruminating animals ; the upper surface of the posterior cheek-teeth accord with those which succeed the anterior deciduous cheek-teeth.

I rather think, says the Baron, that in the species which have but four cheek-teeth on each side, there is but one, the anterior, which is changed ; at least I am satisfied it is so with the Castor, the Porcupine, the Agouti, the Paca, and the Cavy ; but to examine the new tooth in this last species when quite new and unused, the animal must be taken very young ; and what is certainly very singular, in order to see this cheek-tooth in its place, the animal must be inspected some days before its birth. I have satisfied myself that this tooth in the Cochon d'Inde, falls while the fœtus is still in utero ; and by analogy; I suspect that it is the same with all the Rodentia, as to their incisive teeth. These, therefore, would no longer be milk-teeth, but uterine teeth.

In this instance it may be observed, and in many others we may find, though less forcibly, an apparent unwillingness in nature to depart from those general laws to which she submits creation. We are not always able to detect the object intended by the means employed, but when we do so in fact, or fancy we have fully done so, we may sometimes find the general means not dispensed with, in cases where the particular ends do not seem to be contemplated.

As those species which have more than three cheek-teeth, do not, in fact, change the posterior three, we may probably conclude that those which have only three in all never change either of them; accordingly, the Baron could never detect any mutation of cheek-teeth in these latter species, as the Rats, &c.

The Rodentia are totally destitute of canine teeth, and the space between the incisors and the cheek-teeth is void.

The lower jaw is articulated by a longitudinal condyle, so as to confine its movement to a horizontal motion forward and backward.

It seems almost needless to remark, that this dentary formation is in strict conformity with the habit proper to these animals, of dividing hard substances with the incisors, and triturating the pieces with the cheek-teeth.

The head of the Rodentia is flat above, the frontal bone is divided by a medial suture, the parietal is single in some, as the Marmots, Hare, Squirrels, &c.; and double in others, as the Rabbit, the Mouse, and the Dormouse, &c.; the opening of the nasal fossæ is vertical at the end of the muzzle, and is heart-shaped, with the point downward; the orbits are nearly round, and communicate with the temporal fossæ, which are very large. Some genera have perfect clavicles, and others have them only in a rudimentary state; they have the two bones of the fore-arm, but these are frequently attached to each other; the toes vary, and will be found in the generic characters; the posterior extremities are always larger than the anterior, and are sometimes extremely elongated.

The brain is small, and has but few circumvolutions; the eyes in some species are very large, in others, as the Mole

Rat, extremely small and imperfect; the eye pupil is in some species round, and in others elongated horizontally; the nostrils are pierced at the extremity of the muzzle, which is longer than the jaws; the tongue is soft; the ears are in some, as the Hares, very much developed, and in others, as the Mole Rat, altogether wanting. They are in general well covered with hair, except on the toes, extremity of the muzzle, and sometimes the tail, and a few have spines either round or flattened. Some, as the Squirrels, have the toes deeply divided; others, as the Hydromys and the Castor, are palmated, and some are capable of flight by means of lateral membranes extended from the anterior to the posterior extremities, and attached to the sides.

Their locomotion varies, but is most generally plantigrade. Some of them are rapid runners; others, the Gerboas and Gerbilli, great leapers; a few have subterraneous habits and physicalities, as the Mole Rat; and several of them, the Squirrels, &c. are fitted for living in trees. Their nails are proper to each of these varied modes of existence.

The tail, though apparently of inferior importance in the destinies of the species, varies considerably among the Rodentia. The Castor has it naked and flat, the Rat slightly ridged crosswise, the Ondatra compressed, the Squirrel feathery, and the Gerboa tufted at the end.

Some of the Rodentia are more decidedly herbivorous than others which approach the omnivorous character. Their intestines are generally long; the stomach simple, or only partially divided; the cæcum large, except in the Myoxus, a very carnivorous genus, which is totally destitute of this viscera.

Some of the Rodentia are exceedingly destructive to the industry and property of mankind, as the Hamsters, Field Mice, Rats, Mice, &c.; but others administer to our appetite or comfort, as, of the former description, the Hare, Rabbit,

Marmot, Squirrel, &c.; and, of the latter, by their fur, the Beaver, Hare, Rabbit, the Hamster, Chinchilla, Squirrel, &c.

The Rodentia are said in general to be but slightly endowed with intellect; the Beaver and the Rat, however, afford striking exceptions to this observation; for whatever distinction we may draw or invent between instinct and reason, many of the actions of these animals seem to partake considerably of both these characters.

The several genera of this order are spread almost over the whole habitable globe, and some species have become cosmopolite in following man in all his migrations. The Islands in the vicinity of New Holland have, however, hitherto only furnished two species.

The various genera which compose this order will be found not to depart very considerably from the common type. The order is a very natural one, especially when we view the animals which compose it in reference to their manner of breaking and dividing, or triturating their food; there are species and genera in it which approximate the characters of other orders, and the dentary system proper to this order seems first discernible among the Phalangers; it continues in the Kanguroos, and is nearly complete in the Wombat; but these Marsupiata have not the articulation of the lower jaw proper to this order.

This extensive and important order of the Rodentia, the Baron very appropriately commences with the *Castor*, or *Beaver*. There is no animal in the whole order which displays, within several degrees, an equal portion of intelligence. The instinct of sociability, that natural penchant which induces certain of the mammiferous tribes to communicate with each other, and even to enter into means of mutual defence is, doubtless, the result of innate dispositions, which entirely depend on organization, and are not acquired by experience. It is, nevertheless, a great error to deduce, from the same cause, the harmony which usually reigns among these animals; the sacrifice, if we may so express it, which we behold them make of private interest to public good, and their entire forgetfulness of that individual strength which would enable them to hurt one another. If they renounce the right of the strongest, and submit themselves to moral laws, and a kind of consciousness of duty, it proceeds alone from the influence which they exercise over each other, from the education which the young receive from the adult, at an age when they are forced to obey, and perpetually constrained to confine themselves to that circle which circumstances have presented to their association; — thus, themselves, their actions, and their wants, are finally placed under the inflexible control of habit.

An incontestable proof of this fact is, that these animals lose all their social qualities, from the moment that some powerful cause has isolated them from their fellows, and condemned them to live in a state of solitude. The dog himself, an animal both organized for society, and impelled to it by the powerful influence of domestication, is but a ferocious animal, averse from all submission, when educated merely under the influence of inanimate nature, and exposed to no resistance but that which it can escape or overcome. The Beaver, in the same circumstances, exhibits similar phenomena; its instinct remains, but its individual wants being alone developed, place it in a state of open war with its fellows, and destroy all harmony between them.—Brought up together, these animals would have lived in perfect harmony, and laboured in concert; but remove them from such society, and each can live no longer but for himself alone. Many of the Canadian Beavers, shut up in menageries, strongly illustrate this observation, according to the remark of M. F. Cuvier. These being taken very young, and brought up in a solitary manner, in narrow cages, could never habituate themselves to any thing but obedience to their master's will; whenever any attempt was made to unite them, it was productive of nothing but the most violent combats and the severest wounds. It was always necessary to separate them, from the fear of losing animals of so much value, and whose manners were an object of so much curiosity. One of these just alluded to was an extremely mild animal, habituated to the presence of men, and would suffer itself to be touched and carried with the hand from place to place with the utmost confidence; and was brought at last to live familiarly with some dogs; for it is worthy of remark, that certain animals of different species will sooner contract an affection for each other when united by man, than those of the same species; a disposition which, however apparently opposed to the social instinct, is but a result and a confirmation of its existence.

The Beaver just spoken of, was about seven inches in height, measured in front, and about nine and a half in the most elevated part behind; its tail, from the anus to the extremity, about ten inches; the length of the body, from anus to muzzle, about fourteen, French measure. It was an heavy, clumsy animal, without any agility in its motions, and its entire physiognomy was in perfect unison with its physical qualities. In this respect, indeed, all Beavers resemble. But they are fond of water, delight to plunge into it, and steep their food in it. They are all fond of gathering and heaping into a corner all the light substances they meet with, such as straw, the remains of their provisions, in a word, every thing within their reach. Thus was the instinct of building, thus strongly manifested, though in a state of captivity. What was, however, singular, the animal above-mentioned exhibited no propensity of the kind; on the contrary, it remained in a state of the most complete inaction, though at perfect liberty in the midst of a small court, where it might have found every thing necessary for his purpose, had it been induced to build. What, perhaps, was still more extraordinary was, that it shunned the water, and never entered a little basin that was dug for it, either to bathe or moisten its aliments like the Otters. But it was not fond of being alone ; it constantly endeavoured to approach those persons who took care of it, and passed almost the whole of its time in sleeping. It lived thus many months in excellent health, eating little, hardly ever drinking, and giving no manifestation of any sentiment. During all this time it exhibited neither joy, anger, nor desires of any kind ; for its wants themselves had scarcely sufficient power to withdraw it from its lethargic state. Sometimes it emitted a very soft and feeble sound, most usually when it seemed annoyed at being touched, or was desirous of following the persons who quitted it. It replied by some sound when it was called ; for it had learned to recognise its name, or rather the voice of his master, a circumstance which places it far above many others of the Rodentia, which are genuine examples of stupidity. Its want of intellectual activity, and the enfeeblement of its instinctive dispositions, must be attributed to its captivity, but more especially to its *domestication*, in a period of extreme youth. Nothing, in fact, can more completely arrest the development of the instinctive faculties, than the gratification of every want and every desire, and the possession of perfect security; the muscles of the body require motion to develop them, and the powers of the mind exertion.

On the organization of the Beaver it would be superfluous to add much to the Baron's description. The teeth already described, have the faculty of growing by their own power of development during the life of the animal. The hinder toes are palmate, and the fore are free and provided with nails, admirably calculated for digging. This fact alone, were there no other, would in itself be sufficient to prove a superintending intelligence. But proofs on this subject are superfluous.

In walking, the Beaver leans on the ground only the toes of the fore feet; but it rests the entire sole of the hinder. When the feet are raised, the toes approximate, and separate in the contrary motion, and that by the same disposition of the muscles. The animal, in repose, usually is couched upon its tail, which it gathers back between the hind feet. This tail, of an elliptical form, can move vertically and laterally with tolerable force, but the animal usually drags it after him. The eye is small, and the pupil round, and closes almost entirely in a strong light. It never dilates very sensibly, but in a soft and feeble light. The third lid is but rudimental. The Beavers appear to have no necessity of preserving their eyes from the contact of the water. The conch of the ear is simple, and closes when the animal dives; so do the nostrils. Its voice, according to M. F. Cuvier, when the animal is uneasy, consists at first in a little dull sound, which finally changes into a louder tone, similar to barking. Capt. Franklin, in his most interesting account of his journey, relates, that a gentleman, about to shoot a Beaver, was prevented from doing so by the strong resemblance of its cry to that of his own children.

In the same pouch with the parts of generation and the anus, are found two pairs of glands, and the upper pair produces the *castoreum.* The teats are four in number.

So much has been written and promulgated respecting the manners and habits of the Canadian Beaver, that our readers will pardon our not entering into any repetition of accounts so universally known. Few travellers to the New World have omitted to give some notices of this singular animal, and long and detailed histories have been published respecting the dikes and huts which it constructs, and the means it adopts in its operations. These histories have, at times, been carried to such a degree of exaggeration as to set the Beaver next to man in the degrees of intelligence. It would appear, however, that though a very considerable degree of sagacity cannot be denied to this animal, yet that the most wonderful of its operations are referable to instinct rather than to reason. The best proof of this is, that, exclusive of such operations, the Beaver discovers no degree of intelligence at all approaching to that of the higher quadrupeds.

The subdivision ONDATRA, including but a single species, has much affinity with the Castor, particularly in its habits; its organization indeed rather approximates it to the Rats. Accordingly we find that Linnæus, in the 12th edition of his System, associated it with Castors, while his editor, Gmelin, transferred it to the Rats.

M. Sarrasin, in the memoirs of the French Academy of 1725, and from him Buffon, have given us some interesting details of this animal as a species of Rat. Modern systematic zoologists have employed the physical characters, principally of its teeth and tail, to separate it into a distinct division, which Lacépede and Geoffroy have called *Ondatra*, and Cuvier and Illiger, *Fiber*.

The teeth of this subdivision are incisors $\frac{2}{2}$, and cheek-teeth $\frac{4 \cdot 4}{4 \cdot 4}$. The anterior paws have four toes, the posterior five; these are deeply cleft, and are furnished on each side with rough hairs, acting like the web of other aquatic animals, though less perfectly in swimming. The tail is long, and compressed laterally, indicating aquatic facilities.

The *Ondatra* is about the size of a small rabbit, stands very low, and weighs about three pounds. The fur is as fine as that of the Castor, with a thick and soft down underneath. The head is rather short, the eyes large, and the ears short and round.

The long hairs, according to M. Desmarest, are reddish brown on the upper part of the body, and red mixed with ash on the sides, and reddish gray underneath. The flanks are marked with a brown spot, the outer side of the thigh ash-coloured; in general, on the upper parts of the body, the down is ashy near the root, and reddish brown toward the point; the down of the lower part is of a bright shining gray colour.

We have, however, before us three drawings of this animal, neither of which accords strictly with the above de-' scription as to colour; one of these we have figured which has very little of the red tint, but has a colour which appears to result from a mixture of white and brown, or as if a lightish brown surface had been powdered with white or gray; the second is almost uniformly deep brown, and the third is a pure white, with the eyes like the others, and not red.

Whether permanent varieties or distinct species of the Ondatra corresponding with these drawings will eventually be found, future observation must determine. We have by no means sufficient faith in difference of colour, as a specific character, to employ it without further observation.

The Ondatra is an inhabitant of Canada and North America, perhaps' generally. At the period of its amours, the Ondatra, in common with some other Mammalia, gives a strong scent; with it indeed this odour is so strong as to be absolutely pernicious, and the Indians have named a river in Canada, which is much frequented by them, the Stinking River. This odour proceeds from a milk-white liquid; secreted by certain glands situated near the parts of generation in both sexes.

The Ondatra, like the Castor, is a social animal, particularly during the winter; in the warmer seasons it is generally found in company with a single mate. They construct retreats of about three feet in diameter, each of which generally contains several individuals. These retreats or cabins are round, and covered with a concave top, about a foot in thickness; they are made of grass, rushes, earth, &c., and are so compact and hardened, as to be impenetrable by rain or by the severity of cold, even when the surface of the ground is covered with snow.

They do not amass magazines of provisions, like some other social animals, but they leave numerous and long avenues in the vicinity of their cabins, for the ease of collecting roots and other vegetable matter as food, and also for access to the water, where they probably also prey partially on fish. Like other digging animals, they are sometimes exposed to a very hard fate during the severity of winter, and are known occasionally to be driven to devour each other when incapable of procuring their natural food.

They produce five or six young at a time, once in the year. The period of gestation is not known, but the young are generally found tolerably strong at the commencement of autumn.

The voice of the Ondatra is not unlike a sort of groaning; the hunters sometimes draw them from their retreat, and take them by imitating this voice.

Though decidedly aquatic in its habits, the toes not being so completely palmated as those of the Beaver, they do not swim so well or so fast, or continue in the water so long as that animal; they are more frequently obliged to visit the shore; here indeed they cannot run fast, and walk still more awkwardly.

The Ondatra is not a fierce species; if taken early, it may be easily tamed, and fed on the *acorus, calamus nymphæa,* and the roots of aquatic plants generally.

Their pelt or down is made use of in the manufacture of hats; formerly the fur also was in use, but as it cannot be completely divested of its unpleasant scent, it has been

altogether disused. In winter, when they are less offensive,
the flesh is said to be good food.

The subgenus ARVICOLA, or the *Campagnols*, was confounded with the Linnean genus Mus, until Lacepede separated it under the name Arvicola, and Cuvier under that of
Lemming.   Illiger adopted Hypudæus and Georychus as
the generic names of this subdivision.

Of the various divisions created by modern systematic
writers of the Linneań genus Mus, it may well be questioned how far they are really useful in the propagation of
zoological knowledge.   As a subdivision, which the Tyro
in zoology may pass with venial indifference, though the
professor of the science may employ it with advantage,
it may be, in common-place language, all very well.
Close and careful comparison may detect some difference
in the minutiæ of the conformation of the teeth of Arvicola,
and in those of the Ondatra and Lemmings, properly
speaking; but the principal difference after all will be found
to appertain to the tail, which in the Ondatra is scaly
without hair, but in the Arvicola is covered with scales in
conjunction with hair.

In addition to the Baron's specific notice, we shall only
add, that the *Water-Rat* varies by climate.   Its head is
large; the muzzle short and thick; the eyes small; and
the ears short, and hidden in the fur; the tail is scaly, like
that of the Rat, but more furnished with hair; and there
are small scales on the skin of the feet; the fur is long; the
incisive teeth are yellow.   The feet are not palmated,
though the habits of the animal are aquatic.   There are
five toes on each foot, but the thumb of the anterior extremities is extremely short.

The denomination of Water-Rat sufficiently indicates the
habits of the animal.   It is never found like the common
Rat in our habitations, nor does it frequent elevated or dry

soils, but establishes itself on the borders of ditches or rivers, or in wet and marshy valleys. It not only swims with facility, although rather slowly, but dives and runs at the bottom of the water, where, however, it cannot remain much more than half a minute.

It is found all over Europe, in the North of Asia, and in America, but is most common in Russia and Siberia.

It feeds principally on herbs or roots, but takes also young fish, frogs, and water insects ; when unable to flee from an enemy, it will sit on the haunches and defend itself with its fore-paws and teeth with considerable obstinacy.

The female produces six or seven at a birth in her sub-terranean retreat, and several times in a year. Like the Ondatra, these animals are observed at certain periods to emit a peculiar scent.

There are some varieties known, which will be found in the table.

The *Campagnol*, or *Meadow Mouse* of Pennant, (*Mus Arvalis, Pallas*) is very common, says Dr. Shaw, in our island, and is readily distinguished from the rest of the British species by the shortness of its tail. We shall add nothing further in this place to its specific description ; it makes its nest in meadows, and produces a litter of about five or six at a time, twice a year. Its favorite food is grain, which it amasses during harvest. It is frequently found in corn ricks and barns.

This species is a most formidable enemy to the labours of mankind. Wherever the husbandman may have by industry and art overcome the natural sterility of the soil, or directed and fostered its innate fecundity to the benefit of mankind, the Campagnol is quickly found to follow him, to profit by his labours, and appropriate the fruits of his industry. Nor is this destructive little animal content merely with dividing the harvest with man, or even appro-priating the whole of it ; his spoliation extends to the seed in the ground, as well as to that which is not gathered.

In some seasons, the Campagnols have been known to increase in an extraordinary degree, and by biting the straw asunder they lower the grain to within their reach, and by these means will sometimes destroy the whole produce of a field, and then proceed in their destructive office to another. Though they prefer corn to other food, they will nevertheless attack the roots of grass in meadows, and those of plants, as well as nuts, and other fruit in gardens, &c.

Whatever may be the temporary cause of the excessive multiplication of species sometimes so banefully excited, especially among the insects, there is a providential and paternal care by the Author of all things exercised certainly by secondary agency in favour of the whole mass of his creatures; the undue multiplication of locusts and other animals which we happily in this climate only know by relation and history, if continued according to the mathematical ratio of ordinary increase, would soon depopulate the world; but ordinary anticipations and calculations cease when special interposition becomes necessary. Thus we find, in regard to the present species, that when the produce of the summer fails them the quicker, in consequence of their extraordinary multiplication, they are driven to the revolting alternative of destroying one another.

France seems to be particularly subject to what may be called the plague of mice of this description ; and though found in particular provinces at particular times, a square of not less than forty leagues has been not unfrequently known to suffer to a most pernicious degree in one season.

The burrows of these animals, which serve both as retreats and depositories for their stores, are not spacious or deep, but they divide them into two or three apartments, and inhabit them gregariously. The galleries occupied by several families or small colonies are not contiguous ; there is always a space between them. If the inhabitants of one burrow abandon it or perish, others are not found to occupy

K 2

the same, but each colony prefers providing a domicile for itself. These retreats are not in general more than from six inches to a foot from the surface ; but the pregnant females will sometimes deepen the excavation to upwards of two feet by a very small alley or aperture, which, after several sinuosities, terminates in an excavation about as big as a fist, furnished with a soft bed of roots, &c. for the accommodation of the young.

These animals are found all over Europe, nor does the low temperature prevent their inhabiting the uncultivated country in the North of Russia ; they are found in Siberia. They are migratory animals, and like others of a similar description, are not stopped in their journeys, even by deep or rapid rivers.

The *Economic Campagnol* is rendered highly interesting by Dr. Pallas's amusing description of its habits, which we shall abridge in the language of Shaw, referring to the Table for its specific characters.

These little animals make their burrows with wonderful skill immediately below the surface in soft turfy soils, form-ing a chamber of a flattish arched form, of a small height, and about a foot in diameter, to which they sometimes add as many as thirty small pipes or entrances, and near the chamber they frequently form other caverns, in which they deposit their winter stores ; these are said to consist of various kinds of plants, even of some species which are poisonous to mankind. They gather them in summer, har-vest them with great care, and even sometimes bring them out of their cells in order to give them a more thorough drying in the sun. The chief labour rests on the females ; the males during the summer wandering about in a solitary state, inhabiting some old nests occasionally, and living during that period on berries, without touching the hoards which are reserved for winter, when the male and female reside together in the same nest. They are said to breed several times in the year, the female producing two or three young at a time.

The migrations of this little species are not less extraordinary than those of the Lemming, and take place at uncertain periods. Dr. Pallas imagines that the migrations of those inhabiting Kamtschatka may arise from some sensations of internal fire in that volcanic country, or from a prescience of some unusual and bad season. Whatever be the cause, the fact is certain. At such periods they gather together during the spring season in surprising numbers, except the few that reside about villages, where they can pick up some subsistence ; and this makes it probable that their migrations, like those of the Lemming, are rather owing to want of food. The mighty host proceeds in a direct course westward, occasionally swimming with the utmost intrepidity over rivers, lakes, and even arms of the sea. During these perilous adventures, some are drowned, and others destroyed by water-fowl, fish, &c.: those which escape rest a while to bask, dry their fur, and refresh themselves, and then again set out on their migration. It is said that the inhabitants of Kamtschatka, when they happen to find them in this fatigued situation, treat them with the utmost tenderness, and endeavour by every possible method to refresh and restore them to life and vigour. Indeed none of the smaller animals are so much esteemed by the Kamtschadales as these ; since to their labours they owe many a delicious repast ; robbing their hoards in autumn, and leaving there some kind of provision in return, accompanied by some ridiculous presents by way of amends for the theft. As soon as the migrating host of these animals has crossed the river Peuschim, at the head of the gulf of that name, it turns southward, and reaches the rivers Judoma and Ochot about the middle of July: the space thus traversed appears astonishing, on consulting the map of the country. The flocks during this time are so numerous that an observer has waited two hours to see them all pass. Their return into Kamtschatka is in October, and is attended with the utmost festivity and welcome on the part of the natives, who consider their arrival as a sure prognostic of a successful chase and fishery; and they are said equally to lament their migrations, which are usually succeeded by rainy and tempestuous weather.

Wormius gives an accurate description of the *Lemming (Mus Lemmus of Lin.)* It has, says he, the figure of a mouse, but the tail is much shorter; the body is about five inches long; the fur is fine, and spotted with many colours; the anterior part of the head is black, the upper part yellowish; the neck and shoulders black; the rest of the body reddish, marked with small black spots of different figures as far as the tail, which does not exceed half an inch in length, and is covered with blackish-yellow hair; the belly is yellowish-white; the order of the spots, as well as their figure and size, differ in different individuals; about the throat there are several long stiff hairs; the eyes are small and black; the fore-legs very short; the feet have five toes covered with sharp bent claws, the middle are much the longest, the fifth very small, and situated above the level of the rest.

The migratory habits of these animals as related by naturalists exceed credibility, were they not authenticated on very respectable authority. It is to Pallas we are indebted for a particular account of the animal, in common with the rest of his genus Glires Shaw has abridged his details, and we shall adopt his account.

The natural or general residence of the Lemming is in the Alpine or mountainous parts of Lapland and Norway, from which tracts, at particular but uncertain periods, it descends into the plains below, in immense troops, and by its incredible numbers becomes a temporary scourge to the country, devouring the grain and herbage, and committing d'evastations equal to those caused by an army of locusts. These migrations of the Lemming seldom happen oftener than once in ten years, and in some districts still less frequently, and are supposed to arise from an unusual multiplication of the animals in the mountainous parts they inhabit, together with a defect of food ; and perhaps, a kind of instinctive prescience of unfavourable seasons ; and it is observable that their chief migrations are made in the autumn of such years as are followed by a very severe winter. The inclination, or instinctive faculty which induces them, with one consent, to assemble from a whole region, collect themselves into an army, and descend from the mountains into the neighbouring plains in the form of a firm phalanx, moving on in a straight line, resolutely surmounting every obstacle, and undismayed by every danger, cannot be contemplated without astonishment. All who have written on the subject agree that they proceed in a direct course, so that the ground along which they have passed appears at a distance as if it had been ploughed ; the grass being devoured to the very roots in numerous stripes or parallel paths, of one or two spans broad, and at the distance of some ells from each other. This army of mice move chiefly by night, or early in the morning, devouring the herbage as it passes in such a manner that the surface appears as if burnt. No obstacles which they happen to meet in their way have any effect in altering their route ; neither fires, nor deep ravines, nor torrents, nor marshes, nor lakes ; they proceed obstinately in a straight line, and hence it happens that many thousands perish in the waters, and are found dead by the shores. If a rick of hay or corn occurs in their passage, they eat through it ; but if rocks intervene, which they cannot pass, they go round, and then resume their former straight direction. If disturbed or pursued while swimming over a lake, and their phalanx separated by oars or poles, they will not recede, but keep swimming directly on, and soon get into regular order again, and have even been sometimes known to endeavour to board or pass over a vessel. On their passage over land, if attacked by men, they will raise themselves up, uttering a kind of barking sound, and fly at the legs of their invaders, and will fasten so fiercely at the end of a stick as to suffer themselves to be swung about before they will quit their hold, and are with great difficulty put to flight. It is said that an intestine war sometimes takes place in these armies during their migrations, and that the animals thus destroy each other.

The major part, however, of these hosts is destroyed by various enemies, and particularly by Owls, Hawks, and Weasels, exclusive of the numbers which perish in the waters; so that but a small number survive to return, which they are sometimes observed to do, to their native mountains.

In their general manner of life they are not observed to be of a social disposition, but to reside in a kind of scattered manner, in holes beneath the surface, without laying any regular provision, like some other animals of this tribe. They are supposed to breed several times in a year, and to produce five or six at once. It has been observed that the females have sometimes brought forth during their migrations, and have been seen carrying some in their mouths and others on their backs. In some parts of Lapland they are eaten, and are said to resemble squirrels in taste.

It was once believed that these animals fell from the clouds at particular seasons, and some have affirmed that they have seen a Lemming in its descent; but an accident of this kind is easily accounted for, on the supposition of a Lemming escaping now and then from the claws of some bird which had seized it, and thus falling to the ground; a circumstance which is said not unfrequently to take place when the animals are seized by Crows, Gulls, &c.

C. Hamilton Smith Esq. del.

Drawn

London. Published by E.G. Whittaker. Dec.7.1825.

THE RED ECHIMYS OR SPINY RAT OF D'AZARA.

The subgenus ECHIMYS, or *Spiney Rats*, (Loncheres, Illiger,) was established by M. Geoffroy St. Hilaire, principally on the character from which it is named, the back being furnished with spines as well as hair. Their teeth offer some slight variation from their congeners, as may be seen in the text. Their general form approaches that of the Rats ; the body elongated ; the tail varying in length in the different species, but always round and generally scaly ; the anterior paws have four toes and the rudiment of a thumb, those behind have five, and all of them are armed, with bent, digging claws. It does not appear that all the Spiny Rats are included in this subdivision, as the Perchal and the Mus Cahirinus of Geoffroy are referred to the Rats proper.

M. d'Azara has furnished us with some particulars of the *Red Echimys*, and as all that is published of the remaining species of this subdivision is confined to their physicalities, it may be sufficient to refer to the Table for a brief account of their specific characters.

This animal is about eight inches long, with a tail not exceeding three. On the head, sides, body, and flanks, the colour is a mixture of dark brown and reddish ; the under parts are white, and the tail covered with short hair, thick and soft, through which the scaly skin cannot be seen, is of a dark brown colour. The longest spines are about nine lines in length, formed like a double-edged sword ; they are whitish for about two-thirds their length ; whitish-brown toward the point.

In general, the head, neck, and body, are thicker than in the common Rat. The male is rather larger than the female. The Red Echimys is found at Cayenne, Paraguay, and especially between the town of Neemboucon and the river Plate. They dig their burrows in dry and sandy soils, and they are dug so close to each other as to make much precaution necessary in walking over them ; they are about four or five feet long, and about eight inches under the surface.

In addition to the characters given by our author to the subgenus Myoxus, we shall merely add, that their eyes are large and prominent; large ears; mustachios long; anterior paws with four toes, and the rudiment of a thumb, furnished with an obtuse nail; the posterior paws have five toes, armed with sharp nails.

The animals proper to this subdivision inhabit the temperate climates, and live on fruit of all sorts, and reside in trees; hence they seem to be intermediate between Rats and Squirrels. In winter, after having collected a small supply of stores for use when they wake, they fall into a long lethargy.

The *Loir*, or *Fat Dormouse*, (Myoxus glis, Gm.) is the thickest species of this subdivision; its size is about that of the squirrel; the cheeks are covered with whitish hair; the eye is surrounded with a deep brown; the mustachios are long; the upper part of the body is ashy-gray brown, the under part whitish; the tail is covered with long hairs of the same colour as the body, and disposed in a similar manner to those of the squirrel.

The Loir much resembles the squirrel in its manners; inhabits the forest; climbs the trees, and leaps from branch to branch, with less agility, it is true, and lives on the same food as the Squirrel. It is said also to feed on small nestling birds, but it does not make a nest or bed in the trees like the Squirrel. It however constructs a bed of moss in hollow trees, and retreats sometimes into the clefts of rocks; it fears humidity, drinks but little, and rarely descends to the ground.

These animals couple in spring, and the female brings forth in the summer four or five at a birth; they grow quickly, and are said not to live beyond six years. In winter these little animals fall into a torpid state, produced by the coldness of their blood. They possess so little animal heat, that it scarcely exceeds the ordinary temperature of

THE FAT DORMOUSE.

MYOXUS GLIS. Gm.

London, Published by G B Whittaker Dec.ʳ 1824

the air ; this torpor ceases with the cold ; a few degrees of heat above 10 or 11° will reanimate them, and if they are kept during winter in a warm room, they will continue active the whole season ; they will then move about, eat, drink, and sleep at the usual intervals, like other animals. When the cold approaches, they roll themselves into a ball, and in this state may be found in winter in hollow trees, or clefts of rocks, or in holes in walls exposed to the south ; they may be taken and rolled about without rousing them ; nothing, indeed, seems to wake them from their lethargy but gradual heat ; if exposed suddenly before a fire they die ; resuscitation can only be effected by degrees. In this state, however, although deprived of the capability of movement, with the eyes closed, and in an apparent state of death-like indifference, they possess a feeling of pain when sharply inflicted ; a wound or a burn causes them to contract, and to make a slight sort of convulsive leap which they will repeat several times.

It sometimes happens that the Loir wakes in the winter season, when the thermometer rises to 12, 13, or 14° ; the animal on such occasions will quit its retreat, and eat some part of the store of provision reserved for its early resuscitation.

A. M. Mangili, of Pavia, has published his observations on the periodical lethargy of this species in particular, among other lethargic quadrupeds ; those observations were begun in the month of December, and continued till the April following. On the 24th December, the temperature fell to 4° below zero, when a Fat Dormouse confined in the observer's library fell into its hybernating or torpid condition, after retiring to a bed, of slips of paper, placed among the books for it, together with some provisions. On the 27th December, its left side having been exposed, the thermometer placed near it marked 3° and an half ; respiration was suspended, and renewed at certain intervals ; after four minutes of perfect repose the animal respired from twenty-two to twenty-four times together in the space of a minute and a half. When the thermometer rose a degree, the intervals did not exceed three minutes, but the number of respirations continued the same. On the 29th December the thermometer stood at one degree above freezing, when the respirations were twenty-six to twenty-eight, and the interval between the series about six minutes. On the 3d of January the frost became very severe, and the Loir roused itself, passed some excrement, and ate a little, and did not again become lethargic till the temperature became more moderate. When carried to a place where the temperature was kept at from 3 to 5°, its lethargy continued. The intervals of non-respiration became from six to eighteen minutes, and the number of respirations at each series always from eighteen to twenty. He waked again on the 9th of January, when the temperature was at 2°. On the 10th of February the temperature was at 7°, when the number of respirations was from thirteen to fifteen, and the intervals eighteen to twenty minutes. Placed suddenly in a receiver with an artificial cold, 1° below freezing, the animal appeared to suffer, and its respiration became stronger, more frequent, and without interruption ; the cold having been artificially increased to 6° below zero, the animal, after an accelerated and continued respiration, awoke, and endeavoured to escape ; when again put into a box with the temperature at 7°, it quickly fell into the lethargic state, and on the 31st of February, the temperature continuing the same, its respirations were not more than five, six, or seven, after intervals of from twenty-eight to thirty-five minutes ; it continued in this state till the 12th of March, when it awoke.

We may, therefore, conclude, that the lethargic condition of this species is the most profound when the temperature is at from 5° to 7° above zero, and that a more intense cold than this accelerates the circulation, and consequently the respiration, and even revives the animal to activity.

M. Mangili's hypothesis, as to the cause of certain quadrupeds hybernating, seems to be, that the arterial blood necessary to excite and revive the fibres of the cerebral organ, flows less copiously to this organ in the hybernating animals on account of the small number of the arteries he had found in such animals, and of the smallness of their calibre ; this concurring with other exterior causes of debility, diminishes the energy of the fibres of the brain, and produces at first sleep, and eventually, continued lethargy.

The same physiologist took a Loir about midsummer, when the temperature was at 15 or 16°, and put it at the bottom of a large vase, with a bed of hay and provisions. The animal after having tried in vain to escape refused its food, and quickly fell into a state of lethargy ; but, instead of previously rolling itself into a ball, it stretched itself on the back. In this state the intervals of repose were much shorter than in winter, and the respiration was less frequent. It did not wake till the 17th of July, when it effected its escape.

The Fat Dormouse is not extensively located, being confined to the temperate parts of the continent of Europe, but not on the high mountains where the Marmot is found.

This species is still used for food in Italy, and is taken simply by preparing a fit place for its winter quarters in the wood, which is large enough for many of them to retire to, whence they are taken toward the end of Autumn. The Romans were very fond of them as food ; they kept and fattened them for the table in receptacles called Gliraria. Martial tells us that they are fattest after hybernating, when they have had nothing but sleep to fatten on ; on which Buffon observes, that the Loir, at all times fat, keeps itself in condition in winter by waking occasionally, and taking food at intervals.

The *Garden Dormouse*, or *Lerot*, very much resembles the last-mentioned species, but is smaller, the body thicker, the muzzle more pointed, and the tail covered with scattered red-gray hairs, and terminated with a tuft of long black hair ; it is reddish gray above, and white underneath ; the stomach also is neither so large nor so elongated as in the Fat Dormouse, and it differs also in some other anatomical distinctions.

The Garden Dormouse, as its name imports, inhabits gardens, and sometimes finds its way into houses. It makes its bed in a hole, in walls, mounts only the espalier fruit trees, and selects the best and choicest fruit.   It is sometimes also found in the clefts of trees, in old orchards, on a bed of moss and leaves.   It hybernates, and eight or ten are not unfrequently found together rolled up in the midst of a magazine of nuts.

The Garden Dormouse produces five or six at a birth in summer, which grow rapidly, though they do not generate till the following summer.   They are not eatable like the Loir, and give a scent like that of the common rat.   This species is proper to the temperate parts of the European continent.

The *Common Dormouse (Mus Avellanarius Nina, M. Muscardinus, Gm.)* is smaller than the Lerot or Garden Dormouse, and is but little larger than the mouse ; the head is short, the muzzle less elongated, and the eyes larger.   The tail is furnished with hair, ranged on each side like that of the Loir, but much shorter.   This animal is best distinguished from the two preceding by the character of the tail ; all the upper part of the body is brownish yellow mixed with white, the under parts are light, with the throat nearly white.

This **species** is never found in houses, and but rarely in

THE MUSCARDIN DORMOUSE.

MUS AVELLAMARIUS. Lin.

London. Published by G.B. Whittaker. Sep. 1825.

gardens, but is an inhabitant of woods, hybernating in the clefts of trees. Its habits accord with the species already mentioned.

M. Mangili's observations extended to this species as well as to the Loir. A degree of cold below freezing roused the animal; at 4 or 5° of heat, the Dormouse respired 174 times, divided into ten series of movements of respiration, (the most considerable of 30, and the least of 5) in 82 minutes ; at 10° of heat 47 times in 34 minutes, the respirations being 7 or 8 in each period, and the intervals from 4 to 8 minutes. When exposed to the sun, the respiration of the animal was no longer suspended, but went on uniformly and regularly as in a state of natural sleep ; as soon as roused, it ate same pieces of chestnut, again rolled itself up, and fell asleep. It respired for half an hour without interruption, 25 or 30 times in a minute ; then the intervals of perfect quiet took place, and increased progressively. When suddenly placed in an artificial cold of 2° under freezing, respiration became more frequent, as in the case of the Loir.

The result of the experiments on this species seems to prove, that the common Dormouse is of all animals the most disposed to lethargic habits ; that a temperature eitner too high or too low rouses it; that as soon as it is awakened it takes some food, though moderately ; that it passes from its lethargic to its active state in less than half an hour, while the Marmot requires a much longer period ; that the time it takes in waking thoroughly is quick in proportion to the elevation of the temperature, probably because it receives its necessary portion of calorique more rapidly.

M Mangili having exposed a Dormouse when in the lethargic state to an artificial cold of 10°, it died in 20 minutes. When opened, he found a great quantity of blood in the ventricles of the heart, and in the principal vessels which supply and receive from the lungs. He also found the lungs, the veins of the neck, head, and especially of the brain, considerably distended with blood.

The common Dormouse is found, though not very plentifully, in England.

We insert a figure of the Muscardin Dormouse, *(Mus Avellanarius, Lin.)* but shall add nothing here to its specific characters, as stated in the Table.

The subgenus HYDROMYS includes at present, perhaps, but a species sometimes found with a white belly (*Hydromys leucogaster*), and sometimes with a yellow belly (*Hydromys chrysogaster*) ; these, however, are specifically separated by their describer.

The animals proper to this subdivision approach the characters of the Dormouse and the Rat, by their teeth on the one hand, and the Water-Rat and the Beaver, by their aquatic habits, on the other.

The *Hydromys* is about a foot long, and the tail is nearly the same length ; the colour is brown above, and orange-colour or white underneath ; the tail is covered with short rough hair, thick at the base, and of the same colour as the belly ; the colour then becomes blackish as it approaches the end, but the tip is quite white.

We are perfectly ignorant of the habits and manners of these animals, nor is their habitat very accurately known, though they appear to belong to the western coast of New Holland.

We have seen by the text that the *Coypus* of America is placed, though conditionally by the Baron, in the subdivision Hydromys. Subsequent observation has determined that its characters are so far distinct as to separate it into a new subgenus, as proposed by Commerson, which he has named Myopotamus. As the teeth constitute its principal pretensions to a generic separation, and they are noticed in

THE COYPOU.

HYDROMYS COUPUS.

London. Published by G.B. Whittaker, Jan.1825.

the Table, we shall proceed to a short description of the species in this place.

Molina was the first zoologist that noticed this animal as indigenous in Chili. Commerson afterwards procured a drawing of it, and apparently without a sufficient examination treated it as a new genus. Gmelin and his followers in the interim adopted Molina's native name of the animal, and referred it to his comprehensive genus *Mus*. M. Geoffroy, not satisfied with Commerson's presumption on the animal, chose rather to refer it to the Hydromys, or palmated Rats ; but a recent inspection of the teeth by M. F. Cuvier has satisfied that eminent operative zoologist, that the animal required a distinct generic place, and he has therefore adopted the name Myopotamus.

D'Azara has given a detailed account of the Quouiya of Paraguay, which turns out to be the Coypus of Molina, now generally known by that name. It is distinguishable exteriorly from the two species of Hydromys, with which, as we have seen, it has been generically associated by its larger size, and by the colour of the fur, which though subject to vary in different individuals, is generally of a marron-brown colour on the back, each hair on that part being brown and red, with the former colour prevailing; on the flanks, the red predominates, and each hair is brown only at its base. The pelt, hidden under the long fur, is ashy-brown, brighter on the belly. In common with most other aquatic quadrupeds, the Coypus has very little hair on the tail, the naked parts of which are scaly. The colour of the mouth, and extremity of the muzzle are white ; the ears are short and round.

The body measures about two feet in length ; the tail seventeen or eighteen inches.

The male and female are not observed to differ in exterior character.

The Coypus is easily tamed, being endowed with a gentle disposition, nor is it difficult in regard to its food, as it is said to eat almost all vegetable matters. It inhabits the banks of rivers, and sometimes migrates in search of a new domicile for a considerable distance. It is a good swimmer, digs burrows in the earth by means of its strong fore-nails with facility, and dwells in them. The female brings forth five or six at a time, according to Molina, and from four to seven, according to d'Azara, who very soon learn to attend her in all her excursions in the great business of animals, the search of their subsistence.

We now come to the RATS, properly so called, which as a genus we shall forbear to notice; and shall here pass by the *Mus Giganteus* of Hardwicke, (the Bandicote Rat of Pen.?) as being principally remarkable for its size, as big as a Rabbit, and proceed to the *Common Rat*, which, like the Surmulot, or Brown Rat, appears not to be aboriginal in Europe. Nothing indicates any knowledge of this animal among the ancients, and the modern authors who have spoken clearly on the subject, go no farther back than the sixteenth century. Gessner is perhaps the first naturalist who has described it. Had this animal lived formerly as it does at present, among us, and at our expense, it is not probable that all mention of it would have been omitted, especially as we find notices of other animals of a similar kind, less remarkable and less destructive, such as the Mouse, Dormouse, *&c.* Some naturalists think with Linnæus and Pallas, that we have received it from America, and others believe that it is a present of our own to that country, made after we had ourselves received it from the eastern regions. To this question it is perhaps impossible to reply, and with the lights which we possess on the subject, conjecture is but a frivolous amusement. It is certain that the Rat is to be found in all the warm and temperate climates of the globe, that it is wonderfully common in Persia, and multiplied to a prodigious extent in the western islands, where it is not obliged by winter to seek a refuge in the habitations of man, but where the fields during the entire year present it with abundance of nutriment. In all this part of America, accordingly, it has become a perfect scourge, from its ravages and devastations. In fact, the Rat consumes an immense quantity of provision, and destroys or damages still more than it consumes, particularly in the fields, as it cuts up from the roots plants of which it eats but a portion.

With us its favourite abode is in barns or granaries, under straw roofs, or in deserted houses. Sometimes it will burrow in the earth like the Surmulot, or Brown Rat, when it can get no other habitation. Though this last-mentioned species does not mix with the Common or Black Rat now under consideration, and even may sometimes destroy it, yet the natural antipathy commonly supposed to exist between them is an error. The Surmulots do not necessarily exclude the Rats from their vicinity, nay, the two species often live under the same shelter, and in contiguous burrows. This occurs when the place of their establishment affords food in abundance, and excludes the necessity of mutual warfare for subsistence. In the contrary case, we find that the Surmulots not only destroy the Rats, but that the latter, as is well known, will devour one another.

The Rats, like all the Rodentia, with roots distinct from the coronal in the molar teeth, are omnivorous. They are at least as much omnivorous as all the omnivora properly so called. They live indiscriminately on flesh or grain. They will also eat fruit, and all such parts of vegetables as contain saccharine or oleaginous matter. But they lay up no provision for the winter.

These animals bring forth many times in the year. During the season of their amours, they have very violent combats, and utter cries resembling a sharp hissing. They prepare with straw, &c. a nest for their young, which are born entirely naked, and with the eyes shut. They are generally about nine, or sometimes more, in number.

L 2

They are genuine plantigrades, and each foot has five toes, remarkable for their thickness. The thumb of the fore-extremities is visible only by the nail. The teats are six in number. The two lower incisors are narrow, pointed, and developed, in separating from each other. The molars are three on each side; but it is unnecessary to enter into any detailed account here of an animal so well known.

The gait of the Rat is lively; he runs with rapidity, and makes very considerable leaps. Thus he often escapes from his adversaries when attacked by open force. If he is surprised, and obliged to defend himself, he does so with courage, and often with success, by means of his long incisors, which inflict deep and severe wounds. During the intervals between eating and sleep, the Rat is constantly occupied in cleaning and polishing his fur. He can carry his aliments to his mouth with his hands, and drinks lapping.

The Rat is usually about seven or eight inches in length.

The next species worthy of a particular notice is the *Brown Rat,* or *Surmulot,* before alluded to. On this we may remark, that the introduction of a species into a climate that is new to it is usually attended with great difficulties. Man, alone, aided by his reason, can set his industry in opposition to the inclement sky, and create a climate around him suitable to his wants and wishes. But the habits and instincts of animals are never sufficiently flexible to enable them to supply new wants, and adapt themselves to new situations. Unless they are under the especial protection of man, unless his foresight comes to their assistance, they generally perish after a struggle more or less protracted. Some, however, are found which derive from our habits that which the vigour of the climate would have refused them, and from the influence of their constitution place themselves in the very situation in which we would have placed them, had we taken any interest in their preservation.

Such is the case with the Surmulot, or Brown Rat, *(Mus*

Decumamus), which comes to us from the southern regions of Asia, and whose instinct has established it more completely among us than we could have ever done by our intelligence. Vain efforts, indeed, are daily made to naturalize in our climates species that might be useful, and which seem to require much less for that purpose than this animal, whose wants are numerous. Notwithstanding this, it has been introduced and multiplied among us, in spite of every natural difficulty with which it had to encounter, and every effort on our part to expel it. Its multiplication at present is so great that it is impossible effectually to oppose its encroachments and ravages. It finds sustenance and shelter to such an extent in the habitations of man, that he may be considered more as its protector than its enemy.

The Surmulots have found in the burrows which they have dug beneath our roofs that degree of temperature necessary to their preservation. In our cultivated fields, in our granaries, in fact, in all the provision which the foresight of man has collected, they have found an aliment suitable to their life, and favourable to their reproduction. Under other circumstances they must have perished here from the effect of our winters, as they have neither the faculty of lethargizing from the cold, nor the instinct of hoarding up provision. They have profited by the fear with which we have inspired their natural adversaries; and their own natural distrust has preserved them against our attacks and artifices.

The Surmulot is larger than the Rat. They are sometimes found above eight or nine inches in length. The tail is about one-eighth of the body. This animal is less heavy and clumsy than the Marmot or the Beaver, and less light than the Dormouse or Squirrel. Its motions are prompt and lively, and it climbs and swims with agility. It lives underground, as we have said, in deep dens which it digs with a most astonishing facility. Its perseverance in labour produces effects apparently far surpassing the extent of its powers. It penetrates everywhere. It pierces walls and displaces pavements; and, as the Surmulots generally unite in great numbers, when they enter a habitation, they even put the foundation of it in considerable danger. They eat animal and vegetable substances indifferently; grains, roots, and flesh; and though portions of such provisions are constantly found at the bottom of their burrows, yet they lay up no store, at least when they inhabit our dwellings. Buffon says that the old males remain in the country during the winter, and fill their burrows with acorns, &c., which would lead us to suppose that their instinctive propensities varied according to circumstances. But this would be a phenomenon so extraordinary, as to require a very complete and distinct authentication. They make use of their fore feet in eating, and drink much, lapping with their tongues. They bring forth many times in the year, and generally from eight to twelve at a birth. When they are annoyed in their establishments by men or animals, they remove, and sometimes emigrate to a considerable distance. The habitation they then choose for their retreat is in considerable jeopardy. If they are very numerous, it is likely to be overturned. Towards the middle of the sixteenth century, they were observed for the first time in the neighbourhood of Paris, and M. F. Cuvier assures us that in some of the departments of France they are yet unknown. Pallas tells us that they arrived at Astracan in the autumn of 1727, in such numbers, and in so short a time, that nothing could be done to oppose them. They came from the western desert, and transversed the waves of the Volga, which unquestionably must have swallowed up a part of their horde. They have not advanced any further to the North, and are not to be found in Siberia.

The general colour of this animal is a darkish-gray fawn above, and a pale-gray below. The tail is scaly, *i. e.*

covered with small parallelograms of epidermis ranged in circles around it, and underneath the extremity of each lamina of epidermis grow some small gray hairs. The hairs which cover the limbs and the head are short. The mustachios are black, and the soles of the feet, which are naked, are flesh-coloured, as are also the ears and extremity of the muzzle.

No personal description of the *Common Domestic Mouse*, an animal known to every one, seems to be required. The organization of this little species accords in all respects with that of the Common or Black Rat, and the Brown or Surmulot Rat, consequently it agrees with those animals pretty generally in all its habits and modes of life, as far as the disparity of size and strength will permit, except perhaps that the Mouse may be said to have less energy of character, less resolution and determination, than the Rat, even after their relative powers have been duly estimated.

When the parasitical habits of the Mouse, in regard to man, are considered, and he may be said to be exclusively a resident in our common habitations, we seem naturally led to inquire what the habits of the animal were before the progress of human civilization had furnished him with his present means and modes of living. Analogy would lead us to the conclusion that hollow trees, clefts in rocks, and perhaps the burrows of other quadrupeds, may have furnished the Mouse with a residence before the artificial and permanent habitations of mankind supplied him with a more congenial domicile; but we have few facts in support of such conclusion, and so completely is this little troublesome inmate now attached to civilized life, that he has migrated with us to all the parts of the earth to which industry and enterprize have carried us.

Mice, though found in such numbers in a single house, cannot be called social animals; each lives insulated, for himself alone, except when the universal fiat " increase and multiply" induces a temporary association. Gestation is said to last about twenty-five days, and the young are born from four to six at a time, entirely naked, and with the eyes closed ; the mother fosters them about fifteen days, after which time they become independent of her care. They seem to have been from all time creatures of our climate, at least we have no record of their first appearance here, as is the case with the Common and Brown Rat.

The *Wood-Mouse* of Shaw, *Field-Rat* of Pennant, or *Mulot* of Buffon, *(Mus Sylvaticus,* Lin.*)* is in general under five inches long, and the tail is rather less, but it varies considerably in dimensions.   Its colour is very dark, yellowish-brown, whitish on the under part ; the tail is dark-brown above, and dirty-white underneath ; but as it varies in size, and is an inhabitant of a large portion of the earth, so it varies also in colour.   Its head is both thicker and larger, comparatively with the body, than that of the Rat ; the eyes are very large and prominent; the ears are large; and the animal stands higher than the Rat.

The Mulot or Wood-Mouse is found throughout Europe, and, though rarely. in Russia. It is a very destructive little animal, as its habits induce it, like the Squirrel, to lay up a large store of winter provision, consisting of nuts, acorns, corn, *&c.*

These animals multiply occasionally to an extraordinary degree, and become great pests by their predatory and wasteful habits.

The *Harvest-Mouse, (Mus Messorius,* White) is probably the smallest of British quadrupeds, the body not exceeding two inches and a quarter in length, and the tail two inches; and the weight is said to be about one-sixth of an ounce.

Either this species is exclusively British, or it has hitherto escaped the industrious researches of the continental naturalists, for it is doubtful whether it can be identified with the Mus Pendulinus of Hermann.

Mr. White, in his history of Selbourn, a sort of work, by the way, well worthy of imitation, particularly by the Clergy and others, who, with the blessings of liberal education, possess the means of local observation, first made this species known to the public, nor indeed have we any other original account of it. We shall quote his own words, though they have been already transferred by Shaw to the General Zoology.

"These Mice are much smaller and more slender than the Mus domesticus medius of Ray, and have more of the Squirrel or Dormouse colour; their belly is white; a straight line along their sides divides the shades of their back and belly. They never enter into houses; are carried into ricks and barns with the sheaves; abound in harvest, and build their nest amidst the straws of corn above ground, and sometimes in thistles. They breed as many as eight at a litter, in a little brown nest, composed of blades of grass or wheat. The nest is most artificially platted, and composed of the blades of wheat, perfectly round, and about the size of a cricket-ball, with the aperture so ingeniously closed, that there is no discovering to what part it belongs. It is so compact and well fitted, that it will roll across a table without being discomposed, though it contained eight little mice, which are naked and blind. As the nest is perfectly full, how could the dam," asks Mr. White, " come at her litter respectively, so as to administer a teat to each? Perhaps she opens different places for that purpose, adjusting them again when the business is over; but she could not possibly be contained herself in the ball with her young, which, moreover, would be daily increasing, in bulk."

Mr. White informs us, that though they construct nests for breeding above ground, and are found most abundantly in corn-ricks in Hampshire, they nevertheless burrow in winter, and pass the severe season underground.

The character of the sub-genus CRICETUS (the *Hamsters* or *Pouched Rat*) will be found sufficiently noticed in the text and table; we shall, therefore, proceed to some accounts of the species.

Nothing so greatly shows the power and extent of the resources of Nature, as the modes in which she supplies by instinct the want of intelligence, and puts a blind and necessary force in the place of judgment and reason. When this is done, we find those beings which are in reality the most stupid, appear to possess the most extensive intellectual faculties. They seem to approximate to man; to equal, nay, to surpass him in foresight and sagacity. What is most singular is, that these remarkable faculties are usually accompanied by organs the most limited, and physical qualities the most feeble. The circumstance, however, which separates instinct from intelligence, and gives to the latter the most decided superiority, is, that instinct is circumscribed to a small number of actions, out of the range of which it is absolutely nothing. But intelligence, on the contrary, always present, and always ready for action, extends itself to all circumstances, to all times, and to all places. With instinct the world is bounded to it alone, but the reign of intelligence extends beyond the dominion of the senses.

The Hamster, like most of the Rodentia, presents a curious example of extended instinct and bounded intelligence. For man alone the future exists in the present. No other animal is capable of foresight, or of conforming his actions by anticipated knowledge to future contingencies. Other animals exist but in the present, and they appear, in fact, to have little or no perception of time. Nevertheless, the Hamster, an animal feeble and disarmed, subsisting chiefly on farinaceous matters, conceals itself in complicated burrows, which it digs in the midst of the champaign country; and, as if it foresaw the approach of winter, and the period when the fields were to be stripped of their productions, it forms in these burrows considerable magazines, often more extensive than its wants require, and thus it is enabled to wait the return of spring, and the maturity of the harvest. In other respects it is a stupid animal, altogether under the empire of circumstances in which it may be placed. In the solitude of the country it grows timid and ferocious. Brought up in captivity, it familiarizes itself with every thing it sees or hears; but it does not distinguish those who feed it from other persons. The actions of both evidently appear to it nothing but simple motions. The rolling of a stone, the walking of a man, the running of a dog, make upon it but one and the same impression. It can, however, defend itself, and will bite cruelly.

M. Sonnini, however, gives rather a different character to this animal. Although, says he, their principal food is vegetable, they will also devour birds and other weak animals. They also fight furiously whenever they meet, and the vanquished becomes the prey of the conqueror; the fury of their quarrels and combats is pushed to such a length in this species, that both the male and female will sometimes destroy its mate. The same writer relates that the female of a pair confined in one cage destroyed the male, and devoured a part of its viscera. Ferocity, he continues, is the dominating passion of the Hamster; he attacks whatever he meets; neither superiority of strength nor size will deter him; the dimensions of the horse, the address of the dog, not even the appearance of man, will deter him from the attack, and he will rather die than yield or quit his hold.

Fabricius *(apud Gesner, Hist. Quad.)* states that he had seen a Hamster leap on the muzzle of a horse, and hold on with his teeth till he was killed. When preparing for attack, the Hamster voids his pouches, if previously full, and then inflates them so much that the head and neck become larger than the rest of the body; he then erects himself on his hind legs, and darts on the object of his fury, and if he seize it will not quit his hold with life; and Sonnini adds, that he had seen a Hamster confined in a cage, and irritated by means of an iron rod nearly red hot, seize the rod with his teeth, and in spite of the pain he must have suffered still persist in keeping his hold.

The Hamster digs with great facility; each burrow it forms is composed of two entrances, one which conducts to an oblique canal, at the entrance of which the animal throws out and accumulates the earth which it is forced to remove, the other which serves as an exit to a vertical canal, which is the true entrance to the burrow. These two canals conduct to a greater or less number of particular excavations of a circular form, which according to the age of the animal are from about one to five feet in diameter, and communicate together by horizontal conduits. One of these excavations is the retreat of the Hamster; it is furnished with a good bed of dry herbs, and it is here that the females bring forth. The other excavations constitute the magazines of provision.

Each animal has its own burrow. The males have usually but two openings to theirs. The females form several by vertical conduits, especially when they have young ones. The burrows of the old individuals sometimes embrace a considerable extent; they descend to four or five feet in depth, and frequently contain many bushels of corn or other grain. They are sought out with great care, as much to collect what they contain, as to destroy the animals which form them, and which, when they are numerous, cause great devastation in the harvest. In the environs of Gotha, it is said, that in a single year eighty thousand of these animals have been killed; they are discovered by the quantity of earth accumulated at the entrance of their oblique excavation. These Rodentia are not merely granivorous; they will also cat flesh, and often devour one ano-

ther when they meet.   Thus it is, that like the most ferocious animals, they live in a solitary manner, and never seek
each other but at the season of love.   The particulars of their
reproduction are not precisely known.   The rut, it would
appear, takes places many times in the course of the spring,
the summer, and the autumn ; the gestation of the females
lasts four weeks, and occurs three or four times in the
year ; there are usually from six to twelve young ones,
which, after a very short lactation, quit the mother to go
and dig, each for itself, a burrow, and live by their own resources.

Some writers inform us, that the Hamsters pass the winter in a profound lethargy ; others have some doubts concerning the hibernation of these animals.   It appears, however, very certain, that it takes place, and at this epoch all
the entrances to the burrows are closed up.

The Hamster is not much larger than the common rat,
and has the same form of head and general physiognomy.
Notwithstanding, however, the general and numerous relations which approximate these animals to each other, they
differ very essentially, and form very distinct sub-genera.
The Hamsters, like the rats, have four incisors and twelve
molars, equally divided on each side of the two jaws ;
and these teeth have similar relation together.   But in the
rat the coronal is formed of tubercles irregularly disposed,
while, in the Hamster, it is divided by very regular furrows.
The fore-feet have four toes, with a rudiment of a thumb,
furnished with a nail, which, imperfect as it is, may be
considered as a genuine toe ; for under the skin we find the
phalanx of which it is composed.   The hind-feet have
five distinct toes, and those, like the first, are entirely free.
All these toes are furnished with long and sharp nails, proper for digging.   The tail is conical, and about eighteen or
twenty lines in length.   The eyes small, globular, and projecting.   The external conch of the ear is considerably ex-

tended, rounded, and simple. The tongue is thick and soft, and there are large pouches in the interior of the mouth: the upper lip is provided with long mustachios.

The entire body of this animal is covered with long, thick, and soft hair, of a grayish fawn, on the upper parts, and a deep black on the lower. The sides of the head and body, the circumference of the ears and the buttocks, are of a brilliant fawn. There are three spots of whitish-yellow, on each side of the body.

The Hamster is an animal of great neatness and cleanliness It is constantly employed in polishing and cleansing its hair with its paws, which it licks like a cat, and then applies to its body. It climbs with great facility, using its fore-paws like hands; but it walks and runs heavily. As long as it can conceal itself, it remains inactive a considerable part of the day. Otherwise it seems agitated, without, however, being affected by fear. When in its motions it meets with any thing eatable, it stops, takes some part, and casts by the rest for future use. It will eat fruits, roots, onions, and corn, which it prefers to every thing else. It does not drink, urines often, and its excrements are black, of the size and form of a middling bean, somewhat elongated. The Hamster is found in France only near the Lower Rhine, but is common in all the Northern parts of Germany, where the soil is suitable. It is also found in Poland and part of Russia, &c.

The Hamster is one of the best known of the Rodentia. No where has it been spoken of with so much detail and exactitude, as in the special treatise of Sulzer upon it.

We are told, however, that there is a variety of this species entirely black.

The *Canada Hamster* (*Mus Bursarius*, Shaw) was first described and figured in the Linnean Transactions, and subsequently by Dr. Shaw in the General Zoology. We are, however, still uninformed, with sufficient certainty, as to its

CANADA HAMPSTER.

MUS BURSARIUS. SHAW.

THE CHINCHILLA.

CRICETUS ? laniger

Published by G.B.Whittaker. Ave Maria .

dentition; and the allocation of the animal among the Hamsters, must, therefore, still be considered conditional. We have engraved a figure of it from our collection of drawings, but are not able to add anything, either on the subject of its physicalities or habits. Its superficial specific characters are detailed in the Table, as are those of the Chinchilla, a beautiful little animal, presumed to belong to the division of Hamsters.

The *Anomalous Hamster (Mus Anomalus,* Thompson) must be considered also as placed conditionally only among the Hamsters by the character of the cheek-pouches; but another peculiarity equally, or still more observable, connects this species with the Echimys, or Spiny Rats, which is the flattened spines or prickles, particularly on the back.

We are indebted to Mr. Thompson for a first notice of this animal, which was found in the Isle of Trinity, where, however, it is very scarce. It is about the size of the common Rat, but the nose is more pointed; the ears are naked, round, and of moderate size: the pouches are formed by a duplicature of the common tegument, like the pouch of the Opossum, and are of considerable size. The body is covered with fine lance-shaped spines, stronger on the back than elsewhere, intermixed with hair. The Table will supply its other known characters.

It is obvious that this species differs perhaps generically from the other Hamsters, with which, indeed, it seems to have no other relationship than by the cheek-pouches. Exteriorly it has the appearance of the Echimys; nor does it seem to be quite satisfactorily shown that these latter animals are destitute of a similar pouch; the Mus Anomalus may, therefore, in fact, be an Echimys proper, or a pouched modification of that subgenus of Rodentia. In the absence of more certain premises, it seems premature to separate it otherwise than specifically; we have, therefore, in this, as in numberless other instances, followed the stream, and placed it conditionally with the Hamsters.

III.   *

We now come to the genus GERBOA. This genus approximates considerably to the Rats properly so called, by a great number of characters of internal organization, but is sufficiently distinguished by the shortness of the anterior limbs, and the length of the hinder extremities, or to speak more correctly, of the hinder metatarsi, and by the tail, which is covered with long hairs at its extremities.

As to external conformation, the Gerboas exhibit some relations with the Kanguroos. The form of the body is the same in general. The hinder limbs are likewise five or six times stronger than the fore. In both genera the tail is very long; the ears elongated, and pointed, and the eyes very large and round. But though the Kanguroos have so many traits of external conformation similar to the Gerboas, they are infinitely removed from them in most important points, such as the organs of generation, ventral pouch, &c. Erxleben was therefore decidedly wrong when he classed the Kanguroos with the Gerboas, under the name *Jaculus Giganteus*.

The Gerboas have the same teeth as the Rats, that is, they have two incisors in each jaw ; and the lower, instead of being flat and cut scissors-like, as the upper, on the contrary, are conic and pointed. The molars are generally six in number, three on each side. They are slightly sloped. There is sometimes an additional one in the upper jaw.

In the Gerboas the cheek-bones are very prominent, which gives a singular and flatted form to the front part of the head. The muzzle is short, large, and obtuse. A considerable number of stiff hairs extend on each side, and form long mustachios. The nose is naked, cartilaginous, and in one species rather complicated. The ears are long and pointed; the eyes large, and placed altogether on the sides of the head.

The body is a little elongated, larger behind than before, and well covered with soft and silken hairs. The fore-feet are very short and feeble. They have four or five toes according to the species. The thumb or interior toe, where it exists, is very short, rounded at its extremity, and provided with an obtuse nail. The other toes are long and armed with crooked nails.

The hind-feet are as disproportioned as those of the Kanguroos, being four or five times longer than the fore-feet. They are terminated by five or six toes, according to the species, which are armed with short, but large and obtuse claws. The three middle toes are always supported by a single metatarsian bone, terminated by as many articulary pulleys. This arrangement is analogous to what is observed in the Ruminantia and Birds. When there are but three toes, there is in all but one metatarsian bone. When there are five, there are three bones in the metatarsus, one of which alone is very strong, the lateral bones being very short and slender.

Ancient and modern naturalists have both been mistaken respecting the walk of the Gerboa. They have all imagined that these quadrupeds walked on their hind feet only, never employing the fore-feet for that purpose. From this error the genus was named *dipus*, two-legged.

M. Olivier, from a consideration of the structure of this animal's body, puts an end to this error; clearly proving that the Gerboa was incapable of sustaining itself for any length of time upon the tarsi. It usually walks on its four feet; but when frightened from any cause, it endeavours to escape by means of prodigious leaps, which it executes with equal force and activity. When these animals are about to leap, they raise their body upon the extremity of their hind toes, and support themselves upon their tail. Their fore-feet are so closely attached to their breast, that they are scarcely visible. Having taken their spring they leap, and fall upon their four feet; then they elevate themselves again with so much celerity, that it almost appears that they are constantly in an erect posture.

The tail of the Gerboas is as long, or as long and half as the body. It is not very strong at its base, like that of the Kanguroos. Its thickness is pretty nearly equal in the entire of its extent. Sometimes it is perfectly cylindrical, at others quadrangular. It is usually covered with smooth hairs to its extremity, which is terminated by a tuft of long silken hairs. The Gerboas make use of their tails to sustain their bodies when they rise from one leap to execute a new one. It then has the form of an S reversed (ⁿ). M. Lepechin having cut the tails of some of these animals to different degrees of length, observed that the extent of their leap diminished in the same proportion. Those from which he cut it altogether, could not run at all, but fell back directly as they attempted to raise themselves on their hinder feet, wanting the necessary support of the tail.

The female Gerboas have eight teats placed on the whole extent of the belly. The orifice of the vulva seems to be confounded with the anus. The males are generally smaller than the females. The tints of the skin are also less deep.

The genus Gerboa is now composed of several distinct species, one of which is extremely abundant in Barbary, in Higher and Lower Egypt, and Syria, and again in the more northern climates, situated between the Tanais and the Volga.

The others occupy an immense space in Siberia and the north part of Russia, from Syria to the Eastern Ocean, and as far as the northern parts of Hindostan. A late one, recently described by M. de Blainville, has been published, though it would seem erroneously, as belonging to New Holland.

The synonymy of the Gerboas is in a very perplexing state. The generality of naturalists not having observed with sufficient care the number of toes in these animals, factitious species have been the result of this inaccuracy. It will be better, therefore, to reject all testimony concerning such as seem doubtful, and to admit none but what are perfectly authenticated by the most exact observers.

Among the species of this genus, the *Gerbo*, which is extremely common in Egypt, lives in troops, and digs burrows. Without being very wild, it is of an unquiet character. At the least noise it runs precipitately out of its hole. It eats corn, nuts, roots, and all kind of fruits. The *Alactaga* lives in the same manner, but prefers cold and fertile climates to the hot and sandy regions in which the Gerbo chooses its retreats. It sleeps during the cold months of winter at the bottom of its den. But the slightest degree of heat is sufficient to rouse it from its stupor.

All the Gerboas carry their food to their mouth with their hands or fore-paw, and accordingly they are claviculated. They have also under the skin certain glands, analogous to the *Thymus*, with which other quadrupeds, which lethargize at the approach of winter, and pass that season in a state of stupor, are also provided.

It appears difficult to keep these little animals in a state of captivity, and still more to transport them to our climates. They gnaw the hardest wood with an extreme facility.

We shall first notice the species which the Baron has inserted in the Animal Kingdom, and then notice anything that may be interesting in other species mentioned by other naturalists.

The first is the *Gerboa*, properly so called, or *Gerbo, Jerboa*, of Daman and Shaw. The *Mus Egypticus*, Haselquist. *Mus Jaculus*, Linnæus, edit. 10 ; and a variety of other synonymes, or misnomers, with which naturalists have contrived to embarrass the subject as much as possible, but with which we shall forbear tormenting our readers for the present.

The Gerbo, confounded by Linnæus and Buffon with the *Alactaga* or *Mongul*, is, however, totally distinct. Its size

M 2

is nearly equal to that of the Rat. Its head is thick, and large in proportion to the body ; but yet more elegant in form than that of the Alactaga. The nose is smaller, and the ears shorter and larger. The upper incisors, which are generally white, are vertical, square, and divided in their length by a groove, which runs through the middle. The silken hairs of the mustachios are about six inches long. The eyes are large, projecting, and lateral, and separated from each other by about an inch and a half, and have a brown iris. The ears are whitish at the base, the rest gray. The body is rather elongated, and larger behind than before. The general colour of the skin is a clear fawn, and varied with black and zigzag lines, which contrast agreeably with the beautiful white of the under part of the body.

Naturalists do not agree on the number of this animal's toes. Besides the three toes armed with obtuse nails, Sonnini says that he has observed near the heel, a kind of spur, or rather a slight rudiment of a fourth toe. From this circumstance he regards the Alactaga, described by Samuel Gmelin, as apparently belonging to the same species with the Gerbo. Most authors, however, differ on this point with Sonnini. Pallas gives five toes to the *Mus Jaculus*, and but three to the *Mus Sagitta*. Edwards declares that he has not been able to discover the little spurs above mentioned. M. Olivier, and, above all, our illustrious author, coincide with the opinion of Pallas. The only authors of Sonnini's opinion are Buffon, who never saw either the Gerbo or Alactaga, and J. F. Gmelin, who confounds the two species, under the name of *Dipus Jaculus*, while he admits as a distinct species the Mus Sagitta of Pallas, which is undoubtedly the same animal as the Gerbo.

We may as well add here some anatomical points of distinction observed by M. Desmarest.

The tail of the Gerbo has six vertebræ less than that of

the Alactaga, which has thirty-one. The bones of the first are in general stronger. The cæcum is shorter, and scarcely reaches the symphysis of the pubis. Of the three toes of the Gerbo, that in the middle is scarcely larger than the two others. Of the five toes of the Alactaga, the middle is considerably longer than the lateral toes, which are removed back as far as one-third of the length of the metatarsus.

The Gerbos are very common in Barbary, in Egypt, in Syria, and Arabia. But their species is proportionally less numerous as we advance to the north. Nevertheless it does extend in that direction as far as the countries situated between the Tanais and the Volga, as far as 50° north latitude. It was in this country that Pallas discovered his Mus Sagitta. In this country these animals are less numerous than the Gerboas, of which Pallas makes a third variety of the Alactaga or Mongul. But they are found in considerably numbers in the sandy hillocks which border the southern banks of the Irtish.

Egypt is the country in which Gerbos have been observed with most facility. The sands and ruins which environ modern Alexandria, are, says Sonnini, very much frequented by these animals. They live there in troops, and form burrows which they dig with their nails and teeth. It is said that they can pierce through the small stones which are below the bed of sand. Without being exactly ferocious, they are very unquiet animals. The least noise makes them re-enter their holes with precipitation. They can only be killed by surprising them. The Arabs know how to take them alive, by closing up the issues of the different galleries of their burrows, with the exception of one through which they must go out. Their flesh, though not the best of meat, is yet in considerable request among the Egyptians. Their skins, the hair of which is soft and shining, is employed in the manufacture of ordinary furs.

The Gerboas are found represented in an upright posture on the models of Cyrenaica.

Under the name of *Mus Jaculus*, Pallas has joined three animals, differing in size, and the proportion of certain parts, but strongly resembling in the disposition of their colours, their general forms, the tuft of hair at the tail, and other particulars. He even says himself, that in many respects they may be considered as forming distinct species ; and M. de Blainville has divided them, under the name of *Mongul Gerboa, Brachyura,* and *Little Gerboa.*

Samuel Gmelin published a genuine species of Gerboa, under the name of *Cuniculus Pumilio Saliens,* and *Alactaga,* which last word, in the language of the Mongul Tartars, signifies variegated colt. This Gerboa resembled, in all respects, the Mus Jaculus of Pallas, of the large variety, except in the hind feet, which, according to Gmelin, have but four toes, while the Mus Jaculus of Pallas has five. It seems probable, however, that Gmelin was mistaken in reckoning the toes, and did not perceive the internal, or fifth toe. This is particularly the opinion of the Baron, as our readers must have seen.

We shall describe this animal under the name of *Alactaga* alone, the great variety of Pallas, and in the two others we shall follow M. de Blainville, without exactly subscribing to his opinion of distinct species.

This animal is about the size of the common squirrel, and not that of the rabbit, which the name *Cuniculus,* adopted by Gmelin, would lead us to believe. It differs from the Gerbo by the number of toes on the hinder feet ; the Gerbo having but three, and the Alactaga five. In this relation it approaches the two other varieties of the Mus Jaculus of Pallas, of which we shall speak hereafter.

The fur is very soft and pliant, of a yellowish fawn over the body, but varied with a grayish brown towards the crupper. The muzzle is white towards its extremity, and brown above. All the under part of the body is white, as well as the interior of the limbs. The sides are gray. The buttocks are marked with a white spot in the form of a crescent. The tail is longer than the body, covered for two-thirds of its length with similar hair, and terminated by a tuft composed of two ranks of hairs, half white and half black.

The head is oblong, and the muzzle rather protruded, but blunt and thick. The nose is large, flesh coloured, formed like a heart, with crescented nostrils, divided by a partition. There are eighteen teeth, two incisors in each jaw, four molars in the upper, and three below on each side. The mustachios are formed by long and black hairs. The eyes are large, with a yellow-brown iris. The pupil is nearly round. The ears are longer than the head, half cylindrical, oblong, naked, or nearly so, and transparent. The hinder feet are as long as the body, head included. The tarsus and metatarsus are particularly long. In the Skeleton we find only three metatarsian bones, the middle toe of which supports the three principal toes, and is terminated by three articulations pulley-formed. The lateral metatarsians are extremely slender, and but half as long as the middle.

The Alactaga is found in the deserts of Tartary, on the sand-hills which border the Tanais, the Volga, and the Irtisch. This species or variety is not numerous, considerably less so than the *Gerboa Brachyura*. The two, as well as the little Gerboa, inhabit all the region which extends from east to west, from the desert of Crimea, or the neighbouring territories of Taurida Chersonesus, to the countries situated between the Argun and the Onon, and from north to south, from the fifty-fifth degree of north latitude to the tropics. These animals dig very deep burrows, but accumulate no provisions there. This we give on the authority of Pallas, but Gmelin says, that they collect herbs and roots during the summer, form them into different heaps, and transport them by degrees into their burrows, after having left them some time in the open air to dry. Like the Loir and the Marmot, they are in a lethargic state during the winter. They remain in their retreat all day, and sally forth by night to seek their provision, which consists principally in herbs or succulent plants, in roots, fruits, *small birds,* and *insects.* They also devour each other, and always begin by eating the eyes and brain. In the deserts situated to the west of Tartary they feed on the bulbs of tulips, which are there very abundant, and of various other plants, such as the *Chenopodium,* the *Atriplex,* the *Salsola,* and the *Salicornus.* Beyond the lake Baikal, they find bulbs of the *Lilium Pomponium.* In warm climates the females produce many times a year, and the number of their little ones should be considerable, as their teats are eight in number. They dig their burrows with sagacity and surprising activity, scraping away the earth with their fore-paws, and plucking out with their teeth all the roots in their way. They take but a few minutes to make an excavation of two or three inches. Their dwellings are about half an ell deep, Russian measure. The burrow extends obliquely into the earth, and there are many openings above which cut it perpendicularly. These may, with probability, be deemed a kind of breathing holes.

The least degree of cold reduces the Alactagas to their lethargic state, and what is more remarkable, a great degree of heat will produce on them a similar effect. They foresee the cold and rainy seasons, and close their burrows with most astonishing punctuality. In the seasons which are dry, but yet cloudy, they come forth from their retreats in the day-time.

The swiftness of these animals, especially when pursued, is almost incredible. It is so great, that they scarcely seem to touch the earth, and according to the report of Pallas, they cannot be overtaken, even on horseback. Their tail serves as a resting point when they fall to the ground, and a helm during their leap, or rather flight.

They sometimes walk on the four paws, but it is only when they are uneasy, or when they are digging their burrows. They often hop like birds. They carry their provision to their mouth with their fore-paws.

It is difficult to preserve these animals in captivity, unless they receive a sufficient quantity of earth or sand to dig into. They may be subsisted on carrots, fruits, cabbage, bread, &c. They never drink, and yet urinate abundantly.

The Arabs, Tartars, and Calmucs find the flesh of the Alactaga a dainty, and employ themselves greatly in the chase of this quadruped. The Burats and Monguls have fallen into an erroneous opinion that this animal sucks their sheep by night.

The *Gerboa Brachyura, Dipus Brachyurus* of M. De Blainville, is Pallas's second variety of the *Mus Jaculus*. The size is intermediate between the *Alactaga* and *Little Gerboa*. The muzzle is less elongated than in the first of these animals. The mustachios and ears are shorter, and the last are wider. The cylindrical tail, thicker and shorter in proportion, is terminated by a tuft, the hairs of which are not so exactly distinguished, and the white part is less extended. The hind-feet are also relatively shorter, with more robust toes, and the nail of the middle toe less long than those of the lateral toes.

The size of this animal, when adult, and the epiphyses of the bones are completely united, is nearly equal to that of the Common Rat, while the dimensions of the Alactaga approach to those of the Squirrel, and the Little Gerboa is not much beyond the Mus Sylvaticus in size.

In proportion and size, this variety approximates very nearly to the Mus Sagitta or Gerbo, but differs in the number of toes. In this respect it is analogous to the Alactaga and Little Gerboa. It has four distinct toes, and a thumb scarcely apparent on the fore-paws, with a very short nail. Five on the hinder.

The upper part of the body is a pale gray-fawn, varied with brown. The under is white, and the buttocks are marked, as in the two preceding species, with a white transversal band. The muzzle is white at the extremity, and brown above, like the Alactaga, and not the same colour as the back, like the Little Gerboa.

This variety is greatly multiplied. It is found especially in Oriental Tartary, and Siberia. Beyond the lake Baikal, it is the only variety found, and probably in the deserts of Mongolia.

The last variety of the *Mus Jaculus* of Pallas is the *Little Gerboa*, or *Dipus Minutus* of M. de Blainville. It differs from the first in the proportions of various parts, and in its little tail, which does not exceed that of the Mulot. The difference of size does not depend on age, inasmuch as the individuals of this small variety have the epiphyses of the bones as completely consolidated as those of the Alactaga and Gerboa Brachyura.

There are four toes, and a small thumb on the fore-paws, and five on the hinder. The hinder extremities and tail are proportionally longer than in the last variety. The jaws on each side have one molar less than in the Alactaga and Gerboa Brachyura. The thigh-bone, which in the Alactaga is as long as two-thirds of the tibia, is here longer than the tibia. The hairs of the tail are ranged like the barb of a quill, an arrangement less apparent in the preceding species, and the white part of the tuft is less extended.

All the upper part of the body is a yellowish-gray mixed with brown. The under part is a beautiful white, as also are the paws and feet. There is on each side of the buttocks a white transversal band, slightly crescented. The principal characteristic of this variety is, that the muzzle instead of being white is the same colour as the upper part of the body, and that in some subjects there is a white spot upon the body, sometimes a large black spot is found on the epigaster.

The habitat of the small Gerboa is more southern than that of the two preceding varieties. It is found near the Caspian, in regions where the Alactaga frequents, and on the lower banks of the Volga and the Khymer, in regions where the Gerboa Brachyura is also found.

We shall leave to our readers to decide whether the differences now described between these animals, which after M. Pallas, we have persevered in calling varieties of the Alactaga, are sufficient to constitute distinct species. For our own parts, we rather incline to a different opinion. None of the differences above-mentioned are as important as those which exist among the varieties of the human species. But we have already entered our protest very frequently against the undue multiplication of species.

We shall now proceed to other species of the Gerboa, as given by other naturalists, though not actually authenticated by the Baron.

The *Great Gerboa,* or *Dipus Maximus* of M. de Blainville, was first observed by this clever naturalist in London, in 1814, in the Menàgerie of M. Polito, in the Strand. It is about the size of a small Marmot or Rabbit; its head is large and rounded; eyes very large, separated and altogether lateral; the iris is black and the pupil round; the cheeks are wide; the muzzle is short, very thick, with a deep furrow that divides the upper lip in two, and is extended to the partition of the nostrils. The ears are very slender and transparent, not much covered with hair externally; they are large, rounded at the extremity, with a kind of dilatation, equally rounded at the exterior side of their base; the nose is complicated; the opening of the nostrils are oblique and semi-lunar, placed laterally and surmounted with a deep fold, in the form of a V, the branches of which are bifurcated, and the point of which terminates in the furrow of the lip ; the mouth is not deeply cut ; the incisors, two in number in each jaw, are very apparent, long, narrow, and trenchant at the extremity, as in all true Rodentia, and the upper ones have no longitudinal furrow in their anterior face.

M. de Blainville could not observe the number and form of the molar teeth, but he assured himself of the absence of the canines.

The body of this Gerboa is very large, especially in the posteriors. The belly has neither lateral folds nor pouch, like the Didelphes. The fore-paws, very short, are provided with four distinct toes, armed with crooked nails, and no vestige of thumb is apparent.

The hinder paws, on the contrary, are considerably developed ; the thighs exceedingly strong and muscular ; the leg extremely long, as is also the metatarsus, which rests altogether on the ground in a state of repose. The toes are three in number, of which the middle is the longest, and is terminated by a very strong claw. The external one is much less thick, and the internal is the smallest. It is easy to satisfy one's self by the touch, that there is but one metatarsian bone for these three toes   In the individual observed by M. de Blainville, the tail was evidently mutilated ; there remained about two inches of it.

The fur is extremely soft, thick, and well furnished with hair, altogether analogous to that of the Rabbits. The general colour is pretty nearly that of the Surmulot, or rather a brown, a little more fawn-coloured above the long hairs, being black at the extremity ; underneath it is entirely white. A large black spot crosses the eye, and extends across the forehead to the other. The end of the muzzle, or rather the extremity of the nose, is of the same colour ; the mustachios are extremely long, very black, and formed of shining hairs, and on each side of the head are two other brushes or tufts of silken hairs, but much smaller, one above the eye, and the other behind. The hairs which cover the origin of the tail are long, but not tufted. The metatarsus is covered with very short hairs nearly like the Rabbits.

This description proves that this animal belongs to the genus Gerboa. Nevertheless, it differs from all the common species by its larger size, its clumsier body, its more robust hinder limbs, and shorter in proportion.

The individual just described was extremely ferocious and uneasy. It was continually attempting to gnaw the bars of the cage in which it was shut up. Its mode of walking was altogether like that of the Hares or Kanguroos, and it appears probable that it could leap with great vigour. It scratched itself with its hinder paws like the Kanguroos, and licked the front ones like the Rabbits. It was supported with bread, carrots, and other vegetables, which it carried to its mouth with its fore-paws.

Our author is inclined to refer to the Gerboas certain animals of which Desmarest and Illiger have formed a separate genus. The first of these naturalists has given it the name of GERBILLUS, the second, that of MERIONES. We shall give their description after M. Desmarest.

There are six species, four of which belong to the warm climates of the Old Continent, and two to North America.

The characters are not yet fixed in an invariable manner, as the teeth of but one species are known. All resemble in character purely exterior, such as the length of the hinder extremities, the small projection of the cheek-bones ; also in such interior characters as the composition of the tarsus, &c.

The Gerbilli may be considered as long-footed Rats, generally of a small size, with the anterior extremities rather short ; on these there are four toes, and the rudiment of a thumb.

The hinder feet are constantly divided into five toes, all nearly of the same thickness. The metatarsus is remarkably long, and formed of as many bones as there are toes. This last is a remarkable distinction between them and the other Gerboas ; the head also is pointed and elongated, and there is a very small projection of the cheek-bones ; while in the others the projection is considerable, on account of the great curvature of the zygomatic arch. The tail is long and proportional to the elongation of the feet. It is always more or less finely annulated, and scaly like that of the Rats ; sometimes covered with hairs, and sometimes almost naked. Its extremity is not tufted, though in certain species some hairs are longer than others.

Two of these species, the *Mus Tamaricinus* and *Mus Meridianus* of Pallas, have been regarded by some naturalists as belonging to the genus Myoxus. This, from a consideration of characters, would seem to be an erroneous opinion. The Myoxi, or Dormice, live in trees, and are approximated to the Squirrels by characters sufficiently numerous and important. All the subterraneous animals, such as the Rat, the Hamster, the Marmot, and the Gerboa, have common characters, which group and form them into one natural family. All the known species of Gerbilli dig burrows, and many of them accumulate provisions, which circumstance marks a clear analogy between them and the last-mentioned Rodentia.

If the Gerbilli are not real Gerboas, (which we are inclined to believe,) it cannot be denied that the closest affinity exists between them. They resemble not only in that external character which is most apparent, namely, the disproportion between their anterior and hinder extremities, but also in their mode of living. They walk and run only by leaps, and are possessed of very considerable swiftness of motion. One single species, the *Gerbillus* of Canada, has been found in a state of hibernation. M. Raffinesque Schmaltz, in the prodromus which he has published on these animals, mentions the names of ten different species, some of which we shall notice here after M. Desmarest, the others being altogether unknown.

The first species is the *Mus Tamaricinus* of Pallas, commonly called the *Tamarisk Gerboa.* This animal is larger than the common Rat, to which it has a considerable resemblance. The oblong head is terminated by a convex and rounded muzzle; a membranous fold covers the nostrils; long whitish silken hairs; the upper lip is divided into two lobes, and the under is very thick. The external face of the incisive teeth is yellow, and the upper ones are marked by a furrow, and slightly crenulated at their extremities; the lower ones are obtuse.

The Tamaricin has a lively animated physiognomy, arising from its large eyes; the edges of the eyelids are of a clear brown, and without hairs; the ears are almost naked, large, oval, and bordered with a kind of brown down. A transversal fold, somewhat elevated, is observable at the entrance of the auditory conduit; the neck is short, and the proportions of the body are similar to those of the Lerot. The legs are strong, and the hinder longer than the fore, though in proportion shorter in this species than the rest. The fore-feet have four toes, without reckoning a sort of tubercle which stands for a thumb. On the hind-feet are five toes, and the thumb is longer than the external toe. All the toes are naked and wrinkled underneath. There are two callosities on the carpus, and three on the metacarpus. The tail is nearly cylindrical; it is entirely covered with hairs; those on the extremity are the largest, and form a brown tuft.

The hair of the body is softer than that of the Rat, and rougher than that of the Squirrel. On removing this hair, we find a down of a leaden colour underneath attached to the skin. All the upper part of the body is of a yellowish-gray; the sides have a shadow rather less deep, but towards the crupper it becomes brown; the circumference of the eyes and nose is whitish, and there is a spot of the same tint above the eye, and another behind the ear; the sides of the head are of a whitish ash-colour; all the under parts of the body are entirely white, as is the under part of the tail; the upper part is ash-coloured, but cut by about two hundred brown streaks, which form so many rings.

Pallas has observed the *Tamaricin* on the southern coasts and deserts of the Caspian Sea, and this great naturalist is of opinion that this pretty animal also inhabits the warmer climates of Asia. It frequents those departments which abounds in tamarisks and saline plants, such as the *salicornus, atriplex maritimus, &c.*, which form the principal articles of its subsistence. Each individual of this species lives isolated in a profound hole, dug under the roots of trees. This burrow has two galleries, the openings of which will scarcely admit one's four fingers. The Tamaricin never quits this retreat but at night, and is easily taken in the snares which are set at the entrance of its hole.

The second species is the *Gerbillus Indicus*, described in the eighth volume of the Linnean Transactions, under the name of Yerbua. This animal is about the size of the Domestic Rat, but its head is larger in proportion than its body; the ears are large, round, upright, and almost naked; the nose is very round, and furnished with mustachios; the upper jaw is half an inch longer than the lower; the lower incisors are twice as long as the upper; but these last are longer, and divided by a longitudinal furrow; the eyes are large and of a brilliant black; the legs are of an unequal length; the fore ones are shorter than the hind, and have four toes and a small tubercle; the hind-feet have five toes; the three middle ones are twice as long as the one on the fore-feet; the external toe is one-half as long as the others, and the internal is the shortest of all; the claws are white, and of a moderate length.

The fur is of a brown-red, mingled in the upper part of the body with small spots of an obscure brown, arranged longitudinally. The head is of a flaxen hue, particularly around the eyes. All the other parts are white.

The tail is cylindrical, slightly covered, and terminated by a tuft of long and soft hairs, of an obscure brown.

This Gerbillus was discovered in Hindostan, between Benares and Audwan, by Lieutenant-Colonel (now General) Thomas Hardwicke, in 1814. It lives on barley and wheat, and forms considerable magazines of these different grains in the spacious burrows which it inhabits. It cuts the grain near the root, and thus carries off the entire stalk. It never touches its provisions until the harvest is over, and the fields no longer furnish supplies. It only goes out by night, runs very swift, and often leaps; its leaps are often above five yards in length.

The third species is the *Gerbillus* of the torrid zone, or, as it is sometimes called, the African Gerboa, *(Gerbillus Meridianus)*. Some authors have considered this species as the same with the *Mus Longipes* of Linnæus. After the two last it is the largest of the entire genus; its hinder-feet are proportionally more elongated than those of the Indian Gerbillus, but much less so than those of the subsequent species.

Its size is intermediate between that of the Rat and Mulot; its tail is nearly the length of the body, strong, cylindrical, and covered with hairs at the extremity, which form a tuft.

The head is oblong, and the muzzle tolerably advanced; the ears are large, oval; the mustachios very long; the incisor-teeth are yellow, the upper being marked with a longitudinal furrow; the body is thicker in proportion behind; the thighs are thick and fleshy; the feet elongated, large, fit for leaping, and divided into unguiculated claws; the fore-paws have only four toes, and a very short thumb;

the fur on the upper part of the body is of a pale fawn-colour, mingled with gray, and underneath of a beautiful white ; the tail, shorter than that of the Tamaricin, is the same colour as the back, and has neither rings nor spots ; a longitudinal line of a reddish-brown is observable under the centre of the belly.

This animal inhabits the sandy and burning deserts which border on the Caspian, and which are situate between the Volga and the Ural. It digs burrows, and lives on the nuts of the *Pterococcus Aphyllus, &c.*

The next species we shall notice is the *Gerbillus Ægyptius,* sometimes called the Long-legged Rat, and is the same as the *Dipus Pyramidum* of M. Geoffroy.

This species must not be confounded with the Gerbillus of the torrid zone. There is a notable dissimilarity in the proportions of the hinder extremities, they being much more slender and elongated in this than in the other species. There is also a difference of size between these Rodentia, for the Egyptian Gerbillus is hardly as large as a Mouse, while the other may be considered as intermediate between the Mulot and common Rat.

An individual described by M. Olivier had a conical and pointed head, like the Rats; ears oval and moderate ; mustachios long ; and very short neck ; the fore-paws were very short, provided with five toes, the four exterior toes being armed with crooked claws; the internal, or thumb, was very short, and without nail ; the hinder feet were as long as the body, very strong, almost naked, and provided with six unguiculated toes, almost equal in length ; the tail was a little longer than the body, covered above with small brown hairs, and terminated by hairs more elongated ; all the upper part of the body was of a clear yellow, and the under part of a pure white.

This animal was met near Memphis, coming out of the burrow which he inhabited.

The *Gerbillus* of Canada, *Canadian Gerbo* of Shaw, and *Mus Canadensis* of Pennant, was discovered in the environs of Quebec, by Mr. Thomas Davies. It has been figured in the fourth volume of the Transactions of the Linnean Society. It is about the size of a Mouse; its head is small, and shaped like that of a Rat. The ears are very short, and not elevated like those of the Gerboas. The upper jaw is furnished with long mustachios; the fore-paws are proportioned to the size of the animal; it has four toes, armed with crooked nails; the posterior extremities are very long, and terminated by five toes of nearly equal length, and armed with claws like the fore-paws; the tail is longer than the body, almost naked, having but a few scattered hairs at intervals, and no terminating tuft.

This quadruped is found in the meadows and thickets. When surprised, it attempts to escape by leaping with great vigour to distances very considerable for so small an animal. In the winter it retires and falls asleep, rolled up like a ball, in a burrow about twenty inches deep. It places itself then in a sort of little chamber, of an oval form, and never stirs until the middle of spring. No provision is found in this retreat, nor is it exactly known on what substances it feeds.

Another species is the *Gerbillus Soricinus*, very imperfectly known. Its tail is said to be shorter than the body; the latter is a grey-brown in the upper parts; the sides are marked with a red longitudinal streak; the ears are almost naked, and oval; the tail is covered with silken hairs, and of a gray-brown underneath.

It is a native of North America, but its mode of living is entirely unknown.

We ought, perhaps, more methodically to have noticed at the end of the Rats, properly speaking, a new genus of Glires, proposed by Messrs. Say and Ord, in the Journal of the Academy of Natural Sciences of Philadelphia, published in the present year, (1825,) named NEOTOMA. The number of the teeth is 16 ; incisors $\frac{2}{2}$, cheek-teeth $\frac{6}{6}$, with profound radicles. The angles of the ridges on the surface of the cheek-teeth, which, together with the roots of these, seem to found its principal pretensions to a generic separation, are sufficiently marked by the figures. The grinding surface of the molares, say its describers, differs somewhat from that of the molares of the genus Arvicola, as will be perceived by our figures, but the large roots of the grinders constitute a character essentially different. The folds of enamel which mark the sides of the crown do not descend so low as the edge of the alveolar processes ; in consequence of this conformation, the worn down tooth of an old individual must exhibit insulated circles of enamel on the grinding surface. This genus, or sub-genus, must be placed near to Arvicola.

The *Neotoma Floridana,* (Florida Rat), has an elongated snout ; eyes and ears very large, and tail longer than the body ; the ears are large, thin, subovate, clothed with such fine hair as to appear naked ; the whiskers are long ; the body and head are lead-coloured, intermixed with yellowish and black hair, the black predominating on the ridge of the back and top of the head ; the yellow, on the sides, abdomen, and throat ; buff-colour belly, and claws white. It measures between seven and eight inches, and the tail above six. The whole fur is extremely soft and fine.

It was found in East Florida, and its describers conclude their notice by observing, that though the multiplication of genera has become an evil, they have ventured to separate it, from an inability to class it under any of the genera of the systems.

The genus SPALAX, of Guldenstaedt, is another separation from the comprehensive genus Mus, of Linnæus.

FLORIDA RAT.

NEOTOMA FLORIDANA. Say & Ord.

The species included in it, destined to live underground like the two Moles, are remarkable for the cylindrical and elongated form of their body ; the largeness and flatness of their head ; the strength of their incisive teeth, which are cut square in each jaw ; the smallness, if not total absence of external eyes and ears ; the shortness of the paws ; number of the toes, five on each foot ; and smallness, or total want of a tail.

We shall proceed without further observations on the genus to a consideration of the species. The principal of which is

The *Zemni*, *Blind Rat* of Pennant and Shaw, to which Guldenstaedt applied the Greek name, Spalax, has been hitherto referred to the Mole. This singular animal attains nearly ten inches in length, and its cylindrical body is full two inches in diameter. Its thick head, nearly pyramidical, narrower in front, is terminated by a very hard and strong cartilaginous muzzle ; on each side of the head is a salient line, extending from the nostrils to the meatus auditorius. The nostrils are round and narrow ; the opening of the mouth is small. The incisive teeth are extremely prominent and strong, of an orange-yellow colour, those in the lower jaw twice the length of the others; the under lip is shorter than the upper, and does not cover the teeth; the eyes are stated in the text. Aristotle, says Olivier, has observed that externally there are no traces of eyes : if the skin of the head be taken off, a tendinous expansion may be perceived extending over the orbits (and which forms outwardly the salient line above-mentioned,) immediately under which is a glandulous body, oblong, a little flatted, toward the middle of which is a black spot, representing the globe of the eye, and which appears perfectly well or-ganized, though not half a line in thickness. We may perceive in cutting the scelotique, the several substances which compose the eye. Nothing in short appears wanting to constitute a perfect eye, but a greater development of parts.

Whether the Spalax be absolutely blind, or whether it receive any perception of light through the medium of the eye as an organ, does not sufficiently appear by what has hitherto been said by its describers. The presence of what may be called the vestige of an organ, seems perfectly consistent with other instances, in which the application of such imperfect organ is not at all to be traced. On the contrary, it accords with that apparent unwillingness in nature to depart from prescribed laws. The total absence of an accustomed organ is much more anomalous in nature than the complete inutility of an imperfect one. So it seems with the Spalax, which is not without the vestige of eyes, though their application as organs of sight seems doubtful, we might probably say, indeed, nugatory.

It is a common observation among ourselves, that the loss of one faculty is generally in some measure counteracted by the perfection of another, or to speak more methodically, we should probably say, that the absence, or total loss of one sense necessarily calls others actively into operation, which become more perfect by exercise. We have also frequent occasion to observe, that Nature, to a very considerable extent, is ever willing to vary and change the physicalities of a being, in accordance with its circumstances and situation. We need not therefore be surprised to find that the blind Spalax has the organs of hearing in a very perfect state. What is denied on the one hand is prodigally bestowed on the other, and the creature is thereby enabled to preserve its existence. The external ear, indeed, has but a very small outward expansion, but the auditory canal is very large, and the whole organ internally greatly developed.

The neck of this animal is large, short, and muscular, by which the head is capable of considerable strength considered relatively to its size, and the whole animal takes a cylindrical shape ; the feet are short, armed with round trenchant nails, rather larger on the hind feet than on those before.

The whole animal is covered with a short soft fur, the base of which is blackish ash-colour, and the extremity reddish, whence results a general tint of yellowish gray. They are sometimes found spotted with white.

The Greeks, as has been generally assumed, described the Mole, ασπαλαξ, as blind, an error which modern Zoologists have piqued themselves in detecting ; M. Olivier, however, has shown that this wonderful people, whose mental faculties shot forth as it were a meteor through the surrounding density, and anticipated the progress of human art and intellect by many tedious ages, were not so idle in their observations, or incautious in their conclusions.— The ασπαλαξ of the Greeks was, doubtless, the animal now under consideration, which was indigenous in their country or around them, whereas the Mole was an exotic in Greece. The Romans may, however, bear the blame of having led us into this error by rendering the word ασπαλαξ into *talpa*, and applying that word to the Mole of Europe.

The Spalax lives gregariously underground. They bore excavations, which are not far from the surface, in search of food, but dig a hole lower in the earth for personal retreat and safety. They prefer cultivated grounds, and as they subsist principally, if not entirely, on roots, they become serious destroyers of the fruits of agriculture. Their movement are precipitate, turning or running sideways, or even backward with facility, when driven and in danger, and they bite with great force and effect. When on the surface, they almost always carry the head raised, apparently for the purpose more effectually of hearing what is passing around them ; thus relying on their most perfect faculty for a forewarning of approaching danger, which they have not the means of detecting by sight.

RODENTIA.

As the two species included in the genus BATHYERGUS are sufficiently noticed as to their generic and specific differences in the text and table, we shall not repeat or dilate upon them here ; but shall merely insert the short notice of Mr. Burchell, as one of the latest South African travellers, of the *Mus Maritimus*, or Coast Rat of Pennant.

In every part of the Sand Flats, says Mr. Burchell, I observed innumerable mole-hills, and my foot very often sunk into their burrows.   For this reason, it is very unpleasant, if not dangerous, to ride on horseback in such places, as persons are liable to be thrown, by the feet of their horses unexpectedly sinking into these holes.   The animal which make these hillocks is a very large kind of Mole Rat, nearly as big as a Rabbit, with a very soft downy ash-coloured fur, having, in appearance at least, neither eyes, ears, nor tail. It is peculiar to this Colony, and is called *Zand Moll* (Sand Mole.)

From the great softness of the fur, and the abundance of skins that might be obtained, it might possibly constitute an article of some value for colonial trade or for exportation, and it is surprising that no speculative person has hitherto attempted to convert these skins to some useful purpose.

Of the single species, (the *Helamys Cafer, Dipus Cafer, Gm., Cape Jerboa, Pen.*) which composes the division HELAMYS, or PEDETES, of Illiger, we shall merely add in this place Mr. Burchell's account of the animal.   Great complaints, says he, are made against these animals, for the mischief they do to the corn, eating it both green and ripe.   It is nearly the size of a Hare, with long soft fur of a sandy colour, a long tail, black at the extremity, and

THE CAPE RAT.

MUS CAPENSIS.

London, Published by G.B. Whittaker, Dec.r 1824.

hinder legs of twice the length of the fore pair. Its very remarkable gait is occasioned by this disproportion of legs, as it moves, at least when in haste, only by long leaps or bounds. From this circumstance, and its resemblance in several particulars to a Hare, it has obtained from the Dutch Colonists a name signifying *Leaping Hare*. Its ears, however, have more resemblance to those of a Cat than of a Hare; but the two long front teeth in each jaw, and its leaping motion, plainly prove its close affinity to the latter. Its fore feet, which are little more than two inches in length, are provided with very long hooked claws, better adapted for holding its food than for burrowing in the ground, and have every appearance of not being used for the latter purpose: on the contrary, the hinder legs, which are nearly ten inches long, are furnished with extraordinary large and strong nails which might almost be called hoofs, and which seem to be used only for scratching away the earth, for which office they are well suited, although such an application of the hind legs is a singular anomaly, and not easily to be explained, without a more complete knowledge of their mode of life. In this manner it does in fact use them dexterously and expeditiously, making deep burrows, in which it lies concealed all day. As it comes out to feed only by night, it is an animal not so well known from its form and appearance as from its operations. It inhabits the neighbourhood of mountains, whose rocky sides afford them a greater protection than the plains, where they may be easily overtaken by Dogs or other carnivorous animals. No construction can be better suited for ascending, or any worse for descending a steep. There must, therefore, one would imagine, be some singular management on the part of an animal so formed, and at the same time inhabiting such places. It is sometimes, though less frequently, called the *Berg-haas* (Mountain Hare.)

The Genus ARCTOMYS of Gmelin (the MARMOTS), is distinguished principally by the pointed character of the cheek-teeth. We shall proceed to notice the species.

The *Alpine Marmot* is distinguishable exteriorly by a thick inelegant body, short thick legs, large and flat head, short truncated ears, short tail, incapable apparently of elevation, and a general clumsiness of appearance.

Predestined and constructed principally for a subterranean existence, with few other requisites for nourishment than the grass, whether green or dry, which surrounds its habitation ; provided with a sufficient defence against most of its enemies in its burrow, and passing one half the year in an uninterrupted lethargy, the Marmot has little occasion for the sagacity of the Beaver or the Rat, or the agility of the Squirrel or the Hare, or in fact any very positive character to maintain its existence. Accordingly we find its locomotion is slow, it raises itself not without an apparent effort, and though a climber in its natural state, it is slowly, and by pushing against the back, as well as by the feet, that it mounts the clefts and projections of its native mountains.

The Marmots are not found at any distance from their burrow, wherein they live in families, and a very remarkable instinct is observed among them on occasion of quitting their retreats. When they do so, one of the family is placed on an elevated spot near the mouth of the burrow, and within sight of the rest who are seeking food. If an enemy or any new object be observed by the sentinel on guard, he utters a shrill cry, when on the instant the whole company make all haste to their retreat, or if too far from the mouth of the burrow, seek instantly a hiding-place in some cleft or hole. We have here a remarkable instance of the care of Providence over the creatures of this world, among whom universal inequality is not the least notable of the general accidents of their existence. Inequality seems necessarily, in the state of things around us, to generate tyranny, and the world would very soon be left in possession of a very few whose physical powers were indomitable, did not Providence interfere in an endless variety of modes in favour of its weaker creatures.

These interferences, when employed through the instrumentality of instinct, are at once striking and inexplicable. They never seem unnecessarily or prodigally brought into action ; but, on the contrary, are ever essential to the continuation of a race ;—but when employed, their perfection generally outstrips the utmost refinement of reason. This indeed may be explained in the words of the poet :

—— Reason raise o'er Instinct as you can,
In this 'tis God directs, in that 'tis Man.

The burrow of the Marmot is generally in the elevated parts of the southern European mountains, above the limits of the forest, and in the regions of perpetual snow. It is formed of an alley or gallery five or six feet long, sufficient only in size to permit the animal to pass ; at the extremity of this alley is a circular excavation, in which the Marmot retires, and hybernates; sometimes the circular cave has two outlets, forming an acute angle like the letter Y. An excavation is said always to be found in one of the alleys, which is presumed to be made by the animal, in procuring earth to stop the mouth of the burrow previous to its entering on its long winter sleep. The commencement of their lethargy seems to depend on the beginning of the cold, which varies from the middle of September to the middle of October—the newly-formed families then begin their excavations, and provide dried grass to lie on. M. F. Cuvier informs us that they make a spherical bundle of this dry grass, and press it into a state of tolerable consistency, and lie upon it, with the head brought down between the legs; and he adds, that in order to close the entrance of their retreat, they at last enter it backward with a bundle of hay in the mouth, which they contrive to leave at the opening, so as effectually to close it up.

The Marmots passing the whole winter in a deep lethargy make no reserve of provisions, and they become extremely thin during this long period of abstinence, which renders their flesh hard and coriaceous. At the commencement of winter, when they are very fat, the mountaineers seek them for the sake of the meat, which, however, even then would not be very agreeable to refined palates; their fat has the taste and appearance of lard.

The incisors of the Marmot are extremely powerful, so much so that they can only be kept in confinement in an iron-wire cage.

These animals are plantigrade. They have four toes before and five behind. The ears have only the rudiment of an helix on the anterior part, without any other lobe or fold. The eyes are large, with round pupils. The nostrils are naked only on the part which separates them. The upper lip is cleft, leaving the incisors at all times uncovered. The tongue is small, thick, and soft. There are mustachios on the sides of the muzzle, and above and below the eye. The fur is of two sorts, 1. Woolly, which is long, rather frizzled and thicker than the other; of a deep gray for the greater part of its length, and white at the tip. 2. Silky, which is rare, and rather longer than the woolly, black for the greater part of their length, and white at the extremity. On the yellow parts the hairs are entirely of that colour, or with a few narrow black annuli toward their points.

The general colour of the animal resulting from these mixtures on each hair is of a softish sprinkled gray on the back, neck, head, and part of the sides of the body, and

THE BOBAC.

ARCTOMYS BOBAC.

C. Hamilton Smith Esq.r del.t   J. Scott. sc.

London, Published by G.B. Whittaker Dec.1.1825.

limbs ; the under parts are white inclining to yellow, and the latter half of the tail is black ; round the muzzle and the toes the colour is whitish.

The fur is very thick, particularly on the back, flanks, belly, and outer side of the limbs, and especially on the cheeks, where the length of hair entirely changes the apparent proportions of the head.

The *Bobac*, or *Poland Marmot*, very nearly resembles the common or Alpine species. It appears, however, to be larger, and to differ in colour, as stated by the Baron in the text. The specimen we have figured was larger than a Hare, of a pale ochrey colour, with the head, back, and rump brownish. The Bobac, indeed, does not, like the Alpine Marmot, inhabit the most elevated part of high mountains, but seems to require a higher temperature, which it finds in less elevated districts. It prefers dry soils, and is most commonly found in Poland, but extends in parallel latitudes, not exceeding $55°$, perhaps even to Kamtschatka.

The burrows dug by the Bobacs are extremely deep, in which they live in societies of from twenty to forty; they amass so much dry grass in a single burrow, that enough is commonly found to feed a horse for a night. Pallas relates a story which has been attributed to the Alpine species, and also to Rats, but which may well stagger the most credulous ; that when they have occasion to transport a quantity of provision to their burrow, one lies on the back, is laden by the rest in the manner of a cart, and then drawn by the tail to the common magazine.

Of the *Souslik*, or *Variegated Marmot*, it may be sufficient to observe here, that it is much the prettiest of the genus, being spotted, or waved with white, on a yellowish-brown ground. It is partially carnivorous, and in its hoards are occasionally found reserves of birds and small quadrupeds, on which it feeds.

Until lately only two species of American Marmots were admitted into the zoological catalogues, the *Monax* of Linnæus, and the *Empetra* of Pallas; these, however, have been very variously described by different writers, arising, as there seems reason to conclude, from their having, in fact, written from other species, which they had arbitrarily applied to one of these. Our late Arctic travellers have, however, added several to these two species of American Marmots, which, with some others noticed by the French zoologists, have considerably enlarged the catalogue of these animals.

The present division, however, of this work, does not require us to insert here a notice of all the species which have been described, which rather pertains to the Table, to which we refer.

We have engraved, from a very elaborate print done by Lawson, in North America, which appears to be the Arctomys Monax of Gmelin, the Maryland Marmot of Pennant, and the Ground Hog of Travellers. This species is said by Pennant to have the nose and cheeks of a bluish ash-colour, back of a deep brown colour, sides and belly paler, and to be about the size of a Rabbit.

The collection of drawings in our possession strongly evinces the neglect zoology has experienced in this country, by the number of figures it contains, drawn from life, principally at Exeter-Change, the Tower, Liverpool, &c., of specimens decidedly of species not described, no vestige of which specimens now remain. Among the rest is a drawing of a Marmot, by Howitt, which died at Exeter-Change a few years ago, and though unfortunately we have no notes or observations upon it, we have thought it right to engrave the drawing, particularly as if it occur again a description only may be given, without the illustration of a figure. As the plate could not appear without a name, we have affixed Marmot Diana to it, from the crescent which passes from

THE MARYLAND MARMOT.

THE MARMOT DIANA.

London, Publish'd by G & W. B. Whittaker, Feb.y 1824.

one eye to the other. Mr. Cross has no recollection of what became of the animal after its death.

The figure alone, and not the subject, being before us, it is needless to enter into a description, except as regards the colour. The drawing is tinted slight ochrey-brown, whitish on the belly, darker down the centre of the back, with close slightish cross-bars, which appear to arise from the hairs being annulated or party-coloured, and the crouched position of the body, and the consequent folding or ridges of the skin, displaying the two colours of each, with something like transverse regularity. The tail is dark brown, and the lunated mark from eye to eye, and the other from ear to ear are black, except the region round the nostrils, which is pure white.

One of the species brought home by Capt. Franklin, and described by Mr. Sabine in the Linnæan Transactions, may deserve a more particular notice. This is the *Striped American Marmot, Arctomys Hoodii*, of Sabine, and appears to be the same as the Sciurus Tridecimlineatus of Mitchell, (Medical Repository, 1821,) and the Ecureuil de la Federation, of Desmarest. By the examination and allocation of Mr. Sabine, it is now, however, placed among the Marmots ; and though already drawn and engraved by the able hand of Mr. Curtis, we have, on account of its singularity and beauty, inserted a figure of it, from the specimen in the British Museum.

It measures about seven inches and a half from nose to tail ; the top of the head is broad and flat, and obscurely marked with alternate stripes of dark brown and dingy white ; the nose is tapering, and covered with light-brown hairs ; the ears are small, and very short. The whole upper part of the body is marked longitudinally, with alternate dark-brown and dingy-white stripes ; the dark stripes twice the breadth of the light, and dotted at even distances the whole length in their centre, with small spots of dingy white ; there is a dark stripe in the centre of the back, and it is rather broader than the others, of which there are three on each side, but the lowest on each side is not distinctly defined or spotted ; the whole under parts are of a dingy-white, slightly fulvous ; the tail is two inches long, indistinctly banded with dark-brown and dingy-white, the tip being of the latter hue.

Of the *Meerkats* of Southern Africa, Mr. Burchell says, these are a species of Squirrel, of about the size of a common Squirrel. It has no outward ears, and its body is very thinly covered with short coarse hair, which is brittle, and may easily be rubbed off ; but the tail, which is longer than the body, is furnished with long-spreading hairs, as in the European kind. It was seen to live chiefly on the roots of plants, which it scratched up with its fore-feet. It is common in some parts of the colony, and being a pretty little animal, is sometimes domesticated.

The genus SCIURUS, or the Squirrels, includes a great many species and varieties, an enumeration of which, accurately referred to one or the other description, is no very easy task.

Animals extensively located, like the Squirrels, are generally observed to be most subject to deviate into varieties ; nor can we doubt that climate and circumstances are powerful agents in this operation. How many admitted species owe their existence to this cause must ever, in a great measure, be matter of conjecture ; analogy seems to lead to a conclusion that very many do, and when time and succession have, as it were, consecrated a variety into a permanent race, we may be fully warranted in regarding it as a distinct species.

A description of the *Common Squirrel*, in addition to what is said elsewhere, is unnecessary. It is an animal of extreme cleanliness, occupied, without ceasing, in polishing the fur with its fore-paws. When still and awake, it sits habitually, if at ease, with the tail gracefully erect up the back, and curved toward the extremity; but when listening, the body is straightened, and the tail lowered to an horizontal position, in which it may, like that of the Jerboa and Kanguroo, support the body, and prepare it for sudden action. The Squirrel is as aptly fitted for residence in the trees as the quadrumanous animals ; though, on a different principle ; its long, slender deep, cleft fingers, and sharp nails, enable it to mount a perpendicular with more facility than the Monkey, and to seize and hold by the small branches with equal ease. It takes its food to the mouth with the fore-paws while sitting, and drinks in rather a singular manner, by a sort of sipping, though rather by means of the tongue than the lips.

The activity and agility of the Squirrel, when favourably situated, are truly surprising, and its sudden turns, when moving with the greatest rapidity, are almost too quick for the sight to follow. On the ground, however, it has less opportunity of displaying these powers, and though it advances rapidly, it is by means of leaps or gallops ; this arises from the length of the hinder extremities compared with those before.

Squirrels appear to live in pairs, and to confine themselves to a very small range of action, seldom moving by choice from the tree or the immediate vicinity of the spot where they first settle. They build their nest in the fork of the branches ; it is spherical in shape, and so compacted with moss, bits of wood, &c. as to offer an effectual resistance to the rain. Here they sleep during the heat or middle of the day, but toward the afternoon proceed in search of food, gliding along the branches, and leaping to those of another tree with surprising swiftness.

The period of their gestation seems not very precisely known, but the young are born about the month of June. Their affection for, and care of the young is remarkable, and the male partakes with his mate in providing for them. This trait of character seems generally to be observed in monogamous animals, and though certainly attributable to no other origin than that of instinct, naturally impresses us with a favourable notion of creatures, whose actions partake of what among us is attributable to moral character and right feeling.

As soon as the young have attained strength enough to cater for themselves, the cares of their parents are directed to a different object, that of, in effect at least, anticipating the privations of winter, by what with ourselves might be called a prudent appropriation of the superfluities of summer. For this purpose they select the clefts and holes in trees near their nest, in which they stow away corn, nuts, &c. They are observed to have an unhesitating knowledge of the situations of these magazines, even after the snow has reduced almost every thing to one common level, through which they work their way in a direct line to what they seek. There is something in actions like these so nearly allied, not only to memory, but association, that it seems very difficult to distinguish the difference between them, as blind impulses or as reasonable acts.

The blindness of the instinct of accumulation, however, is sufficiently evinced in the tame Squirrel, which, after being captured in the nest, and removed from the influence of example or instruction, is equally impelled by its instinct, though no longer necessary. If an inferior food be given to a tame Squirrel, and another more agreeable to its palate offered shortly after, the animal will not carelessly drop or throw away that which it is already possessed of, but will endeavour to conceal the one before it receives the other.

There is another common action of the Squirrel which, assumes the appearance of a result of reason and deduction. When disturbed, or hunted in a tree, it never fails to keep as much as possible on the opposite side of every branch it may fly to, so as to have the shelter of the branch between itself and its adversary ; thus rendering it difficult for the hunter to get at it with any missile. This effect, indeed, however useful on the occasion, may not be contemplated at all by the animal, whose only object may be to keep out of sight of its enemy. Other actions of a very elevated character, in an intellectual point of view, have been attributed to the Squirrel, particularly by the German naturalists, which need the most unequivocal confirmation before they can be admitted ; in the absence of which, the rehearsal of them may be dispensed with, though Linnæus himself may have sanctioned by recording them.

Toward the end of winter the Squirrels moult, when much of their beauty is suspended till the middle of summer, at which time the new fur and soft feathery tail are in full splendour—the new fur first appears at the end of the tail, and tips of the ears.

They emit a sharp, but soft whistle, like that of the Guinea Pig, which frequently betrays them when in a tree. When uneasy, they also make use of a sort of grunt.

The regularity of action of the Squirrel is remarkable in a state of captivity, dancing, as it were, in regular time, in which it seems to delight, and if kept in a circular moving cage, and allowed occasionally to go out, it will frequently return of its own free will to run round its wiry tread-mill. It is said also, that Squirrels are fond of the contact of any thing cold, and will lie with their belly on polished marble for the pleasure of the sensation.

The common Squirrel, in high latitudes, seems subject to vary to a gray tint, and in that dress has been erroneously confounded with the Gray Squirrel of America. There remains, however, still considerable doubt, whether the Gray Squirrel of the Old World be in fact only a variety of the common one ; or whether, though the common species be subjected occasionally to become grayish, there be not still a

O 2

distinct gray species. M. F. Cuvier, who describes the Gray Squirrel in question, observes, in conclusion, that his object is not so much to prove that the Petit-Gris does not belong to the common species, as to show that the true relationship of these animals is not sufficiently established, and that further observations are indispensable to that end.

The *Alpine Squirrel (S. Alpinus*, F. Cuvier) has been lately described by M. F. Cuvier. If it has been previously noticed by Gesner, Aldrovandus, or Klein, it has been treated as a variety of the common breed. It is certain, however, that the characters which distinguish it from the last-mentioned species are not accidental, nor are they the effects of age, sex, or season—they have all the constancy of specific characters, the young have them at their birth, and the adults never change them, nor are they, according to the describer of this animal, of that nature, as to be considered like the characters of races which generation may reproduce, but which are attributable, nevertheless, to fortuitous causes, and not to the primitive organization of the animal, at least we have neither sufficient examples nor analogies in this genus to abstract the differences which distinguish the common Squirrel from that in question, whence to infer their specific identity.

Nature is so rich, so prodigal of her powers, so infinite in means, and we are so much accustomed to see the several species vary from the influence of surrounding circumstances, that we may well be surprised at finding so few species of Squirrels in this part of the world, in countries which seem quite favourable to the nature of these animals, and to their perfect development; while in America, on the other hand, under parallel degrees of latitude, a comparatively large number of species of these animals is to be found.

Moreover, when we perceive, or think we perceive, any

break in the general harmony of nature, it usually turns out that our assumption arises from imperfect observation. It is therefore much safer to suspect anomalies than to attempt any explauation of them by hypotheses, however ingenious or probable.

Upon reasoning of this nature, and upon the actual differences to be observed between the two, M. F. Cuvier proceeds to treat the Alpine Squirrel, which he names the Squirrel of the Pyrenees, S. Alpinus, and which is found equally on the Alps and Pyrenees, as a distinct species from the common Squirrel.

Its colour is a very deep brown, sprinkled with yellowish on the upper part of the body, and pure white underneath, insides of the limbs gray, sides of the mouth bright yellow, and the edge of the lips white; the feet are yellow, and a yellow band separates the white of the neck and chest from the gray part of the limbs. The tail, viewed sideways, seems quite black, the latter part of each hair being black, but toward the root annulated black and yellowish. In its proportions and size, this species resembles the common, except that the head is rather less.

The *Gray*, or *Carolina Squirrel*, (*S. Cinereus*, Lin.) is rather larger than the common species. It seems, as M. F. Cuvier has observed from a number of specimens, to present no fixed character as to colour, except within certain limits some whitish-gray, others with a mixture of yellow, &c. Indeed no species, perhaps, occurs more subject to differ in this respect, particularly when, like that in question, it is not extensively located. We have no doubt that such-like differences, which in this instance have been so observed upon by M. Cuvier, are, in very many other cases, neither more nor less than what constitute the specific separation of many enumerated in zoological catalogues; nor do we, for an instant, exclude our monograph from the application of this observation. To cleanse this stable of Augeas would indeed require the labours of a zoological Hercules. We cannot hope alone to turn a river through it; the only chance of effecting the object is, by divided labour and judicious industry.

Zoology, while it includes the pursuit of the physicalities of animals, without which, any comprehensive methodical view of the subject is hopeless, has also a second great branch, in what may be called the study of their moral characters and impulses ; the invisible, though limited operations of their humble minds, are better calculated to enlarge our sphere of thought and reflection, and to teach us soberly to estimate our own relative situation, than their material differences of external form, and internal organization, while the harmonious adaptation of these two branches displays the being, the providence, and skill of the great Maker and Preserver of all

As this species differs essentially, at least but little in its specific character, from others, we shall merely insert M. Cuvier's observations on its manners and habits. It has all the gait of the common Squirrel ; eats, seated on the haunches, and holding its food in the hands, &c. Nothing can be more lively or rapid than its actions. While running with the utmost speed, it stops in an instant, turns and returns by the same path, without any apparent cause or excitement. In captivity, it suffers itself to be taken and handled without the slightest resistance. It nevertheless seems to have no knowledge of persons, and never answers to its name, however familiarised it may be with the voice that calls ; but it understands the sound occasioned by cracking a nut, and comes readily when any food is offered to its sight. It is very fond of warmth, and lies and basks in it as a great luxury. Toward evening, it collects together all the hay and straw in its cage, forms it into a ball, and retires to rest in the midst of it till morning.

The *Masked Squirrel*, or *Capistrate*, is larger than the

THE MUZZLED OR MASKED SQUIRREL.

S . CAPISTRATUS .

London Published by G & W.B. Whittaker. Sepr 1824.

European species, and though considerably subject to vary
from gray to black, is ever distinguishable by white round
the muzzle, and on the extremity of the ears. The specimen described and figured by M. F. Cuvier, was entirely
black, with the exception of the ears, muzzle, toes, and
end of the tail. He also mentions another like the former,
but which had the sides of the body principally on the crupper and flanks, under the tail; and toward the scrotum, a
great number of silky hairs tipped with white.

The gray variety, described by M. Desmarest as the most
common, has the head, with the exception of the ears, end of
the nose, and lower jaw, covered with hairs, entirely black,
with as light reddish tint within the ears and at the corners
of the mouth. The back blackish, sprinkled with white, each
hair being annulated with these colours, the flanks gray,
and the under-parts white, and the tail is of the same character.

The specimen from which we have engraved, accords
with M. Cuvier's description, rather than with that of M.
Desmarest's; but it differs from both, in having a distinct
white spot on the forehead. The Baron, we have seen,
treats this Sc. Vulpinus of Gm., and the Black Squirrel of
America, as varieties. It seems really very probable that
black and white, variously marked as to the disposition of
these two opposite colours, may be of common occurrence
in this species.

Of the *Jeralang*, or *Leschenault's Squirrel*, we present a
figure, from the specimen in the Paris Museum. The body
of this species is rather more than a foot in length, and the
tail is about the same. The hairs on the back of this animal are brown from the skin, and yellowish toward the
point, whence results an ochrey-brown, but the flanks are
dark dull brown; the tail, which is flat, has the upper part
dark dull brown, with the under-side yellowish; the nape of
the neck is gray; the head, throat, belly, anterior and in-

ternal sides of the limbs are white ; the paws are yellowish brown.

Dr. Horsfield says, that the external covering of this species is subject to greater variations than any of the Indian Squirrels, some being pale yellow, and others deep brown. In several the colour is uniform, in others it is distributed in irregular patches of different shades ; but the separate hairs are not variegated or banded, as in many other species. The darker specimens, he states, have some resemblance to the S. Bicolor ; but the examination of numerous specimens has convinced him, that these two species are decidedly distinct. The tint in the Bicolor is of the deepest black, and the tail in the adult specimens, is often gray or yellowish, and it has a different form and termination.

The tail, therefore, seems a much more correct test of this species than its general colour. The specimen brought by M. Leschenault de Latour from Java, and deposited in the French Museum, (the same we have engraved,) is remarkable, and would be with aptness named the Whiteheaded Squirrel, and, in fact, it was at first so named by the French; but the species being subject to vary, so far as to lose this distinctive character, it becomes necessary to distinguish it by another name. It is the Jeralang of the Javanese.

The disposition in the species of this genus to vary, already alluded to, is equally apparent in the common species of our own part of the world ; in those, particularly, of a gray or dark colour in America, and in the larger species of Asia and the Eastern Isles. On the last of these, including the Jeralang, or Leschenault's Squirrel, we have lately had some very valuable observations from the pen of Dr. Horsfield.

The *Sciurus Bicolor* is found on the continent of India, and in Cochin China, almost entirely black above, and golden yellow underneath. It measures about three feet

LESCHENAULT'S SQUIRREL.

S. ALBECEPS. Geoff.

Hamilton. Smith. Esq. del.

J. Scott. sc.

London. Published by G.B. Whittaker Dec.r 1. 1825.

from the nose to the end of the tail, which is about half that length. In the eastern parts of Java it is generally found with the upper parts of the head and neck, the entire back, the sides of the body, and the limbs dark-coloured ; but the first varies from intense brown to tawny, and often passes into yellowish-gray. The separate hairs composing the fur have either a uniform dark tint, or are dark only at the base, and yellowish at the extremity. According to the distribution of these hairs, the external coat of the animal receives its character. The surface is either uniformly dark-coloured, or it is marked with irregular tawny discolorations, of different shades of intensity, spreading in broad transverse bands, or in patches of various extent. The sides of the neck, the shoulders, the upper parts of the legs and feet, the tip of the nose, and root of the tail are dark, but varying from blackish-brown to chestnut and reddish-brown, and in general a ring of the same colour also surrounds the eyes, but this is not found in the specimen in Paris, whence our figure was taken. Sometimes there is white across the head and neck, giving the animal the appearance of the white-headed variety of S. Leschenault's.

The under parts of S. Bicolor are generally yellow ; but this again is subject to vary from light fulvous to Isabella yellow. The dark colour of the upper parts is frequently separated from the light under-parts by a darker streak, slighty approximating to the peculiarity in that particular of the striated species. The tail is dark at its base only, and in the remainder of its length agrees with the under parts.

In several specimens, the tawny colour predominates throughout, and only small patches of brown appear on the shoulders, and sides of the neck, and the base of the tail, and in a few instances, the colour of the upper parts is Isabella yellow with a grayish cast, while the under parts are pale yellow, and scarcely any distinction of colour is apparent above and beneath.

The ears are acute, of moderate size, and covered with delicate hairs, without any brush-like appendages. The mustachios consist of numerous long, stiff, bristly hairs, arising from the sides of the nose, and the upper lip, and diverging from the head. A small tuft of separate, short, stiff, black whiskers, parting backward, arises from the cheeks about midway between the angle of the mouth and the ears.

The front teeth have a gall-stone-yellow tint, inclining to orange, the upper lip is deeply divided. In its general habits, as well as in the form of its head, and in the proportions of the neck and limbs, the Sciurus Bicolor agrees with the other large Indian Squirrels, and like these, it also has a short, broad, obtuse nail on the thumb, which has been aptly compared to that of several Monkeys. The thumb its lf is not lengthened or separated from the toes, but consists of a thick fleshy tubercle supporting the nail. The claws on the other toes of the fore feet, and all those on the hind feet, are acute, and greatly compressed, as in other species of this genus.

The hairs of the upper parts are coarse. At the base, the separate hairs are supplied with down, but they are ridged and somewhat bristly at the extremity, and are not regularly applied to the skin; on the breast and abdomen, the fur is softer, and the arms and paws are bordered with a beautiful series of hairs, generally of a deep fulvous tint, and extending laterally from the shoulder toward the ears. From the outer margin of the fore-arm, near the foot, arise several long, stiff, straggling bristles.

We have availed ourselves thus largely of Dr. Horsfield's observations on this species, not so much for the purpose of describing it individually, as to illustrate the general history of the large Indian species. Dr. Horsfield's researches

THE JAVAN SQUIRREL OF PENNANT.

SCIURUS BICOLOR. Sparman.

milton Smith Esq.ʳ del.ᵗ

J. Scott sc.

London Published by G.B. Whittaker Dec.ʳ 1.ˢᵗ 1825.

in Java, and the valuable collection at the India-House, have enabled him to dilate upon them, with great benefit to the subject, and we shall subjoin the result of his latest investigations.

The varieties of Leschenault's Squirrel, and the S. Bicolor, are in many instances so strongly marked, that they appear to be distinct species, until the gradual passage from one to the other becomes apparent, by the examination of a series of specimens:—the name of Albiceps was first given to the former of these; but as its application could not be general, M. Desmarest applied the name of its discoverer as a specific appellative. One variety in the India Company's Museum has the upper parts of a testaceous colour, and all the under parts white. There is a specimen of this variety also in Paris, of which we have a drawing by Major Smith: it was treated as a distinct species, under the name of S. Hypoleucos.

Of the large Indian Squirrels, which are the subject of these remarks, three species remain, which appear to have clear distinguishing characters—the Sciurus Bicolor, the Sciurus Leschenaultii, and the Sciurus Maximus. They have all a dark colour above, varying from brown of various shades to black, a pale tint underneath, and a large obtuse nail on the thumb of the anterior extremities; but the Sciurus Maximus has distinguishing characters, in a reddish brown tint above, in a very large tuft of hairs arising from the ears in several coloured bands on the cheeks, and in a tail more full and bushy than in the other species. The Sciurus Leschenaultii is distinguished by a more uniform colour above, inclining to chestnut, a nearly white under-side, and a grayish tint on the nose and anterior part of the head. The most common dress of the Sciurus Bicolor is black above, and yellow underneath; in this it is easily distinguished from the two other species.

The manners of the Sciurus Bicolor present nothing peculiar. It is by no means scarce in any part of Java; but it is by far less prolific than the Plantain Squirrel. It rarely approaches the villages and plantations, and the cocoa-nut trees suffer but little from its depredations. It retires into the deepest forests, where its food is abundantly supplied by wild fruits of various kinds. The natives sometimes keep it in a domestic state, and sometimes eat its flesh.

The *Sciurus Maximus*, or *Great Squirrel* of Sonnerat, received the superlative epithet from Gmelin, and one cannot but observe an impropriety in applying any relative term as a specific name, when the discovery of a new species may at any time destroy its application; thus the Sciurus Maximus, it is now discovered, is at least equalled in size by Leschenault's Squirrel, if not also by Sciurus Bicolor.

This species is said by Dr. Horsfield to be reddish-brown on the upper part of the neck, back, and sides; the legs and feet reddish-brown; the shoulders, rump, and the hinder part of the back and thighs are black; the nose and lips are of a dirty flesh-colour; the cheeks and circumference of the eyes are dirty-brown; the fore-part of the face is very dark reddish-brown; the ears are short and round, and covered with bright reddish-brown hairs, which form a tuft larger than the ear itself; between the ears is a broad pale yellow band; from the ears down to the cheeks there runs a vertical reddish-brown line; the throat, inside of the fore-legs, the breast, belly, and inside of the fore, and inside of the hinder, thighs are yellow; between the upper lip, and before and behind each ear, it has very long and strong black whiskers; the irides are dark-brown; the tail, on the upper part and sides, is covered with long close hair, which may be made to stand nearly erect; they are black everywhere except the lip, which is dirty yellow or white; the under side of the tail is covered with short hairs ; the hair on the body is long and harsh.

This species was first described by M. Sonnerat in his travels, and appears to be the same as the Malabar Squirrel, the Barbary Squirrel, and the Great Squirrel of Pennant and Shaw.

The *Barbary Squirrel, Sciurus Getulus,* Gm., next mentioned by the Baron, is one of the striped or lineated species, and is about the size of the Common Squirrel. The upper parts are brown, mixed with reddish and ash-colour, with four longitudinal white bands, two equidistant on each side the spine ; the internal edge of the white bands are bordered with black ; the lower parts are yellowish-white. Gmelin refers the habitat of this species to America, but it appears to belong to Northern Africa, particularly Barbary and the adjacent parts of the Asiatic Continent.

The *Palm Squirrel* has a ground-colour nearly corresponding with that of the Barbary species, but is distinguished from it by having the forehead more arched ; the ears are short, but large and hairy; there is a white stripe down the dorsal line, and one parallel with this on each side. Pennant says, there is a yellow stripe on the middle of the back, another on each side, and a third on each side of the belly ; the two last at times very faint ; the hairs of the tail are rather short ; the upper side of it is of the same colour as the back, and the lower is reddish in the middle, with two lateral lines of a deep brown, and a black border, each hair being reddish at the base, then annulated reddish and brown, and finally terminated with white. Mr. Pennant observes, that authors describe this kind with only three stripes, but the specimen he describes had five.

According to Clusius and Ray, this species does not erect its tail, like other Squirrels, but has the faculty of expanding it.

It is obvious that the Palmiste of Brisson and Buffon

differs considerably from the Palm Squirrel of Pennant, that is, in external markings. The peculiarity of the tail, however, seems probably to limit these differences to those of variety, and not of specific distinction. Judging from three drawings in our possession, all of which exhibit longitudinal stripes, though varying in number on the under side of a fan-like tail, we should incline strongly to the opinion, that this species, like those already noticed from Dr. Horsfield, and in conformity with the general observations on this genus, varies greatly in non-essential characters.

We have engraved one of these from a specimen in Mr. Bullock's late museum, which is pretty uniformly of a slight ochrey tint, lighter underneath and round the eyes. The prominence of forehead, largeness of the ears, neither acuminated nor pencilled, and the fan-like construction of the tail is observable in this specimen; but the dorsal stripe, and one on each side of it, is broad, and of a very dark-brown colour, sprinkled with single hairs of a lighter tint, and corresponding with the colour of the stripes on the lower, or reverse side of the tail. The hairs diverge from the central line of the tail, as if parted; this central line, for a short way up, is dark, but soon becomes lost in the general ground-colour, which surrounds it; beyond this on each side, are three dark stripes, which meet at the top, and the edge is fringed with hairs of the general ground-colour.

We have never been able to ascertain the fate of much the greater part of the novelties and curiosities in zoology that disappeared on Mr. Bullock's sale, except some that were bought by M. Temminck. It is, however, much to be regretted, that an extensive selection, at least, was not thought worthy of being made public property.

We have affixed the name of Palm Squirrel var. to this species, with a mark of doubt, which future observations may perhaps remove.

THE PALM SQUIRREL. *var*

SCIURUS PALMARUM. var

STRIPED AMERICAN MARMOT.

ARCTOMYS HODII.

Griffith sc.

London, Published by G. B. Whittaker, March 1 1827.

Of the species with round tails, or flattened only at the
end, we shall first notice that described by Dr. Horsfield
and M. F. Cuvier, the *Lary* of the latter, and the *Sciurus in-*
signis of both.

This species is the *Bokkol* of the Javanese. It is grayish-brown, inclining to tawny above and on the sides, and
white underneath, with an intermediate streak of a ferruginous tint, extending from the angle of the mouth to the
posterior extremities, with different shades of intensity,
and diffusing itself irregularly over the thighs and flanks.
But a distinguishing character is afforded to it by three
black lines, about one-fourth of an inch in length, extending from the neck to the rump. One of these is placed in
the middle, and follows the course of the spine ; the others
are parallel to it, at a distance equal to the breadth of the
lines. The upper parts are delicately variegated, in consequence of the alternate bands of gray, tawny, and black,
with which the separate hairs are marked. The muzzle
and forehead are nearly uniformly gray, with a slight diffu-
sion of a blackish-brown colour. It is white on the throat,
neck, and abdomen. The intermediate ferruginous streak
begins on the lateral parts of the head ; on the throat it is
obscure and partial. It is diffused over the shoulders, and
mixes with the gray of the sides, and the white of the
lower parts. The feet have the same tint as the muzzle and
forehead. The tail, is deeper in colour than the body, and
is obscurely undulated with brown and black, having gray
hairs irregularly scattered over it. The teeth are reddish-
brown, and the mustache is black. Towards the extremity of the tail, the hairs are loosely disposed in the Bokkol,
but they never separate into two rows, as in the common
Squirrel.

The animal measures, from nose to tail, about seven
inches and a half ; and the tail is as long as the body and
neck together.

The Bokkol is a rare animal in Java, and probably so also in Sumatra.

Another species, the *Sciurus Affinis*, is described by Sir Stamford Raffles. The upper parts are fulvous-brown, with a cast of gray ; it is variegated with delicate transverse bands. The white of the head, the under-parts throughout, and the extremity of the tail, are gray. This latter organ is cylindrical, somewhat distended in the middle, and then tapering to a point ; the transverse bands are irregular between the sides and abdomen ; a stripe of reddish-brown intervenes, which is continued to the neck. The whiskers consist of black and white hairs intermixed. The ears are rounded above, and without any brush-like appendage. The length of the body and head is nine, and of the tail, seven inches.

Sir Stamford Raffles discovered this species in 1819, in the woods of Singapore.

The *Sciurus Tenuis* is described by Dr. Horsfield. The entire length of the body and head is five inches and a half, and of the tail, five inches. It differs from the Affinis in size, and in having generally a darker tint. The colour above is uniformly and delicately variegated with dark, tawny, and blackish-brown ; towards the sides, the tint becomes fulvous, with some lustre, but no defined line or streak is apparent. The sides of the head and neck have the same colour. The under part of the neck and abdomen, and the extremities interiorly, are pale yellowish-gray, with a slight tint of fulvous. The tail above is gray, with obscure black bands, and irregular tawny specks ; it has the same form as the Sciurus Affinis, being narrow at the base, broader in the middle, and terminated by a long attenuated grayish tuft, which is obscurely banded with black. The separate hairs covering the tail have a band of a fulvous colour at the base, of black in the middle, and gray at the extremity.

THE GINGI SQUIRREL?

S. GINGINLANUS — Shaw.

T. Landseer sc.

London Published by G. B. Whittaker, Dec.r 1.st 1826.

The *Gingi Squirrel* has the head, neck, back, and sides of a testaceous colour, not unlike the Hare; but each hair is short, close, and harsh, and generally annulated, black and brown, and white near the skin ; on the under part, the hairs are very scanty, almost altogether wanting on the belly, and entirely white. On each side of the animal, and before the termination of the testaceous-coloured hair, there runs a white stripe from the articulation of the fore leg to the fold of the thigh ; this stripe is not perfectly straight or equidistant all along with the spine, but is slightly crescented, with the ends tapering to a point, and inclining upwards. The tail, near the insertion, is the same colour as the back, but in the rest of its length each hair is white at the base, succeeded by a broad ring of brown, and finally terminated with white, whence the tail appears to be covered with dark brown and white hairs intermixed.

This species is an inhabitant of Southern Africa. We have engraved a figure of it, from a skin in our possession.

In addition to the figures of species in this genus, already noticed, we have selected two from the interminable collection of our liberal friend, Major Smith, for engraving, as presenting much novelty as well as beauty ; one belongs to the section of these animals with round or cylindrical tails, and the other to that which is distinguished by the hairs of this organ being arranged in a flatted form. They are drawn from specimens in Mr. Peel's museum in Philadelphia, and were brought there by the American Missouri travellers, Messrs. Lewis and Clarke, from whom, respectively, they take their specific names.

Clarke's Squirrel has the back, upper part of the head and neck, cheeks and tail, of a delicate silver gray colour ; the shoulders, flanks, belly, and posterior extremities, both within and without, are white, with a slight ochrey tint; on the sides of the nose and the fore-arms this tint deepens in

intensity; the head is rather flatted and thick, the ears small
and round; the eyes black, and situate on the sides of the
head very far distant from each other, leaving a wide ex-
panse of forehead; the nostrils are semilunar in shape; the
upper lip is cleft, and there is a black spot on the chin; the
tail, which is flat and spreading, is very beautiful, not so full
near its insertion as toward the middle, and again diminishing in breadth till it terminates in a point.

Lewis's Squirrel has the upper part of the head, neck,
shoulders, fore-arms, to the articulation of the arm, back,
flanks, the posterior moiety of the thighs, and a band round
the belly, of ochrey gray colour; all the under parts, the inside of the limbs, and the paws are pure ochrey; the ears
are small, round, and far back; the eyes are black and surrounded with the same colour as the back; the nostrils open
at the extremity of the muzzle, forming a denuded black
snout, the upper lip is white, and the whiskers very long;
the tail is very beautiful, extremely thick or bushy, cylindrical and annulated, with seven black and six white bands,
with the termination black.

This appears to be the *Sciurus annulatus*, described by
M. Desmarest, *Encyclopédie Méthodique*, article *Mamma-
logie*, from a specimen in the museum at Paris, whose habitat is unknown. His specific characters, however, are fur
of a bright greenish gray above, without lateral white
bands, white underneath, tail longer than the body, round,
annulated, black and white: of the size of the Palm
Squirrel. These differences of colour may be sufficiently
accounted for, to reconcile the probability of the identity of
the species of these two specimens.

The subgenus PTEROMYS of our author, or the Flying
Squirrels, seem to differ physically in nothing from the
Squirrels, properly speaking, except the flying apparatus or
parachute, and the additional base which supports it. The
skin of the sides is greatly extended, and is hairy on both

LEWIS'S SQUIRREL.

S. LEWISII.

C.Hamilton Smith Esqr.
delt. Mus. Philadelphia.

London, Published by G.B.Whittaker, Dec.r 1st 1825.

CLARK'S SQUIRREL.

S. CLARKII.

Hamilton Smith Esq.
Mus. Philadelphia.

London Published by G.B.Whittaker.Dec.r 1.st 1825.

sides, and attached to both the anterior and posterior extremities ; there is an osseous appendage to the hind feet, which furnishes an additional support to this membrane in its destined office of supporting the body in the manner of a parachute, in the extended springs made by the animal from one tree to another, and is analogous to a similar apparatus, with which the Galeopitheci· are provided·; the Flying Squirrels are therefore to the Squirrels proper, what the Galeopitheci are to the Monkeys.

The *Siberian Flying Squirrel*, or *Polatouche*, the *Sciurus Volans* of Linnæus, Schreber, 223, has been confounded with the American species ; but Linnæus distinguished the former under the specific addition Volans, and the latter under that of Volucella.

This species is one-third larger than the American, the head is rounder, and the muzzle larger and slenderer, the eyes also are larger and nearer the nose, the tail is shorter, the fore feet are also shorter and the hind feet longer. The colour is whitish-gray on the upper sides, the belly white, with an under down of a brownish tint, the lateral membrane is bordered near the body for its whole length by a gray hair band ; the long hairs of the tail are ashy-gray on the upper sides. The iris of the eye is black.

This animal feeds principally on the young shoots of the pine-tree, and Pallas states, that its excrement acquires so resinous a quality from the nature of its food, that it will burn with a bright flame, giving a resinous scent.

It is entirely an inhabitant of trees, springing from branch to branch, and from tree to tree, with great agility, aided by the powerful assistance of the lateral membranes. It frequently sits like the Squirrel, taking its food to the mouth with the fore paws, with the tail sometimes vertical, or nearly so, at others horizontal. It seldom quits the cleft of a tree selected for its retreat, except during the night, and its eyes, like those of many birds, are furnished with a nictitating membrane. It bites when irritated, with considerable effect, and is not often tamed.

The female brings forth in May, from two to four at a time; the young are born incomplete, without hair, and with the eyes closed, which do not open for about a fortnight. Maternal instincts, in common with all others, however blind, however involuntary in the creature, are always proportioned to the necessity in which they originate. In proportion as the young of these animals are helpless and imperfect, in the same ratio, we may fairly conclude, are the affections and maternal cares of its parent increased; thus we find that the mother warms and covers them in the ample folds of the lateral membrane, which at this period become employed in an object secondary, perhaps, to its original destination, and when she quits the nest to find food for them or for herself, she covers them with the greatest care with the moss accumulated in her bed.

The fur of this animal, though it is an inhabitant of very high latitudes, is said never to change in colour with the season, like that of many other arctic quadrupeds.

It may be proper to observe, that this species has been named *Sapan* by Vicq d'Azyr and by Desmarest; but as the word was decidedly borrowed from a Virginian root, which was properly applicable to the American species, the name cannot be retained.

The American species, *Sciurus Volucella*, Lin. is described by Buffon, under the name of Polatouche, which, as already stated, is applicable to the last. F. Cuvier has modified the name Sapan, applied improperly, as already stated, to the last species, and given it to this. Catesby and Edwards have described the present species under the name of Flying Squirrel, but as this has rather a subgeneric than a specific allusion, it may be better to distinguish the present under the name of Assapan, from the last under that of Polatouche.

All the upper parts of the body of this animal are grayish brown, each hair being black for the greater part of its length, and yellowish brown at its point, the under parts are white, separated from the darker colour by a black line, the under part of the tail is bright ashy gray ; the hairs of the tail are disposed like the barbs of a feather on each side, and consequently the tail is flat. It measures about six inches from nose to tail, which is about five inches long.

With the exception of sight and hearing, the senses of this animal appear rather blunt; the eye is remarkable for its large size, convexity, and extreme sensibility, and the external ear is remarkably large, and the internal canal extended ; the nostrils open laterally at the extremity of the muzzle, which is naked and extends beyond the jaws ; the tongue is soft, and the upper lip cleft; between the eye and the extremity of the muzzle are long mustachios, forming, doubtless, says M. Cuvier, a particular organ of touch.

This animal is weak and timid, passing the day in its hole in a tree, and seeking its sustenance only at night. It feeds on grain and buds, and seldom descends to the ground. When frightened it utters a weak soft cry, and voids its urine. It is altogether inoffensive, and endowed with a very limited intelligence: all its resources are confined to its agility and retired habits.

Most American travellers have spoken of this animal, though under different names ; its organs of flight, and its general description are, however, sufficient to determine its identity. We have, indeed, but little information on its natural habits, though M. F. Cuvier has observed them in confinement. In this state they continue hidden in their bed during day, and when they do quit it at night to take food, the slightest noise or any strange object, will quickly drive them back to their retreat.

Another species of the Flying Squirrels is the *Sciurus Sagitta*, of Pennant, which is an inhabitant of Java, in which Island, according to Dr. Horsfield, there are four distinct species, two of which he describes as new to systematic catalogues. We shall abridge his account of one of these last, the *Pteromys Genibarbis*, or *Kechubu*, of the Javanese, in which he compares this species with the *S. Sagitta* of Pennant, sufficiently to serve as specific descriptions of both.

The specific epithet Genibarbis is derived from a numerous series of vibrissæ, disposed on the cheeks in a radiated manner, on the sides of the head, above the upper lip, and on the extremity of the lobes of the ears, which seem to distinguish it from all others. These on the sides of the head are longer than the head itself, spreading, and of a dark colour; those on the cheeks occur from the posterior canthus of the eye, towards the jaws, consisting of above twenty separate bristles, closely applied to the sides of the head, about an inch in length in the middle, and gradually decreasing at the upper and lower margin ; those again on the ears arise from the base of the posterior portion of the lobe of that organ, and constitute a fascicle of long and slender bristles, partially concealed by those on the cheeks, spreading far from the head, and exhibiting a character very different from the brush-like appendages, which constitute the pencilled ears of several species of Squirrels.

The structure of the ear of this animal is stated, by Dr. Horsfield, to present several peculiarities which distinguish it from the other Javanese species. The interior of it is large, naked, and disposed transversely near the extremity of the head, divided in the middle by a continuation of the concha, which separates an extensive meatus auditorius externus from the superior cavity. The lobe is short, linear-oblong, with an inflected margin, and surrounds only the superior portion of the ear opposite to the continuation of the concha ; it descends abruptly, and from the base of its inflected margin arises a thick tuft of silky white hair. The naked interior portion is partially covered by the bristles of the cheeks.

The length of the anterior cartilage of the flying membrane qualifies that organ for a more complete expansion in this and in other Javanese species, than in the remaining Flying Squirrels. The tail is nearly two-thirds the length of the animal, with the hairs spreading laterally from it like the barbs of a quill, these are long, delicate, silky, slightly undulated and numerous, affording to this organ a degree of fulness which is highly ornamental.

The general covering is very beautiful. The fur consists of long hairs, downy at the base, closely arranged on the neck and back, but more distant on the flying membrane. For an animal inhabiting a hot climate, this fur is uncommonly thick ; solitary, bristly hairs project in many parts from the downy covering.

The general colour is gray on the upper parts, and white underneath ; on the neck, the back, and the tail, this colour has a brownish tint inclining to tawny, and the intensity of the colour varies in different parts. The hairs are whitish, closely arranged, and delicate ; along the cartilage, by which the membrane is expanded at the extremity, they form a close fringe, which is continued along the entire lateral border of the membrane. The separation between the upper and lower parts is strongly marked between the neck and shoulder. The animal is about the size of the common Squirrel.

The Kechubu is an inhabitant of the forests of Pugar, one of the most sequestered districts of the eastern portion of Java, and like the S. Sagitta is extremely rare. Like other species of this genus, it lives on fruits. Its retreat is generally found by a shrill sound the animal emits in the night. It is nocturnal in its habits, like the other species.

The Sciurus Sagitta is mentioned as a native of Java by Linnæus, in the 12th edition of the Systema Naturæ, on the authority of Nordgrew, and a detailed description of it is given after the specific character. Professor G. Fischer has arranged the Sciurus Sagitta as a synonym of S. Petaurista. Pennant also has formed the opinion, that they were the same animal ; but Boddært admits its existence as a distinct species ; and Pallas enumerates, with his usual precision, the characters which distinguish it from S. Petaurista.

To distinguish the Sciurus Sagitta, as far as it is yet known, at once from the Kechubu, it is sufficient to state that it is described as having a ferruginous brown colour, that the flying membrane extends from the head to the anterior extremities, and forms an acute salient angle behind the wrist, and that it is only one palm in length. The appearance which is expressed by the specific name Sagitta is not observed in the Kechubu. The S. Sagitta is deep brown on the upper parts, and white underneath, the tail bright brown.

For the other species of this subdivision, we shall merely refer to the table, observing only, that they are marked by a difference in the tail from those already mentioned, which has induced their separation into a section distinct from the rest. This is, that the tail is round with full fur, but shaped like that of the generality of quadrupeds, whereas in those already mentioned, as has been observed, the tail is flat or horizontal, and the long hairs of it spring from each side in the manner of the barbs of a common feather.

The genus CHEIROMYS, of the Baron, including the single species *Aye Aye, (Lemur Psilodactylus,* Gm. *Long-fingered Lemur,* Shaw,) is inserted by our author as a sub-

C. Hamilton Smith Esqr del^t

THE AYE-AYE.

CHEIROMYS MADAGASCARIENSIS. Cuv.

London. Published by G. B. Whittaker Sepr 1825.

genus of Sciurus ; but whether the Baron did so on a sufficient consideration of the bony parts of the animal, we are
not able to say ; M. Geoffroy has made a distinct genus of
it, under the name of Daubentonia ; and Shaw has referred
it, as have most of the subsequent systematic writers, rather
to the Lemurine than the Squirrel tribe, among which
Gmelin had placed it. It appears, like many other species,
to partake of double analogies; but is, perhaps, more
nearly allied to the Lemurs, or the Tarsiers, close to which
we have placed it in the table.

The upper jaw of this curious-looking animal has four
cheek-teeth ; the lower jaw but three ; the head is thick,
spherical, and large; the muzzle is short and pointed, and
not arched, as in the Rodentia in general ; the nostrils are
terminal ; the upper lip is entire ; the eyes are large, directed forward, and not laterally, as in the Squirrels; the
eye-lids are cleft; the ears are large and naked ; and the
mouth is large. By a longitudinal section of the head, it
appears that the area of the brain is much larger than that
of the face ; the contrary of which is the case with the
Rodentia generally, and the orbits are distinct.

The fore-legs are very short, compared with the hind ;
the bones of the arm and the hand are similar to those of
the Lemur. All the feet have five toes, which, on the anterior feet, are extremely long and thin; with these the animal obtains the larvæ and insects that lie under the bark
of trees. The hind-feet have the character of hands, the
thumb being opposable to the fingers.

Most of these characters bespeak a strong analogy with
the Tarsiers ; but the teeth are those of the Rodentia. The
animal, therefore, seems strictly intermediate between the
Quadrumana and the Rodentia ; and, therefore, to require
the inconvenient increase of another genus, which our
author has named Cheiromys.

The fur is of a bright fulvous or yellowish down, marked

on the back by long, rough, brown hairs, some of which are white at the tip; the limbs are brown, and the tail black. It is about the size of a Cat.

The Aye Aye appears to be of a timid disposition, and of nocturnal habits; it is very slow in all its movements. M. Sonnerat kept several for two or three months; it was very difficult to rouse them during day their food consisted of boiled rice, which they took to the mouth with the long fingers, in the manner in which the Chinese use their eating-sticks for the same purpose.

It is an inhabitant of Madagascar, and takes its vulgar name from the noise it makes.

We omitted to notice, either among, or immediately after the Marmots *, an animal, which for its rarity is engraved from Major Smith's drawing, from a specimen in Mr. Peel s museum, Philadelphia.

Of the correct allocation of this animal in systematic arrangement there seems to be some doubt, which we are not able to remove, by stating the particulars of its organic analogies. It is named in the museum the *Prairie Dog*, and is the type, we believe, of Raffinesque's genus Cynomys.

The traveller Le Rey says, an animal is found in the plains beyond the Missouri, called *le Prairie Chien;* it is smaller than the gray Fox; it digs holes and burrows in a light loamy soil, and in the same holes a small speckled Snake takes shelter, which, the Indians say, is the Dog's guard. The Indians have many superstitious notions respecting these Dogs; the Ay-oo-wars, or Nepeirce nation, have a tradition, that the human race sprang from the Dog and the Beaver. All the other nations hold them in great veneration.

* The Marmots and their consimilars are not very satisfactorily ascertained; they have lately been revised, and the result at present will be found stated in the table.

C. Hamilton Smith Esq.r del.t

THE PRAIRIE MARMOT.

ARCTOMYS? Ludovician.

Published by G.B.Whittaker Dec.r 1826.

The specimen in question was brought from Louisiana, (where the animal was known by the name of the Prairie Dog,) by Messrs. Lewis and Clarke; we must, therefore, take it for granted to be of the same species as that mentioned by Le Rey.

The colour of this animal is pretty uniform, varying only in intensity; on the back, however, it may be said to be red, mixed and lowered with white ; the tail, particularly towards the end, yellowish ; the rest of the animal light ochre colour, darker on the head and back of the neck. The head is altogether more spherical in shape than that of the Maryland Marmot, with the nose more truncated ; the ears very small and circular, with a black edge ; the eyes are large and round; the nostrils cleft laterally ; the upper lip cleft ; the space from the nose to the lip whitish, and furnished on each side with a few long whiskers ; the toes are five on all the feet, with rather long, sharp, unguiculated claws; the tail is about one-fifth the length of the body, which measures, from the tip of the nose to the anus, about sixteen inches.

In James's Expedition to the Rocky Mountains, we are informed that these animals occupy burrows in particular districts of limited extent, and that these dwellings are usually denominated by the hunters the Prairie-dog villages. These villages are various in extent, some confined to a few acres only, while others are extended over as many miles.

In the progress of excavating their burrows, these animals form truncated conical mounds, raised about six or eight inches above the surface, and almost two or three feet in circumference at the base; the whole face of these mounds, but more especially the top of them, is trodden down, and rendered solid and compact; the top is perforated by a comparatively large hole, which descends vertically for one or two feet, and then goes off in an oblique direction downward. The burrows are usually about twenty feet from each other, and they have generally several occupants, as seven or eight may be usually seen upon one mount, where they delight to sport in fine weather. On the approach of danger, they retreat to their dens; but if their cause of alarm be not too close, they remain barking and flourishing their tails on the edge of their holes, or sit erect to reconnoitre. From their habit of burrowing, it is extremely difficult to take them.

They pass the winter in a lethargic state; but lay up no provisions. They protect themselves from the cold by closing up the entrance of the burrows. Within their retreat, they construct very neat globular cells, of fine dry grass, having an aperture at the top large enough to admit a finger, and so compactly formed that they may be rolled about without receiving injury.

Of its dentition and internal structure, we can add nothing ; it is therefore not possible to determine, whether it belongs to the Marmots, to which (always remembering that genera with single species are ever to be suspected) we should be strongly inclined to refer it.

The genus HYSTRIX is thus characterized. The clavicles are imperfect, there are two incisive teeth in each jaw, and four cheek teeth on each side in both jaws ; these have flat crowns, surrounded with a line of enamel, which enters more or less deeply into the external and internal edge, and appears to cut the tooth into two parts ; there are also little radii of enamel, which wear down by detrition ;- the muzzle is thick and truncated ; the lip cleft; the tongue furnished with spiny scales ; the ears short and round ; the tail various in the different species; the anterior feet with four toes; the posterior with five, all armed with thick nails.

The spines, commonly called quills, distinguish this

genus most obviously, and at a glance; but they are not exclusively peculiar to it; they are found also in the Hedgehog, Tenrec, Echimys, and Echydnes. The Echimys, however, which alone of these belongs to this order, have the spines in shape like a sword-blade, and not cylindrical or tapering, as in the genus in question.

The variations in organized creation are so numerous that their mere observation becomes trite and useless; the business of natural history is not merely to find out and observe them, but so to class and divide them by factitious boundaries, founded on natural demarcations, that the memory may be enabled the more readily to recal them at command. The majority, however, of these countless variations seem mere modifications of a much smaller number of given types. Thus the different modifications of the integuments and excrescences in animals, which are almost as various as the genera, or even the species, may all be referred to a very few sorts; hair, feathers, and scales are found in the different classes, modified in an endless variety of modes; and when the generality of these integuments is considered, we are naturally surprised to find something of a perfectly different character, but limited to a very small number of unimportant creatures, though differently modified in each species to which they are proper.

When the destinies of any given animal require particular capabilities in any one direction, we generally find some one organ fitted for the intended object, either modified or greatly developed; thus the mowing or cutting of grass, the tearing or dividing of flesh, and the pounding or triturating of grain, are all performed by the same organs, the teeth; nor in general, and but with a few exceptions, is one order of quadrupeds furnished with any particular organ, which may not be found, though differently modified, in the other.

This observation, however, does not apply to the spines with which a few Mammalia are furnished ; nor does nature appear, in this particular, consistent with her general operations ; a new creation seems called into an unusually limited action : and the Porcupine, Hedgehog, Tenrec, and Echimys, are furnished with spines differently modified, but which appear to be of no other use than as weapons bestowed on animals not otherwise endowed with courage or ferocity to use them, or particularly exposed beyond others to hostile attacks, or playing any very important or prominent character in the great theatre of the world.

It is in this genus, and more especially in the species of the Common Porcupine, that the spines are most completely developed. Their nature, however, more particularly belongs to the specific description of the animal, though they are too well known to need much detail.

The natural history of the *Porcupine* has been long a great desideratum, until M. F. Cuvier, with his usual attentive observations on many of the Italian specimens, was enabled, in the year 1821, to add many interesting facts to its history, and to strip it of many of those errors with which simplicity and credulity had invested it.

This species has been the subject of numerous misrepresentations. Buffon indeed corrected some of these, and principally the belief, that these animals had the faculty of darting their spines ; but others of these errors were not detected by him, for want of means of observation, and these have survived to a much later period.

We are informed by Agricola, that the Porcupine of Italy was an exotic in that country, brought either from Africa or India. It has, however, long been naturalized in the South of Europe. The only difference observed between the Porcupine of Italy and that of Africa is, that the former is rather less than the latter, and that the spines are not so strong. That these differences, however, result from difference of climate and locality, there can be no doubt: but for the historical information of Agricola, however, they would probably have been separated by zoologists into distinct species, as many others have doubtless been, on grounds not more substantial.

The European variety is found principally in the kingdom of Naples, and in the southern parts of the Roman States. It avoids populous parts, and selects stony and dry situations, exposed to the south-east or south, in which it digs deep burrows, with several entrances, where it lives in solitude and security. Its extreme timidity seems to induce it to continue in its retreat, and to seek its sustenance only in the night. At the end of twilight, it approaches cautiously to the principal entrance of its retreat, and does not risk itself entirely outside, without first assuring itself that all is safe. As soon as it has gained sufficient confidence, it proceeds in search of fruit, bulbs, roots, &c. &c., or, at the accustomed period, of its mate. It hibernates in its retreat, but its lethargy appears to be by no means so deep as that of some other genera, as on the first fine day it may be found awake again, and in search of food.

Thus peaceably and quietly does the Porcupine pass its allotted period of existence, which extends to fifteen or twenty years ; and it is remarkable, that an animal so well prepared for defence should been dowed with so gentle a disposition. Other genera of this order, as the several species of hares and rabbits, may be said to be possessed of a means of defence from the rapacity of almost every animal of prey, in the timidity of their disposition, in conjunction with a certain degree of celerity of motion; without these, indeed, they are almost utterly defenceless : but this is not so with the Porcupine. The ordinary physical powers of inferior animals are in general insufficient to avail against the spines of the Porcupine; and, to render these more effectual, it has, like the Hedgehog, the power of rolling itself into a ball, and thereby presenting an uninterrupted surface of spears. It has therefore, in fact, few enemies to fear—but man. As their flesh is esteemed good, they are frequently hunted, or rather sought for in their retreats. It is not, therefore, apparent why they should be endowed with so great a degree of timidity and distrust.

Although the Porcupine has not the power, so long attributed to it, of darting its quills at its enemies, it is nevertheless not always merely passive when attacked. If closely pressed, it will throw itself with impetuosity on its opponent, and always sideways; and it is on the sides of the body that the spines are longest and strongest. It seems most anxious, when attacked, to protect its head, and yet no animal perhaps can bite harder, for the largest carnivora cannot inflict deeper or more dangerous wounds with their canines, than the Porcupine can with his incisors. The thickest and hardest boards soon yield to the gnawing operation of these teeth; and they have hardness enough to act on iron wire, hence it becomes necessary to line the cage with sheet-iron wherein they are confined. Its natural inclination for a solitary existence can only be partially overcome. It will habituate itself in time to occasional action, but will never become familiar, or any thing like affectionate.

Reasoning and presumption long supplied the place of observation, in regard to the *modus copulandi* of these animals. Aristotle's statement could not have resulted from observation, as is now sufficiently ascertained. The spines are erected by means of muscular action; without this, therefore, they lie level with the skin, and the presumed difficulty alluded to, as in the case of the Hedgehog, is completely removed. The period of gestation is sixty-six days, and the young are born with the eyes open, and covered with spines, which are then about seven lines in length;

the colour of these spines is the same as that of the parents, which they resemble also in all parts of their organization.

The Porcupine is one of the largest of this order, measuring from nose to tail nearly three feet, and the tail four inches. Its physiognomy is thick, and its plantigrade walk is slow. Its obtuse muzzle, thick nostrils, and very small eyes, give it an unpleasant appearance. It belongs to M. F. Cuvier's division of Omnivorous Rodentia—that is, the cheek-teeth have their roots distinct in form from their crowns, or the protruded part of these teeth. Their dentary system is, however, stated elsewhere, as well as the toes and nails.

With the exception of smell, most of the senses of the Porcupine are obtuse, judging at least by their organization. The eye is extremely small, with two lids only. I think, says M. Cuvier, the pupils are round ; the ears are but little developed and simple, as to the number of folds ; the nostrils are open at the two extremities of a sinus, in form of a cross, which traverses the muzzle uninterruptedly, edged with thick lips or fleshy sort of pads, invested with a peculiar integument, and some short hairs ; the upper lip is cleft, and the tongue is but little extensible, and furnished with horny papillæ, large in the middle, and narrower, and similar to those of the Cat on the sides. Independently of the spines, the body has two sorts of hair, one long and bristly, and the other curly and woolly. The mammæ are six, three on each side, placed on the flanks, and not on the abdomen.

The head and the neck are furnished with very long hairs, which the animal can erect at will. The muzzle, the sides of the neck, the throat and anterior part of the shoulders, the limbs, chest, and belly have only short hairs ; the spines cover the posterior part of the shoulders, the back, sides of the body, thighs, and crupper ; the largest, as before stated, are on the sides, and on the anterior part of the back ; those on the tail are open tubes attached to the skin, like the other spines, by a slender solid pedicle. All the parts which are covered with spines are black, the spines are annulated black and white, and the tubes are all white. The woolly hairs are reddish, so that the general colour of the animal is dingy.

The ancients knew this animal under the name *Hystrix*, on account of its harsh bristly hairs, like those of the Hog. This name has been translated into most modern languages, and Linnæus has added cristata, from the sort of mane on the neck and back, as a specific addition to this species.

The *Canada Porcupine (Hystrix Dorsata*, Lin., and *Urson* of Buffon), is remarkable for the length of the hair on its body, and the shortness of the spines. The male is of a deep brown, nearly black ; the female of a lighter brown: the spines are visible through the hairs only on the head, crupper, and tail, without close inspection ; they are an-nulated, dark brown and white ; there are no spines under-neath ; the eyes are round, and the ears short, and nearly hidden in the fur. It measures nearly two feet six inches, from nose to tail, which is about nine more. The spines are strong and sharp, are said to be so formed, as to ap-pear, when magnified, as if barbed at the tip with nume-rous small reversed points or prickles, and are so slightly attached to the skin, as to be loosened with great ease : and the animal will sometimes purposely brush against the legs of those who disturb it, leaving several of the spines stick-ing in the skin. The specimen engraved is a female, from a drawing from life, by Major Smith, in Canada. While the Major was taking his sketch, it struck several of the quils of its tail into the hand of a man who was simple enough to attempt to caress it. The animal appeared fearless, and walked round him several times, till it sud-denly paid him the compliment, which produced a few drops of blood, and many ejaculations.

C. Hamilton Smith Esq.t delt.

THE CANADA PORCUPINE.

HISTRIX DORSATA.

It makes its retreat under the roots of trees, and is very unwilling to be wetted. It sleeps a great deal, and lives principally on bark. It drinks in summer, and in winter swallows the snow. It makes a slight mewing kind of noise.

Its flesh is said to taste like that of the pig, and is eaten by the American Indians, who also use the fur, after having plucked out the spines, which serve in the stead of pins.

The Baron includes, among the species selected for specific characters in the text, the *Hystrix Fasciculatus* of Linnæus; the *Brush-tailed Porcupine* of Shaw, and *Malacca Porcupine* of Buffon. A more complete description of this species is a desideratum. It has been referred to the division of Spiny Rats, instead of that of Porcupines, by M. Desmarest after M. de Blainville; but as Cuvier places it with the Porcupine, we shall insert a short notice of it here.

The head of this species is described by Buffon, as more elongated than that of the Porcupine, consequently it is more allied to the Rats in this particular: the muzzle is covered with black skin; the eyes are black and small; the ears small and round; the moustache five or six inches long; the back and sides covered with long strong spines of a flat form, with a flute or furrow along each side. Most of these spines are white at the point, and black about midway of their length, others are black on the upper side, and white underneath; the lower parts of the animal are covered with whitish hairs, but the legs have a few that are black; the tail is moderate in length, round, naked and scaly, for the greater part of its length, but toward the point is a bundle of long flat excrescences, similar to strips of parchment. The head and body are about eighteen inches in length.

The specimens of this species which have been seen in captivity were untameable. They erected their spines when irritated, in the manner of the Common Porcupine. They

Q 2

were observed to be nocturnal, and fed on fruits in preference to nuts.

It is an inhabitant of India, and the Peninsula of Malacca.

We are not able to add any thing to the short notice of the Baron, at p. 88 of this volume, on the species of Hystrix Macroura, which appears to be allied to this.

The *Prehensile-tailed Porcupine*, or *Couendou (Hyst. prehensilis*, Lin.) is as we have seen, included in the general genus Hystrix by the Baron. By the amateurs of generic divisions, two species of the American prehensile-tailed Porcupine have been pointed out, and both have been separated into a distinct genus, and also into a subgenus of Hystrix, by the prehensile character of the tail. The synonima are stated in the Table, by which it will be seen, that the species described by Buffon is presumed to be distinct from that of Herman and of D'Azara, who has given us some account of the manners and habits of the animal. The description of Buffon's species, with the habits of that of D'Azara, shall be abridged. Their specific descriptions are too nearly allied to admit of any wide difference in their modes of life.

The nose of the Couendou, like that of the Common Porcupine, is thick and obtuse, and covered with brownish hairs ; the ears are naked, covered only with some short spines on their edge ; the face is furnished with long black whiskers : the body is covered with spines upwards of two inches long on the back, one inch and a half on the forelegs, and not half that length on those behind, each spine being white at the base and at the point, and black in the middle ; one half the tail is also covered with spines ; the tail itself is long, tapering, and pointed, blackish, and covered with scales from the the middle to the point ; the under part of the tail, from the point to the termination of the spines, has a few hairs of a bright brown colour, the rest being furnished with scales, as is the whole upper

side, and on the legs are a few hairs interspersed among
the spines of those parts.

The habits of an animal may, in almost all cases, be anticipated by a consideration of its make and organs. That
this is true, in a general way, is beyond all doubt ; but it
is also true, in most cases, to an extent not so obvious at
first sight. It must be admitted, indeed, in regard to the
spiny quadrupeds, that there are few organs of animals,
few incidents to their physicalities less apparent in their
object, and perhaps less influential in their habits,
than the spines. They seem of no use in enabling the
animal to procure its proper food ; and if they are material to it for self-preservation from external violence, we do
not observe any corresponding deficiency of self-protection,
or excessive degree of exposure, to call for any extraordinary means of protection. When it is said, therefore, that
the organs of a species generally proclaim its habits, the
spiny races may be considered as forming an exception to
the observation, in regard at least to these excrescences.

But the prehensile, or rather subprehensile tail of this
particular species, which, as we have observed, methodical
writers have employed to distinguish it as a genus, bespeak
its habits of living in trees ; nor are we disappointed in
this expectation, the Couendou is in fact a tenant of the
forest, climbs with great facility, by means of its claws, but
is said to avail itself of the holding power of its tail only
when descending. In common with other inhabitants of trees,
the Couendou, when on the ground, is awkward, and as it
were out of his element ; its motions are then slow, and apparently painful, nor does the animal seem willing to move,
till the imperious dictates of hunger rouse him to action.
The sustenance of the animal is also found in trees ; it consists of fruit, leaves, and flowers, and even the wood when
young, which it cuts with ease, by means of its strong incisors, common to the whole order. It has, however, none

of the propensities of the omnivorous Rodentia to eat flesh, but confines itself entirely to the food already stated.

The period of gestation, and other incidental particulars are not known, but it is once only in the year, and toward September or October, that the female brings forth, and the young are said to be very few at a birth. The slowness, uncertainty, or difficulty of reproduction, may afford a solution of the question, why these animals are furnished with extraordinary means of protection and self defence; but we do not know, that the circumstances of all the spiny quadrupeds will warrant such a conjecture generally.

The Prehensile Porcupine is a native of many parts of South America.

The genus Lepus includes many species, but these have so many points of identity or similarity, or, in other words, are so nearly alike in most particulars, that it is difficult to distinguish whether their relative specific characters are those of affinity or analogy. This difficulty is increased when it is considered, that they are widely spread over the earth's surface in the new world, as well as in the old, and that a diffused allocation is a great promoter of those differences which are decidedly attributable to variety. The common Hare, the Rabbit, and the Variable Hare, are natives of Europe; one species, the Tolai, is very common in Siberia. The northern and the southern extremities of the peninsula of Africa produce Hares very nearly allied to one another. In North America we have a species very like our own, and in South America is found the smallest of them all, the Tapiti. These species appear to be the best known and admitted by zoologists, though several others have been mentioned by travellers as distinct.

All the species are alike under the continued influence of fear, and as their eyes are presumed not to be perfect during daylight, and their lateral direction prevents the animal seeing directly forward, they rather rely on their hearing, the organ of which is very perfect, to warn them of approaching danger. Perfectly defenceless, indeed, and exposed to countless enemies, they have no chance of safety but in the expedition of their flight, and unless forewarned by the acuteness of one or more of their senses of the approach of an enemy, they would invariably fall the victims of surprise. Their Creator, while he has left them a prey to so many other animals, has provided them with one mode of self-defence in a rapid locomotion, rendered more efficacious by a quick susceptibility of danger. Lively timidity, however, must be attended with pain ; and if there be any disparity in the distribution of good and evil to inferior creation, all, except sportsmen, must pity creatures which exist constantly under the excitement of acute fear.

Ordinary conclusions, drawn by analogy, are frequently erroneous when any of the great objects of Providence are the subjects of our contemplation ; this seems exemplified in the providential preservation of the several races of animals. The doctrine of the special interposition of Providence, at least in ethics, is beside our subject; but the existence of beings around us, whose speedy extermination would seem the ordinary consequence of their situations and relationships with other creatures, evinces an extraordinary suspension of common consequences in physics in a very forcible manner.

The heart is said to be larger, relatively to the other parts, in these than in most other animals, and it has been noticed, that Pliny observed generally, that all animals of a fearful disposition have the heart of considerable size.

The several species of the genus Lepus, or the Hares, properly speaking, exhibit a very close approximation to a common type. In noticing the common Hare, therefore, we shall state several particulars equally applicable to all the species, and shall refer to the Table for the short specific characters which distinguish the rest.

The several species of this genus, which, although they are so analogous as to compose one of the most natural groups of Mammalia, are, nevertheless, spread over a very wide surface. This fact should, perhaps, make us very cautious on the subject of specific appropriations. The great and astonishing changes wrought in the physicalities of animals, by change of place and circumstances, are, to a considerable extent, well known to us ; but the limit of this operation is by no means defined by us ; hence, it becomes very difficult to distinguish between analogy and affinity in animals. The Lions of Barbary, of South Africa, and of Asia, differ not inconsiderably from each other ; but their affinity is not doubted, and but one species is acknowledged. The Hyænas of northern Africa and those of the south differ, perhaps, but little more than the varieties of the Lion externally; in one race they are striped, with the hair elongated on certain parts ; in the other they are spotted, with the hair equal, and the osteological differences consist rather in degrees of development than in essential characters; yet the affinity of these has hardly been suggested, they have always been treated as distinct species. And in the species of the preceding order *(Hystrix,)* we may notice, that there is a difference between those located in Africa and Asia, and those of Europe, though these differences have never been ventured upon as other than characters of variety.

These different results, from premises so much alike, seem greatly to be attributed to accident. When the natural history of an animal, as in the common example of the Hog, is sufficiently known, when its first transportation from one climate to another is matter of unquestioned notoriety, whatever may be the differences that have sprung up in the race before our eyes, and however widely the exotic race may vary from their ancestors, we dare not to affirm that there is any specific difference; but when the allocation of an animal is involved in the general obscurity of past unrecorded events, and the consequences only are known to us by inspection for the first time, it is easy to anticipate a great probability, at least of the descendants from the same root being treated as distinct.   Hence, the great caution necessary in editing a species as new.

These observations, however, have no other than a general application when applied to the genus before us, resulting from the great similarity of the several species which compose it, and which, at the same time, we may observe, are supported by the sanction of the greatest names in zoology.

Mindful of the wide field before us, of the necessity for abridgment, and of the simple elementary pretensions of these essays, we are unwilling to dilate on the external physicalities, at least, of a species so well known as the *Common Hare ;* and though the teeth would well deserve a particular description, every one may acquaint himself with them by the much more effectual mode of observation. The palms of the feet, being covered with hair, is a remarkable character for its singularity in this species. The eyes have no accessary organ, and the pupil is elongated horizontally.   The nostrils are circular, and almost hidden in a fold, by which means they are capable of being closed. The upper lip is cleft. The tongue is thick and soft. Their very long ears, like the nostrils, are capable also of being closed; it appears, therefore, that the excessive development of these organs is not without a corresponding perfection in the senses to which they belong, a perfection which may, at times, when the animal is in a state of repose and safety, not only be unnecessary but irksome to it had not its Creator provided against this evil, by enabling it to suspend the action of the organs. They have ten mammæ.

Hares are capable of reproduction at a year old ; their gestation is thirty days, and they bring from two to five at a birth.

There is one very remarkable physical character of the Hares, as also of the Rabbits, the females are provided with a double matrix, and the consequence is, that two contemporaneous fecundations can proceed together ; this does not arise from any real superfetation, but by the fecundation of a second distinct organ. This accounts for the extraordinary prolific quality of the species of this genus.

Hares do not, like Rabbits, dig themselves, or seek a subterraneous retreat, they merely look for a convenient hollow place in a furrow, which is called their seat, and in which they pass quietly the greater part of the day, till the approach of evening rouses them to seek their sustenance. They vary this seat according to the season, and select the most favourable positions for receiving the benefit of the sun when necessary. In severe weather, they betake themselves to the shelter of the woods. Their sexual appetites are very strong early in the spring, at which time their actions, from this excitement, have led to the observation of the madness of a March Hare.

The great length of their hind legs, compared with the others, disqualifies them from any other than a leaping motion, or an interrupted gallop. Like the Kanguroos, &c. they will sit on the tarsi, and use the fore feet to convey food to the mouth, and clean the fur, &c. They drink lapping, and can bite with much severity if driven to it.

The intelligence of the Hare is considered as very limited, nor have its instincts much of a prominent character. Its doubles, however, in coursing are very notorious; but if this be the result of intelligence, it does not accord with its conduct on the same occasion, in returning to the spot from which it was at first disturbed. The timidity of its disposition is noticed by its specific epithet *Timidus*

The Hare is placed, by the Levitical law, among the ruminating animals. These, however, have all distinct stomachs, which is not the case with this animal, nor does there seem any reason, grounded on observation, in verification of the statement. It is true, that the stomach, at first sight, does appear double, owing to a particular fold within it; that the cæcum also is so very large, that it may be regarded as a second stomach, and that the habit the animal has of frequently moving the lip, though the jaw remain stationary, gives it the appearance of ruminating. These, however, are but semblances, and not the real character of ruminating, and some other explanation would be more satisfactory. Interpolation in the text has been presumed, but in the absence of a more decided elucidation, the adversaries of the Old Testament must make all the advantage they can of the presumed zoological ignorance of the Jewish lawgiver.

The voice of the Hare, though never perhaps exerted except when the animal is irritated or wounded, is not a sharp cry, but is rather strong, and like the human voice. Although solitary and silent, they are not altogether so wild as their habits appear to indicate. Their disposition is gentle, and if taken young, they are capable of training and education. M. Desmarest had one a considerable time about his house; it lost all its natural wildness, and its habits had become quite familiar, at least to all it knew, but of strangers it was still fearful. In winter it sat before the fire between two large Angora Cats, and a sporting Dog, with whom it lived on the best of terms ; at table it was generally close to its master looking for food, and if thwarted in its expectation, would beat with its fore paws in rapid succession on the hand or arm of the person so treating it. It acquired an excessive degree of fat, a common consequence with these and many other animals in a domesticated state, which not unfrequently die of an accumulation of fat.

As a further proof of the subserviency to the powers of man of the most predominant moral quality in this animal, its timidity, we may refer to an exhibition which has been common about the streets of London, certainly in one individual, and probably in several, of a Hare which moved fearlessly about upon a table in the midst of the surrounding multitude, the tones of a hand organ, and the mixed noise and confusion of a public street. The Hare was taught further to beat a tambourine, which it did with great rapidity, and in the manner of that described by M. Desmarest when soliciting a boon from its master; and as a still further proof how completely its fears were neutralized, it was accustomed to pull the trigger and discharge a pistol, rather large in dimensions and calibre, and commensurate consequently in report, and when it is considered that this was an affair of recurrence perhaps almost every half hour, it can hardly be supposed that the animal was taken by surprise, as to the consequence of pulling the trigger. There are few animals so completely stupid as not to learn by reiterated practice the immediate consequences which invariably follow from any particular act, nor did the animal in question exhibit the least alarm or shock on making the report, which would, in all probability, at least without similar training, have the effect of turning a Lion.

A Hare, a Cat, a Rat, a Mouse, and a Sparrow have also been exhibited together about the streets, confined in one common cage.

The *Rabbit*, in all its physicalities and relative proportions, is extremely assimilated to the Hare. The habits and instincts of the two form, perhaps, their greatest differences. Although provided with similar organs, clothed in the same dress, and inhabiting the same countries, they seem to have a natural aversion for each other; a hatred, observes M. F. Cuvier, which nothing can soften. Love, which unites the Dog and the Wolf, the Goat and the Sheep, the Horse and the Zebra, cannot conciliate the Rabbit and the Hare. However violent their sexual desires each for its own species, and however nearly the two may be allied, they will under no circumstances approach each other ; or, if by chance they meet, a combat generally follows, which not unfrequently terminates fatally to one; hence Hares are not found where Rabbits are plentiful.

The Rabbit differs very materially from the Hare in its habits. Unlike the latter, which is contented with a mere seat, a slight concavity on the surface of the earth, the Rabbit digs deep and tortuous burrows for retreat and security ; for the facility of forming these retreats, it generally selects dry and sandy soils. The burrows have usually several entrances, and are common to many individuals, all, as it is said, of one family. Where Rabbits are plentiful, their burrows are separated by very slight intervals, and sometimes even ,communicate with each other. These retreats have no determinate or fixed forms, but the galleries intersect each other in various directions.

When a warren is established, so rapid is the increase of these animals, that its continuance is only limited by want of food. They produce at or even before six months old. The period of their gestation is twenty or twenty-one days ; but as, like the Hares, they have a double matrix, they are sometimes known to produce two distinct litters at an interval of only a few days. The young are brought forth at the bottom of the burrow, in a bed prepared by the mother. They are born covered with hair and with the eyes open, but they do not venture from the burrows for two months, when they begin to eat ; and the father, who has not seen them before then, adopts them, and assists in finding young and tender herbage as food for them. Eight or nine years appears to be the ordinary term of their existence.

Rabbits, when confined, lose some of their natural qualities, and acquire some others, nor are they so esteemed for the table. It appears also, that races of these animals which have been long domesticated lose altogether the instinct for burrowing, nor do the sexes pair monogamiously as they are presumed to do in a natural state ; the males in particular in a domestic state not unfrequently destroy their offspring, though they do not eat them, whence it seems probable that domestication has the effect of eradicating from their nature the instinct of protection of the young, as well as the inclination for digging, and probably other instincts. The females nevertheless in this state seem still more prolific than when wild; they will sometimes produce twentysix young in sixty days. It is said, however, that after a particular race of Rabbits has attained its maximum of development in confinement, its prolific powers altogether fail.

Domestication has produced various varieties in these animals, black, silvery white, and some with long silky hairs, called Angora Rabbits.

All the particularities of its organization have a close analogy with those of the Hare, and as the animal is so completely known, any additions on this subject may, perhaps be spared.

The Rabbit is said to be originally from Spain, but it has been for ages common in the rest of Europe, and is now transported into Africa and America.

We are assured on the authority of those who have paid great attention to the subject, that Rabbits live in a social state, and take an interest in each other, and even have something like respect for the right of property. In their republic, as in that of Lacedæmon, old age, parental affection, and hereditary rights are respected ; the same burrow is said to pass from father to son, and lineally from generation to generation ; it is never abandoned by the same family without necessity, but is enlarged as the number of the family increases, by the addition of more galleries or apartments. This succession of patrimony, this right of property among these animals, has been long observed, nor have the modern investigations in zoology disproved its existence. La Fontaine thus takes notice of it—

> Jean Lapin allégua la coutume et l'usage.
> Ce sont leur lois, dit-il, qui m'ont de ce logis
> Rendu maître et seigneur, et qui, de père en fils,
> L'ait de Pierre à Simon, puis à moi, Jean, transmis.

In the absence of facts to establish the contrary, Buffon very naturally conjectured that the *Variable Hare* did not differ specifically from the common species ; the mere fact of the change of colour with the season being by no means uncommon in several species. Conjecture, however, must give way to the result of observation and experience, which have now established the distinctness of this species.

The Variable Hare is about a fourth larger than the Common Hare, but the head, relatively with the body, is not so large, and the ears are considerably shorter ; the eyes also are situate nearer to the nostrils ; the legs, also, compared with the body, are not so long, and the tail is shorter, having less vertebræ. The summer dress is brown, varied with whitish and reddish-gray ; but in winter it is all over of the purest white, with the exception only of a slight border of black round the ears, and a yellowish tinge on the soles of the feet. The iris of the eye is at all times of a yellow brown, and the tail is always white, but in winter it becomes slightly tufted with a woolly kind of hair.

We have already had occasion to notice, that few operations of nature have hitherto more completely escaped detection than this periodical mutation of colour in animals. The manner and the object are alike obscure, and the mere fact that it occurs principally, perhaps entirely, in the natives of high latitudes, while it naturally refers to climate for a rapid solution of the problem, still leaves us quite in the dark as to the mode of working it.   However climate may excite this change, there must be a predisposition in the nature of the animal to the excitement, for the Variable Hare, either domesticated or confined in a house, in an uniform mild degree of temperature, will nevertheless assume its white dress with the change of season, only a little later than if exposed to all the severity of the cold. It must be observed moreover, that in autumn, and before the actual operation of the cold can have had much effect, the fur seems preparing for the change, and that instead of resulting from the cold, it actually anticipates the season, and becomes white when the cold is not greater than it is in spring, and at the time when the opposite mutation to its summer dress is in progress.   It is said also that the Emperor of China preserves them in a warm climate, where they change yearly, as well as in higher latitudes.

Pallas informs us that this animal has very warm blood, and that during the severest frost the internal heat is from 103 to 105 degrees of Fahrenheit.

In many parts the Variable Hares are observed to change their haunts with the season, but not with regularity, or in troops, or in any preconcerted manner, like many migrating animals ; but they are occasionally observed, on the approach of winter, to quit the elevations of the northern mountains, and to return there at spring.   It is, however, thought not to be the cold, but want of provision which drives them to this change.   In Greenland, however, this has not been observed, but the white Hares of that country are said to continue of that colour even in winter, and some doubt therefore may remain whether the species be in fact the Variable Hare or the common species become permanently white by climate. The fecundity of the Greenland species seems by no means prejudiced by the severity of the cold, a fact, which confirms the statement of Pallas as to their great internal heat. The female brings forth sometimes as many as eight at a time.

In the deserts of southern Russia, toward the fiftieth degree of northern latitude, where the Variable Hare ceases to be common, there is said by Pallas to be a numerous race, to which he applies the name *Lepus Hybridus*, and which he suspects to be the issue of the common and variable species, and to be steril. Notwithstanding, however, the high respectability of the authority, we may reasonably doubt the fact, in the absence of repeated observation. There seems, however, every reason to conclude either that varieties of these two acknowledged species occasionally approach each other, or that there are distinct races intermediate between the two.

We shall add nothing here to the descriptions of the other species of this genus, except the *Tapeti*, or *Brazilian Hare*. This is the smallest of the known species: it was mentioned by Marcgrave, and the earliest writers on the American quadrupeds; but not being sufficiently described, it became almost forgotten, or was treated merely as a variety of the Common Hare. Marcgrave stated that the animal was without a tail, but the figure which accompanied his description represented one ; his text, however, prevailed, and Linnæus, Erxleben, and Brisson treated the presence of the tail in the figure as an error, and named the animal *Lepus Ecaudatus*. Johnson and Gessner referred it to the Guinea-pig; and subsequent zoologists have treated it as a variety of the common American species; but D'Azara has more recently described and established it.

The general form of the body is that of the Hare or Rabbit. From the tip of the nose to the insertion of the tail it measures about eighteen inches, and the tail itself with the hair upon it, which makes it round, does not exceed ten lines. The fur is varied brown-black, and yellowish above, with the upper part of the head red-brown, without any sprinkling of yellow; the cheeks are grayish; a lightish line passes round the eyes; the lower edge of the nose, the lips, and the under part of the head, the chest, and belly, and insides of the legs, are white.

There is a small specimen in the Paris Museum not larger than the Guinea-pig, which has the fur browner than in the adult state; and the ears are short, compared with the rest of the body; behind, the neck is red, and there is a white collar; the tail is so short as not to be perceivable.

The Tapiti does not burrow in the earth, but lives in woods, and sits on the surface like the Common Hare; when hunted, he endeavours to hide himself under the trunks of trees, or in the high grass. The flesh tastes like that of the Rabbit. The female is said to bring forth but one litter of three or four in the year.

The sub-genus LAGOMYS, of our author, differs very slightly in generic character from the Hares; the teeth and toes are alike, and it is principally in their diminutiveness, and in their habits, that they differ from the genus Lepus. They are Hares in miniature; or, as the word Lagomys imports, Hares like Rats, fitted for their Siberian habitat, and endowed with a strong instinct of accumulating provision for winter use; an instinct essential to their preservation in the severe and inhospitable climate they inhabit.

The *Pika*, or *Alpine Hare* of Pennant, has the body thick; and the head, compared with it, rather long; the nose is hairy; the whiskers are very long; the eyes are small; the ears round; the feet short; on each side of the os coccygis is a hardish greasy lump, or substance, which, when the animal sits, forms a thickish tubercle, about the size of a nut; the body is thick, and short.

The colour of the species is reddish-yellow, varying in

THE TAPETI HARE.

LEPUS TAPETI.

intensity in different parts of the body ; it does not seem at all subject to change with the season. The ears are black, with a whitish edge ; the nose is brown, and the whiskers black ; the fur is shorter, and more harsh than that of the Hare, and more allied to the fur of the Marmot. There are six mammæ, two pectoral, two abdominal, and two unguinal.

The Pika is an inhabitant of the highest mountains of the extreme north of Europe and Asia, in the thickest and most sequestered and humid forests. Sometimes it is said to live alone, and at others in small societies. It digs burrows between the pieces of rock, but sometimes sits in the manner of the Common Hare in the clefts of rocks, or of trees ; from these it issues generally after sun-set ; or if the weather be very dark, during day, to graze.

The instinct of amassing provision for winter food is very remarkable in this species, and its mode of doing it is still more so. About the month of August, the Pikas cut and collect large parcels of grass, which they spread and dry, and, in effect, convert into hay ; this they collect into stacks of about seven feet in height, at a convenient distance from their subterranean retreat ; having done this, they excavate a way from their burrow, which opens under the stack, and which they use as a road to their provision, while the snow of a Siberian winter buries almost every thing under one common surface. These poor animals are frequently plundered of their winter store by the hunters, by whom these hay-stacks are much prized as food for their horses.

Actions like this, which do, in fact, anticipate remote, but certain evils, and wear all the semblance of doing so by the powers of reason and deduction, equally surprise and astonish us, while the mental facultiesof these creatures are evidently so limited on all other subjects as to render it probable that such faculties differ from our own not merely in degree, but in kind. It is not merely the prescience of the remote evil of starvation, or even the direct means of providing against it by amassing a store of provision, but it is something equivalent to the knowledge that such provision requires preparation for the purpose, and the best and only means of preparing it that is thus displayed to our observation and astonishment. Little as we know of the mode in which this instinct is excited, it is obviously nothing short of the special interposition and care of Providence for His creatures—the agency—the mode of operation—the final causes may be the legitimate objects of inquiry, though these have hitherto eluded as well the scalpel of the anatomist as the philosophy of the metaphysician. Invited by almost every species, as it occurs, to enlarge on this subject, we feel it impossible to do so with the hope of explaining that which has hitherto been perfectly inexplicable; a future opportunity may, however, lead us into some further observations on this phenomenon of Nature—the instinct and mental faculties of animals ; at present, we shall merely acknowledge the finger of God, as displayed by the unconscious Pika for its own preservation, and pass on.

Although well known by the Siberian hunters, this species almost escaped the notice of travellers, till the time of the indefatigable Pallas, who first gives us the details of its description and history. It is known by the name Pika or Peika, which Pallas adopted as specific, though it is now become a generic name by the Tungouse people dwelling beyond Lake Baikal. It has, however, various names among the different northern Asiatic tribes.

The size of this species is said, by Pallas, to vary considerably, though in reference uniformly to its immediate habitat. Thus the largest are those of the Altaic mountains, which are about as big as the Guinea-pig, and weigh upwards of a pound and a quarter ; those of the Daniria, and the environs of Lake Baikal, are less in size and weight, and the smallest are found beyond the Janisea, near the town of Krasnogan; the weight of these last is under six ounces.

The head resembles that of the Rat, and also of the Pika Pusillus, (the Calling Hare of Pennant,) but it is larger, and not so broad, and the muzzle is less obtuse.

In order to ascertain the particular plants which the Pikas selected and dried for their winter provision, Pallas examined minutely one of their collections ; they appear to be selected very carefully, to be cut just at the proper period of ripeness, and dried so slowly that they form a fodder both green and succulent; neither thorns, nor hard or ligneous stems, and even very few flowers, were to be found among it, but a few acid or bitterish herbs are mixed, by way of giving relish to the rest ; small parcels of large leaves seem to be carefully separated from the rest, though kept with the general stock

We have already stated that the half-civilized sable hunters, instead of imitating the provident labour of the Pika, appropriate to themselves the fruits of its industry, and leave the poor animals to starvation and death. The life, indeed, of this interesting and laborious species is one of continued danger, "sufferance is the badge of all the tribe," exposed to the privations consequent in high latitudes, and long frosts ; they have also many enemies to dread. If their little stores escape the rapacity of man, they themselves are exposed to the attacks of their carnivorous compatriots, particularly of the Weasel tribe, and are subjected to less acute, but more protracted torments, from the larvæ of a species of insects, (Œstrus, Lin.,) lodged under their skin, and living on their very substance: such, says a describer of this animal, such is the common lot of imbecility; the weak are ceaselessly exposed to every kind of oppression, and their labour, industry, and talents frequently become the sources of evils and persecutions; they are robbed, abused, and considered happy if allowed only the sad alleviation of complaint.

Erxleben was of opinion that the *Ogotone*, or *Gray-Pika*, was the same as the species last mentioned ; Pallas, however, has sufficiently marked its specific characters as distinct from that, as well as from the Sailgan or Dwarf Pika, the Calling Hare of Pennant. It measures little more than six inches in length, and is of brownish-gray colour above, and white underneath.

The habits of the several species of this newly-formed sub-genus of our author, tend to the same end, but by different means : their providence and industry is variously employed. The Ogotone, so called by the Mongole Tartars, is very common in their deserts, and beyond Lake Baikal. It is principally in the heaps of stones lying here and there, that these animals are found ; they select a sandy soil, for the facility of digging deep burrows, having two or three entrances, and the bottom well furnished with a thick and soft bed of leaves. The vicinity of these burrows is easily found by the dung of the animals heaped together. In winter, when they quit the earth less frequently, similar heaps are found deposited in one part of the burrow.

These animals wander about commonly during the night, and seek principally in the defiles of the mountains, and the banks of rivers, the tenderest bark, and young shoots. In spring, they graze on the scanty herbage growing on the sand, and they transport so large a quantity of this to their retreats, that the galleries of the burrows are almost stopped by it. The inhabitants consider this as a sign of an approaching storm.

The same plants compose the winter reserves of these animals ; but these are not formed within the burrows, for want apparently of space. Numerous small heaps of about a foot in height, and of an hemispherical form, may be seen near their burrows, from about the month of September. In spring, when the snow melts, these little heaps of forage are no longer seen, and nothing but the dispersed refuse remains.

These animals have much vivacity and quickness of motion, but they are extremely timid, and are not easily to be tamed; their cry is a sharp sort of hiss. Still less in dimensions and powers than the Pika, properly speaking, the Ogotones are even more exposed to a host of enemies; the birds of prey watch and seize them by day, the Lynxes principally subsist on them by night, and the Marten Polecat, and others of the Weasel tribe, wage war against them at all times.

Pallas gives us little information on the subject of the re-production of these animals; they must, however, be very prolific, as in spite of the number and voracity of their enemies, they continue very abundantly in their native regions.

The *Calling Pika, Lepus Pusillus,* Pallas, very much resembles the last species, but is still smaller, and weighs only from three ounces and a quarter to four and a half, and in winter two and a half. The head is rather larger than in the other two species, and thickly covered with fur. It is extremely pretty, covered with long soft hair, brownish-lead colour, or gray, with a yellowish tinge on the sides; the ears are triangular, and edged with white. It is also an inhabitant of Tartary, and especially the banks of the Volga.

These little animals dig burrows, which are deeper than those of the last-mentioned species, in places covered with underwood, and abounding in herbage; they generally remain earthed during day, and quit their retreat on the approach of night, in search of food, returning to it again before sun-rise; they sleep like the Common Hare, with the eyes open, and appear to see equally well by night and day. In the evening, and again in the morning, they call to each other by a lengthened and reiterated cry, not unlike that of the Quail, and which may be heard at half a German mile distance ; when exerting their voice, they stretch out the neck in the manner of the dog. They cry but seldom during day, and then only in tempestuous weather. In winter they form, on the grass beneath the snow, small avenues, the more easily to procure their food, which consists of grain, leaves, small branches, and tender bark. In summer they select leaves and succulent plants ; but, at times, during the severity of winter, they are reduced to the necessity of feeding on the dung of large herbivorous quadrupeds, as the horse and sheep : often situated far removed from water, they have nothing in common, but the dew for drink ; but if they have the opportunity they drink rather frequently. They are very cleanly, and particular in depositing their own excrement in heaps, near the entrance of their burrow, which frequently betrays their place of refuge to the hunters.

The female brings five or six at a time ; the young are born with the eyes closed, and the skin naked and blackish, but the fur appears about the sixth day, and their subsequent growth is extremely rapid. When touched, and also at the approach of the mother, the young ones utter a cry similar to that of young birds.

These animals generally lead a solitary life ; but no animal is more gentle in its disposition, or more ready to associate with man. Pallas states, that their fears on captivity will abate in a day, and that they will then almost immediately become tame.

When sitting, the body of this animal is nearly rolled into a ball ; if placed in the hollow of the hand, it fills up the whole concavity ; but it sleeps with the body stretched on the ground, and the eyes reclined backward. It moves by little leaps ; but as its legs, especially those behind, are very short, its motions are neither very light nor active, and

Its leaps are inelegant. It rarely sits on its haunches, but is fond of rubbing the face with the fore-paws, and scratching itself with the hind; a habit induced, probably, by the number of parasitical insects which infest its fur.

From the small size of this animal, it might naturally be suspected that it would be incapable of the intensity of cold to which, in high latitudes and its elevated habitations, it is obnoxious; but that it is fitted for the station and circumstances in which we find it, is apparent, from the great heat of its blood, which will raise Fahrenheit's thermometer to 104 degrees. In common with the other species of this sub-genus, it is not known to lethargize during winter, nor does its colour vary with the season.

D'Azara, with unusual forbearance in a zoological writer, describes, with minuteness and simplicity, the various species of Mammalia which the province of Paraguay offered to his notice. Classification and arrangement are not his object. Amid all the barbarous proper names he uses, by which the several species are known in their native countries, we have no coining of classical compounds, the further multiplication of which is much to be deprecated. His essays, which include descriptions of many animals new to Europeans, are highly illustrative, useful, and unaffected.

But let us not be misunderstood. Divisions of the subject are necessary evils in zoology, which must be tolerated for their balance of utility, and divisions must have names; it is only when philological learning is displayed rather to illustrate the describer, than the things described, that it becomes obnoxious to censure and detrimental to the cause of science.

To speak well, however, of D'Azara, without qualification, conveys no small censure on Buffon. The great object of the former seems to be to detect, and sometimes almost maliciously to expose, the errors of the latter. The errors of fact in Buffon are, perhaps, even more venial than his errors of imagination. With all his originality, he was, necessarily, from the extent of the subject before him, in no small degree, however judicious, however discriminating, a compiler, and we may be surprised rather at his general accuracy than at his partial mis-statements. Zoology owes, perhaps, more to him than to any man; by his fascinating work he rendered it an elegant and fashionable pursuit. Ray and Linnæus introduced it to the study; Buffon fitted it for the drawing-room. We should do better, therefore, to emulate the useful labours of these great men, than to criticise their want of that information which time and the progress of science have brought to light.

These observations arise from the singular occurrence of introducing a new genus, as yet unnamed, for the *Viscache.* The animal intended, seems, by the description of D'Azara, to possess qualities of an exclusive genus, though heretofore treated as a hare. The Spanish zoologist was not the first to describe this animal, but he has done it more minutely than his predecessors: by these it has generally been referred to the genus Lepus. Desmarest, in the *Encyclopédie Méthodique,* gives a marginal description of it from D'Azara, as an appendix to the genus Dasyprocta, or Agouti, which almost amounts, in fact, to moving it from Lepus, and erecting it into a sub-genus of Dasyprocta, or making it a genus intermediate between these two; but our business is rather with a description of the animal, which must be from the text of D'Azara.

The Viscache is about the same size as the Common Hare; but the tail is longer; the head is thick, but swelled or inflated, so that the eye appears sunk behind the jaw; the muzzle is acutely truncated, and hairy, with the nostrils opening like a cleft; the ears are very long, large, and elliptical in shape; the neck and body, like the head, are thick, large, and cylindrical; the tail is nearly nine inches long, round, and thickish at the insertion; toward the tip, for nearly two inches, it is naked, the rest of it being hairy; the hairs on the upper side of the tail are long, erect, and very stiff, and give the organ the appearance of being compressed laterally; the anterior feet have four toes, which are, compared with each other, nearly equal, with sharp thick nails, calculated for digging; the hind feet have only three toes, of which the middle is the largest, armed with nails similar to those before; the internal side of the middle toes has a large gland, furnished with stiff hair, or rather bristles; it is plantigrade in its motion.

The general colour is dirty-white; the fur of the body being of two sorts, one entirely white, and the other, which is largest, black, except at the root, where it is white; the bristly hairs on the upper part of the tail are dark, but the sides of the tail are bright brown; the sides of the head are black, the hair on these parts being long, harsh, and stronger than those of the Hog; the mustaches are upwards of seven inches long; a whitish band proceeds from the extremity of the muzzle, and passes between the mustache and the eye, and terminates behind the latter; round the eye is brown; the throat, belly, and insides of the limbs are white. The female does not differ in colour from the male.

The softness of the general fur, and the harshness of the long hairs of the tail and whiskers, are very analogous to those of the Chinchilla; an animal at present considered to belong to the Hamsters, but whose real characters are not very satisfactorily described. One might conjecture that the Viscache should belong to a distinct genus, and that the Chinchilla should be associated as a congener with it.

The Viscache is an inhabitant of Brazil and Chili, and not of Paraguay; it excavates burrows in the earth, which contain many individuals, and which have various ramifications; these burrows occupy a circular space, the diameter of which is upwards of fifty feet, and the openings into it are extremely numerous.

Its motions are quick, by running, and not by leaps, like that of the Hare or Rabbit; it seeks its food almost entirely by night, and lives exclusively on vegetable matter. It is not esteemed as food.

We have now arrived, in conclusion of this order, at a few remaining species, which have been separated into almost as many genera; separations either made or adopted by our author.

The first of these is the genus HYDROCHÆRUS of Brisson, Erxleben, Geoffroy, Illiger, and Cuvier. It includes at present but a single species, placed by Linnæus, first with *Mus*, afterwards with *Sus*, and finally with *Cavia*. Brisson separated it generically, and applied the name Hydrochærus to the genus; this was adopted by Erxleben, who applies the same name to the Tapir. Pallas, and after him Gmelin, replaced it with the Cavy, referring the present generic name to the Tapir only. Finally, Illiger and our author have re-established the genus, as founded by Brisson. The Guinea-pig was indeed included in the same classification with it, but M. F. Cuvier has also separated that generically, as we have seen in the text.

In common with the three following genera, the *Capybara*, (by which name from Marcgrave we shall distinguish the only species included in this genus,) in common with the three following genera, has only the rudiments of clavicle.

It differs from the Agouti in the teeth, which have flat crowns, instead of enamelled scales. It is still farther removed from the Paca, by the same character of dentition, and by the number of the toes. It is more nearly allied to the Guinea-pig in the teeth and number of the toes, but differs widely in relative size and strength, and in the semi-palmated feet, and consequent aquatic habits.

The Capybara is the largest species of its order, with a thick pig-like body. The fore feet have four toes, and the hinder feet but three, all with large, obtuse, strong nails, almost covering the last phalanx, and approaching the character, in that respect, of the solidungulous and pachydermatous quadrupeds. These toes are all partially united by an intermediate membrane. The head is particularly large and thick, with the muzzle truncated. The teeth are, incisors $\frac{2}{2}$, canines $\frac{0\,0}{0\,0}$, cheek teeth $\frac{4\,4}{4\,4}$, Desm. Dict. says $\frac{4\,4}{3\,3}$. The molars are formed of scaly lamina, united by the cortical substance deposed between them; the fourth, or last in the mouth, is as big as the other three together, composed of twelve oblique parallel laminæ. The ears are round and moderate. There are twelve mammæ, both pectoral and ventral. No tail. The hairs are thin and bristly, and the whole appearance is pig-like. It measures about three feet from the muzzle to the anus.

The hairs on the upper part of the head and body and external sides of the legs are black; for the greater part of their length from the root and at the point, the rest is yellow; round the eyes and the insides of the limbs are yellow, so that the whole animal is of a dirty dingy colour, lighter underneath, with black mustachios. The female is rather smaller than the male.

This animal, as may be presupposed from its palmated feet, is an inhabitant of watery places, and dives and swims with much facility; it often sits on its hind feet. It seldom cries, except when frightened or hurt, when it utters a full-toned sound, of which *a*, sounded broad, and *pa*, may convey an idea. It seldom quits its retreat but at night, and then generally in company, without going any great distance from the water. It is a bad runner, and to avoid danger, generally has recourse to diving, rising again to the surface at a very considerable distance from the spot where it first plunges, and continuing for a very long period under water ; its flesh is good, and the Spanish Americans eat it on fast or meagre days. It is of a tranquil, gentle disposition, and neither seeks to injure nor quarrel with other animals, is easily tamed, answers to a call, and willingly follows those it knows. The female produces, in general, four or five at a time, on a bed of dry grass, &c. They are said to live in families, and never quit the vicinity of where they are born. They are common in all the lower parts of South America.

The genus Cobaya or Anoema of F. Cuvier, is one of those, which, until quite lately, comprehended but a single species ; the existence of a second, however, has been recently indicated. How far the few species which come at the end of this order may require the several generic appropriations into which we see them separated, whereby the larger and more important divisions of zoology are applied to very limited purposes, we shall not even hazard an opinion.

The oft-repeated difficulty on the subject of distinct species or of varieties, occurs in this limited genus. Organic affinity would identify the *Wild Aperea*, which we shall first slightly notice, with the menial Guinea-pig ; superficial dissimilarity, however, would unhesitatingly separate them into distinct species.

The characters which distinguish this division as a genus will be found sufficiently noticed in the text and table ; the

THE APEREA or WILD GUINEA-PIG

COBAYA APEREA. — *Cuv.*

London Published by G.B.Whittaker Sep.1.1826.

form of their cheek-teeth, and toes unconnected by any intervening membrane, constitute the principal traits which separate them from the neighbouring genera.

In the superficial peculiarities of a thick and heavy body, large, erect, naked, and transparent ears; round, large, and prominent eyes; head and muzzle similar to those of the Rabbit; want of tail, and character of the fur, the Wild Aperea and the Guinea-pig are alike; but in colour they are very dissimilar. The Wild Aperea is pretty uniformly of a dingy reddish-gray, or hare-colour, on the upper parts, and lighter underneath. The colours of the Guinea-pig will be stated in the subsequent account of that animal.

One difference between these animals in a wild and domesticated condition would lead to a strong suspicion on the propriety of their specific identity. The Wild Aperea is said by D'Azara to bring forth but one at a birth, and that rarely: while one pair of the Guinea-pigs, according to Buffon, may by possibility produce a posterity of a thousand individuals in one year.

The Wild Apereas lie concealed in the clefts of rocks, in which it is not difficult to surprise and take them; they also frequent thickets and underwoods, without penetrating much in the wood, or forming a burrow. They eat plants of almost all kinds, and seek their food principally during the night. It is gentle in disposition, and is easily tamed. Its flesh is esteemed as food, and the animals are hunted as game in their native country.

The history of the metamorphosis this animal has undergone by its domestication is involved in the general obscurity of past events, as we shall have occasion to observe presently; and we cannot, in conformity with the practice and opinion of the best zoologists, identify specifically the Wild Aperea with the Common Guinea-pig, without the expression of a certain degree of scepticism on the propriety of so doing.

The captivating language of Buffon was employed with all its effect and elegance in describing the weakness, insensibility, and insignificance of the *Guinea-pig;* and M. F. Cuvier has been not less energetic on the same subject than his great precursor in zoological science. The latter of these writers, however, by limiting his publications to what he has been enabled to see and study in nature, is less excursive and hypothetical, and more implicitly to be trusted in all his statements; whenever, therefore, we find him pursuing the path of Buffon, we may be assured that it is not merely because the track was beaten to his feet, but because he was satisfied from observation that it was the correct one. We shall proceed to abridge his notice of the animal in question.

Gentle in disposition, docile through weakness, almost insensible to every thing, the Guinea-pig has the appearance of an automaton endued with fecundity, but created only to represent a species. These are the traits, says M. Cuvier, with which Buffon terminates his article on this animal, and which characterize it with the utmost propriety. Every animal receiving from nature the instinct of self-preservation, receives with it within certain limits the means of obeying it. All endeavour, in the first instance, to avoid their enemies; but if escape be ineffectual, they employ some one or more of their organs as weapons of self-defence. The Monkey will endeavour to wound with his nails and his teeth; the Dog limits his exertions to biting; the Cat tears with her talons; the Ruminantia exert their horns; the Horse will kick; the generality of the Rodentia employ their formidable incisive teeth to bite, and sometimes their bent nails to tear—the Guinea-pig alone cries out, and endeavours to fly. Whether nature has so constituted this animal, or whether domestication has rooted out its natural instinct, it appears to have no notion of opposition, to escape from harm; and its long incisive

teeth, which are capable, if brought into action, of inflicting severe wounds, are equally unemployed with its nails,
in all purposes of warfare and self-defence. When seized,
it utters a sharp cry, makes some efforts to recover its
liberty, and to escape from the hand that holds it ; but it
is to this only that its efforts are limited. We should probably not find a second example of this extreme inoffen-
siveness in the whole class of Mammalia. Similar instances in either of the lower classes would be rare.

Notwithstanding, however, the extreme imbecility of the
individuals, the species maintains its existence ; for nature,
while she has left them an easy prey to creatures of a more
energetic character, has made them peculiarly prolific.
This, indeed, is very generally the case with all the weaker
animals, whose chances of premature destruction are so
much multiplied ; if the span of their existence is frequently
abridged, their means of reproduction are proportionally
increased, and the most complicated of the Creator's sub-
lunary works are thus preserved from those casualties which
might otherwise in some degree counteract the power that
called them into life. It may, indeed, be said that species
and genera, to an extent limited only by the powers of imagination, have existed and perished by casualties, or the
want of sufficient powers successfully to contend for their
preservation ; and that there are others, as the animal in
question, whose powers are nearly balanced by its casualties,
though the former prevail.

Were this hypothesis, however, correct, there can now
scarcely remain a doubt that facts would have been brought
to light to support and confirm it. Philosophy has penetrated the bowels of the earth, and man's insatiable appetite for knowledge and discovery has revelled in the artificial mine as well as in the natural cavern. Facts have accumulated upon facts in geological science, destructive of
preconceived notions as to the theory of the earth and of

its numerous inhabitants, but nothing has occurred to confirm the idea that modern contemporaneous races have perished from want of sufficient powers to keep their place.

That genera and species once existed on the surface of the earth which now no longer enjoy life there, is certain from their fossil remains; that many of these were rather tyrants than the victims of power is most probable, for some of them certainly exceeded in strength any which now remain. It is not therefore to casualties of this description that their present nonentity is to be attributed—the calamity which destroyed them was of a general overwhelming character, as the circumstances and situation of their remains sufficiently evince. But we must not be led away into the wide field of fossil osteology and geological science; our present object is simply to observe that the power which created was able by its providence to preserve, during the period prescribed for the existence of the creature; and that there are no facts to warrant the supposition that creation has ever been rendered abortive in a general sense by the excess of power in one race of animals working the total destruction of another. Man, indeed, who is in some respects the Vicegerent of Heaven here, might be able to do much in this respect. Wherever human civilization has advanced, the savage races of animals have receded. Our own island was once pestered with Wolves, but we proscribed them, and circumstances favoured their disappearance. Whatever man may be able to effect, there is no reason to suppose that he has in fact done much in the case in question; and we may well suppose that special interposition would not be wanting to check even in him whatever would be prejudicial to creation in the mass, and to the will of the Creator, if ordinary means became ineffectual for the purpose.

The female of the Guinea-pig is capable of gestation at

six weeks or two months old; the period of it is uncertain; the number at a birth are four, six, eight, or twelve, according to the age and strength of the mother; thus, as Buffon states, several hundreds may be bred from a single pair in the course of one year.

These animals are born covered with fur, and the eyes open, and in eight or nine months acquire their greatest degree of developement. Lactation does not last more than twelve or fifteen days, when the young are driven to shift for themselves by the male, and the female becomes again ready to breed.

The Wild Aperea does, in all probability, present traits of character more or less differing from those of its domesticated descendant; and as the modifications the animal has undergone in person by domestication are so very striking, it is probable that its intellects may have been equally modified. The domesticated race has doubtless been tamed for a long series of years, and that by the original inhabitants of South America; for M. F. Cuvier states that it appears by the original paintings of Aldrovandus, that toward the middle of the sixteenth century, the Guinea-pig had the white, red, and black patches, which distinguish it at present; it must at that time also have undergone all the modifications of which its nature was capable, since two centuries and a half have induced no more.

The colour of the Guinea-pig is varied with white, red, and black, in large irregular patches, differing in their relative situations in different specimens. The fur is very short, close, and shining; the naked parts are flesh-coloured, as well as the parts of the skin covered with hair; the circle of the iris is brown. The animal is too well known to need the illustration of a figure.

Although naturalized among us, the Guinea-pig is not inured to our climate; it is only by keeping it from cold and wet that it can be preserved. When the temperature is low, and many of these animals are kept in one place, they huddle together to increase the warmth ; but this resource is weak, and if the cold continue, and they continue to be unprotected from it, they quickly perish. Curiosity is the only motive to keep them, for they are not eaten, and their fur is of no value.

The mode of living of these animals is the same as all the other Rodentia which have cheek teeth without distinct roots ; they feed exclusively on vegetable substances, unless a perversion of appetite, caused by confinement or by hunger, induce them to eat flesh, like the Cows in Iceland, which feed on dried fish for want of their more natural food. Although not provided with clavicles, they convey the food to the mouth with the fore feet. They drink lapping, though but seldom, whence originated the idea that they never drink, and the common practice of keeping them without water; this they endure very well, so long as their vegetable food is fresh, but when fed on dry matter, there is no doubt they suffer from this unnatural privation.

When they are pleased, they make a continued gentle murmur ; if frightened, they utter a sharp cry, and their desires they express by a slight grunt.

This grunting has doubtless induced the very ill-applied vulgar name by which they are known.

They measure from eight inches to a foot in length.

" The Aperea," says M. F. Cuvier, " was the only species of Anoema hitherto known ; but another remained hidden, and which is one third larger than the Aperea or Guinea-pig. The museum of Surgeons' College in London has had the skull of it for a long time ; it was there, in 1814, that I had the first knowledge of it: since then, it has been sent from Brazil to the Museum in Paris, where M. Geoffroy St. Hilaire has named it *Hilaria.*" We have

THE AGUTI OR OLIVE CAVY

CAVIA AGUTI. GM.

London Published by C. B. Whittaker June 1825.

inquired for this skull at the College, and have seen a mis-
cellaneous collection there, but without the good fortune of
finding it. This, however, is one of the countless instances
to prove that our general communications with all parts of
the world have, in fact, though unsolicited and unsought
for, produced us very many interesting novelties in zoology
which our national inattention to the science has caused
us to neglect and forget.

We shall not repeat here the characters which have
induced the separation of the AGOUTIS into a distinct genus.
The Common Agouti and the Acouchi have been long
known, and have been referred, by different naturalists, to
different genera, till the modern reformations in zoology
induced the generic separation of these animals. Ray and
Linnæus placed them in their comprehensive genus *Mus;*
Gmelin, Erxleben, and Boddaert referred them to *Cavia;*
Brisson, to *Lepus,* or, rather, *Cuniculus;* Illiger proposed
the genus *Dasyprocta* to distinguish them, and F. Cuvier
that of *Chloromys.*

The Agouti is about the size of a Hare; but the head is
more assimilated to that of the Guinea-pig, by the thickness
of the muzzle, and flatness of the top of the head; the ears
are large, short, and nearly naked; the body is thicker
behind than before; a mere conical tubercle, without hair,
stands in the place of a tail; the anterior feet have four
distinct toes, and a fifth, distinguishable only by its nail;
the hinder feet, about a third longer than those before, have
but three, but they are thicker than the others, armed
with flattish triangular nails; the animal has them almost
always half bent; the hairs are moderately long and close to
the body; on the crupper they are rather longer than elsewhere; the colour on this part is bright orange-yellow;
the upper part of the body is coloured with yellow-brown
and blackish; the under part is yellow, inclining to gray

and reddish; the dorsal line is blacker than the rest of the body, and the legs are nearly black. The incisive teeth are of a deep yellow colour; the cheek-teeth, four on each side in both jaws, have their crowns perfectly flat, oval, notched on each side, and crossed by some regular ridges. There are twelve mammæ.

The Agouti is an inhabitant of Guiana, Brazil, Paraguay, and some of the Antilles. D'Azara says, expressly, that there are none in Rio de la Plata; nor does it appear that they are found in Mexico: it is extremely common in Guiana; but has been driven off such of the Antilles as are well cultivated: they are not to be seen in Martinique, but are found in St. Lucy, and seem to be extremely rare in St. Domingo. It is an extremely voracious animal, and eats almost all sorts of vegetable food, but its favourite is nuts of all sorts; it will not, however, refuse flesh.

It generally feeds sitting on the haunches. When the supply is plentiful, it conceals its superabundance in clefts, or holes in the earth, and is said sometimes to leave it there for six months together. It drinks lapping, and its urine is extremely fœtid.

Its motions are very rapid, particularly on rising ground; but it is subject, like the Hare, to roll over in descending a hill, and from the same cause, of the length of the hind legs. It is active during day, and troops of twenty and more are often seen in Cayenne running together. It sits frequently on the heels, in the manner of Squirrels, and then rubs the head and ears with the fore-paws. It prefers woods and covered places, and generally lives in the cleft of a tree. The female prepares a bed of leaves for her young, which are born in a perfect state, rather more than six inches in length; she carries them about from place to place. The period of lactation and nonage is short.

It appears that the Agouti can be easily tamed; but this is not often attempted, on account of the mischief the animal

C. Hamilton Smith Esq.ᵈᵉˡ.ᵗ

THE PATAGONIAN CAVY, of Pennant.

London Published by G.B.Whittaker Sep.1.1826.

does; it will cut, in a few seconds, any cord with which it may be fastened; gnaw itself a way through the door which is closed upon it, and in general escape with ease from every confinement. When frightened, in a natural state, it stops to listen, and strikes the earth with its hind feet, like the Rabbit and the Porcupine; if much irritated, it utters a cry like a young pig, and erects the hair on the crupper. D'Azara even states that if its fears be excessive, the contraction of the skin is so great, that the hairs fall off in handfuls. We know that something similar happens to the quills of the Porcupine when the animal erects them too suddenly.

Sight seems the most acute of the senses of this animal, which otherwise, in general, are obtuse, and its intelligence is very limited. It is one of those animals whose natural qualities are only to be known in a state of liberty; as soon as they are domesticated, they become almost immoveable, eat if food is given them, but without avidity—distinguish nothing, neither the person who feeds, nor the accustomed hand at which they receive their food, neither gentle treatment nor rough, not even the stick that strikes them.

The specific characters of the other species, being almost all that is known of them, will be found in the Table.

The *Agouti des Patagons*, or Patagonian Cavy of Shaw, is described at considerable length by M. D'Azara, in his valuable Essay on the Natural History of Paraguay. We shall lay before our readers the substance of his account of this animal.

M. D'Azara informs us that this Agouti is not precisely known among the natives of Paraguay, but that he has seen and caught many of them between the thirty-fourth and thirty-fifth degrees of South latitude in the territory of the Pampas, to the South of Buenos Ayres. The domicile of

the animal extends over the entire country of the Patagonians.

The people of this region call it a Hare; but it is more
fleshy and larger, says D'Azara, than the Hare of Spain,
and differs from it also very materially even in the taste
of its flesh. Two of these Pampa Hares, a male and a
female, are almost invariably found together, and they run
with a wonderful degree of force and velocity for a little
time; but they are speedily fatigued, and a well-mounted
horseman can overtake and catch them with very great
facility. This is generally performed either by entangling
them in a net, or striking them with a ball.

The voice of this animal, heard during the night, has a
very singular, and by no means an agreeable, effect; it is
a loud, sharp, and unpleasant cry, which may, says M.
D'A., be thus expressed, *o, o, o, y:* when taken, it also
cries in the same manner. The independent Indians eat
its white flesh, as likewise do the labouring people among
the Spanish South Americans; but they find it excessively
inferior to that of the various kinds of Tatous.

M. D'Azara was informed that this Hare generally produced the young in the *vizcachères,* or burrows, dug by the
Vizchacha, or *Cavia Acuschi* of Linnæus, and that when
pursued it always took refuge there: but, according to M.
D'Azara's personal observation, who hunted many of them,
none of them ever confided for its safety to any thing but its
own swiftness, although it might have availed itself of the
resource of many of those *vizcachères,* or burrows. He never
either found them in their form in any other position than
couched after the manner of a Stag, and like that animal,
they generally run to a considerable distance.

When taken young, they are easily tamed, will suffer
themselves to be scratched and patted, receive bread from
the hand, eat of every thing, are suffered to quit the house

with impunity, and will return to it with equal freedom.

Their length is about thirty inches; the tail an inch and
a half, without hair, thick, and as hard as a piece of wood;
it is incapable of motion, cylindrical and truncated, and
slightly curved towards its origin.

The anterior height of the animal is about sixteen inches
and a half; the posterior about nineteen and a half; and
the circumference measured, under the cheek, about fifteen
inches and a half.

There are four toes on the front feet; the nails are sharp,
black, strong, and adapted for digging; one toe is larger than
the rest, and the internal toe is next in size; the others are
nearly equal; on the sole of the fore-foot is a callosity,
which is bare, and not very hard, about the size of a wal-
nut. There is another similar callosity on the hinder foot,
but of a larger size; both resemble, in form, a top, the
axis of which is perpendicular to the base, which base is
situated in the sole of the feet; the tarsus is about seven
inches long, comprehending the nails, and is exceedingly
callous from the heel half way; it is this tarsus which
rests upon the soil, and supports the animal, for it cannot
employ the remainder of the foot for this purpose, in consequence of the callosity above-mentioned, which occasions
considerable inconvenience.

There are three toes on the hind feet, and they are
longer than those upon the fore; the middle one is the
longest of the three, and the two lateral ones are of an
equal size.

The limbs are slender, but sufficiently nervous and muscular; the head has a slight resemblance to that of- the
Hare, but rather more compressed at the sides. The upper
jaw has more height than width, and is embellished with
long and black mustachios; there are also some above the

eye, and the upper lid is furnished with very handsome lashes.

The mouth is exactly similar in form to that of the Aperea; but the upper teeth are narrower than the lower; the eye is large, its aperture being considerable, and the two nostrils are divided on the same level, and separated by a groove.

The ear is raised about three inches above the head, and in its greatest breadth is about two inches; it is not sharp at its point, where there are a few hairs; its anterior edge is folded towards the conduit, and the posterior edge, on the contrary, is folded from the base to about half its length the contrary way.

M. D'Azara found, on dissecting a female, two young ones, without hair, and about an inch and a half in length; the sexual part appeared to be in the anus ; there is one pair of teats towards the middle of the belly, and another about three inches and a half more forward. What is most remarkable in the fur, is a white and narrow band, which commencing on one of the haunches, proceeds to the other, round by the tail; but this colour is also introduced between the legs, and occupies the entire lower part of the body, including the lower part of the breast; between the forelegs, the colour is a clear cinnamon, and also under the throat; the side of the head is of the same colour, as well as the exterior part of the fore-legs, the lower part of the sides of the body, and the lower part of the buttocks, and tarsus; under the head, the hair is white, as well as the inside of the ear, which is brown on the outside; all the rest of its covering is brown hair, with small white points, except that which is on the crupper, in the neighbourhood of the white band, where the hair is quite of a dusky cast.

M. D'Azara has seen carpets made of this hair, and they

are much esteemed for their softness, and the agreeable effect they produce to the eye.

The Hares which Buffon says are to be found towards the Straits of Magellan, are of this description; but they are extremely different from the Hares of Europe, to which he compares them; for, independent of all that has been already observed, the animals in question, when they are not running, proceed by steps, and not by jumping.

The Brown Paca has been known since the time of Marc-grave, who has described it, and given an imperfect figure; it has been generally referred, by systematic authors, to the genus Cavia, and has been considered as a single species. M. F. Cuvier, however, has distinguished two species, the Brown Paca, Cælogenus Subniger, and the Yellow Paca, Cælogenus Fuscus; these he has separated into a distinct genus, under the name CÆLOGENUS. The specific description of one of these, from that naturalist, will sufficiently convey the generic characters of both.

The *Brown Paca (C. Subniger)* has the thick heavy appearance of the pachidermatous animals; its legs are heavy; the neck short; the head thick; and the body round. This animal, and its congener, are among the omnivorous rodentia, what the Capybara is among the herbivorous division of the same order. The numbers of the teeth are, incisors, $\frac{2}{2}$; canine teeth, $\frac{0}{0}\frac{0}{0}$; cheek-teeth, $\frac{44}{44}$; of the latter, those in the upper jaw are about equal in size; but, in the lower, they diminish in size, gradually from the last to the first; each tooth, before it is worked down by trituration, has on the crown four tubercles; but when the animal is aged, and the teeth have consequently been much used, the tubercles become worn down, and the intervening riges of enamel widen; these also by further use become nearly obliterated, and the triturating surface of the tooth becomes nearly smooth. All the feet have five toes, but in the hind feet the little toe and the thumb are extremely short; the nails are conical, thick, strong, and fitted for digging; the tail is little else than a naked tubercle perfectly motionless.

The external ear is round and moderate; the eye-pupil is round; the tongue is short, soft, and thick; the upper lip is cleft; the inside of the mouth is furnished with pouches, and the zygomatic arch is very much enlarged lengthwise; the skin of the cheek folds under the bone, and presents a deep concavity, which is unexampled in any other quadruped; the object of this it is not easy to see, unless it be to preserve the pouches more completely; strong moustaches proceed from the sides of the muzzle, and behind the eyes.

The fur is very short, and harsh, of a dingy brown colour on all the upper parts of the body, but varied with four ranges of parallel white spots, commencing at the shoulders and terminating at the buttocks; these spots are so close together, that when viewed in certain positions, they have the appearance of an uninterrupted line, and the lowest series is nearly blended with the colour of the belly, which is white, as are also the under jaw, a part of the internal side of the limbs, and the nails. It measures, from nose to tail, about two feet.

The Paca is very like the Agouti in the general make of its body, as well as in its organs. These two genera have the same system of dentition; nearly the same organs of sense, both are without clavicles, and though they differ in the number of toes on the hind feet, this is no very important character. It is in the cheek-pouches, and the fold of skin under the zygomatic arch, the general form of the head, and in the fur, that these animals differ essentially.

In its natural or wild state, the Paca digs burrows in the earth, in the neighbourhood of forests, and provides for its necessities by nocturnal excursions. In a state of confinement, few animals exhibit more obtuseness of intelligence ; it sleeps a great deal ; is, however, extremely cleanly in its habits ; collects all the straw, hay, and herbage it can procure for a bed, and is particularly careful to keep it clean. Like most other animals deficient in intelligence, it is strongly endowed with instinct.

The Yellow Paca seems to differ from the Brown in nothing but the ground colour of the fur, which is reddish-yellow. It has not, like the other species, been observed in Paraguay.

FIFTH ORDER

OF THE

MAMMALIA.

THE EDENTATA,

OR quadrupeds without incisive teeth, will form our
last order of unguiculated animals. Although associated by characters merely negative, they have,
nevertheless, some mutual positive relations, namely,
very large nails, which embrace the extremities of
the toes, and approach more or less the nature of
hoofs; a certain inertness or want of agility, arising from the obvious organization of their limbs.
There are, however, certain intervals in these relations by which the order may be divided into
three tribes.

THE TARDIGRADES

will form the first. They have a short face. Their
name is derived from their excessive slowness, the
consequence of a construction truly heteroclite, in
which Nature seems to have amused herself by the
production of something imperfect and grotesque.
The only existing genus, or

The Sloths (Bradypus, L.),

Have cylindrical molars and sharp canine teeth, larger than the cheek-teeth ; two mammæ on the breast; and fingers united together by the skin, and marked outwardly only by enormous compressed nails, always bent toward the inside of the hand and the sole of the foot. Their hind feet are articulated obliquely on the leg, and lean only on the external side ; the phalangers of their toes are articulated by a close ginglymus, and the first become attached at a certain age to the metacarpian or metatarsian bone, which also become in time attached to each other for want of use. To this inconvenience in the organization of the extremities is joined another, not less considerable, in their proportions. Their arms and fore-arms are much longer than their thighs and legs, soth at when they walk they are obliged to draw themselves along on their elbows. Their basin is so large, and their thighs directed so much on the sides, that they cannot approximate the knees. Their locomotion is the natural effect of so disproportioned a structure *. They live in the trees, and never quit one until they have completely stripped it of all its leaves, so painful is it to them to mount another.

* Mr. Carlisle has observed, that the arteries of the limbs commence by infinite ramifications, which finally unite into one, from which the ordinary branches proceed. This structure being found also in the *Loris,* whose walk is not less slow, it is possible it may have some influence on slowness of motion. The Loris, the Orang-otang, the Coaita, all very slow animals, are all remarkable by the length of their arms.

We are told that they let themselves fall from a branch, in order to avoid the trouble of descending otherwise. They have but one young at a birth, which they carry on the back.

The viscera of these animals are as singular as the rest of their conformation. Their stomach is divided into four sacs, analogous to the four stomachs of the Ruminantia, but without leaves, or other salient parts interiorly, while their intestinal canal is short, and without a cæcum.

The Aï (Bradypus Tridactylus, L.), Buff. XII. v. and vi.

Is the species in which inertness, and the details of organization which induce it, are carried to the highest degree. It has three toes, or rather three nails on each foot; the thumb and the little toe reduced to mere rudiments, hidden under the skin, and attached to the metatarse and metacarpe; the clavicle, also rudimentary, is attached to the acromion. The arms are twice the length of the legs; the hairs of the head, back, and limbs, are long, thick and without elasticity, almost like dried grass, which gives it an hideous air. Its colour is gray, often spotted on the back with brown and white. Many individuals have between the shoulders a bright fulvous spot, which traverses a black longitudinal line. We are ignorant whether it forms a distinct species. Its size is that of a Cat, and it has a very short tail. It is the only

mammiferous animal known which has nine
cervical vertebræ.

The Unau (Bradypus Didactylus, L.), Buff. XIII. 1.

Which has only two nails on the fore-feet, and
no tail; is rather less unhappily organized
than the Aï. Its arms are not so long; the
clavicles complete. There are not so many
bones attached in the feet and hands; its
muzzle is elongated, &c. It is less by one-half
than the Aï, and of an uniform gray-brown,
which sometimes assumes a reddish tint.

These two animals are originally of the hot parts
of America. They would probably have been destroyed long since by the numerous Carnassiers of
this country, if they had not some defence in their
long nails*.

Dr. Shaw, Gen. Zool., has described under the
name of *Bradypus Ursinus* (Prochilus, Ill.) an
animal indigenous in India, brought alive to
England, of the size and nearly of the form of
the Bear, with five toes armed with nails on all
the feet, without incisors, with canines and
molars; but they are unequal among them-

---

* It is singular that the Two-toed Sloth was not known before
Seba, and that it was for a long time obstinately referred, on the
authority of that ignorant collector, to Ceylon. Erxleben has
maintained that it belonged to Africa, because he took the Poto of
Bosman, which is a Galago, for it (See this last genus). The fact
is, that the Unau comes only from South America.

selves, which appears to indicate a generic difference with the Sloths. It would be very interesting to have the anatomy of this singular animal *.

The second tribe includes

THE COMMON EDENTATA,

with a pointed muzzle. Some have cheek-teeth. There are two genera of them.

The Tatous, or Armadillo, (Dasypus, L. †)

are very remarkable among the Mammalia, by the scaly and hard helmet composed of compartments, like little pavements, which cover their head, their body, and oftentimes the tail. This substance forms a buckler over the forehead; a second, very large and very convex, over the shoulders; a third, similar to the preceding, on the crupper, and between these two last, several parallel and moveable bands, which give the body the faculty of bending. The tail is sometimes furnished with successive rings,

* Buchanan (Travels in the Mysore, t. ii. p. 198,) assures us, that it is a true Bear, which lives on white Ants, fruit of Sorgho, &c.

The Translator has inspected the skull of the specimen described by Dr. Shaw, in which the alveoli of the incisive teeth are complete, and the remaining teeth are perfectly ursine. According to Dr. Shaw's description, the animal must have been deprived of its incisive teeth during life.—*Ed.*

† Tatou is their Brazilian name. They are also called *Quir-quincho*. The Spaniards call them *Armadillo*, on account of their armour; the Portuguese *Encouberto*, for the same reason. Dasy-pus (hairy feet) was one of the names of the Hare or Rabbit among the Greeks.

sometimes only, like the legs, with divers tubercles. These animals have large ears; sometimes four, sometimes five great nails before, but five always behind. The muzzle is pointed; the cheek-teeth cylindrical, separated from each other, seven or eight in number, everywhere without enamel on the interior side; the tongue is soft, but little extensible; some separate hairs between their scales, or on the parts of the skin which are without scales. They dig burrows, and live partly on vegetables, and partly on insects and carcasses; their stomach is simple, and they are without cæcum. They belong to the warm, or at least the temperate, parts of America.

The species may be almost distinguished by the number of their intermediate bands combined with the form of the compartments; the bands nevertheless are subject to vary one or two, according to the individuals.

The Three-banded Tatou or Armadillo. Tatou Apara, Marg. Apar, Buff. Mataco, d'Azara. (Dasypus Tricinctus, L.), Schreb. LXXI. *A.*

With three intermediate bands; the tail very short; the compartments regularly tuberculous; five toes on all the feet. It enjoys the faculty of rolling itself up, shutting the head and the feet between the bucklers, and thus forming a complete ball. Of Brazil and Paraguay. This is one of those found most to the south. Its dimensions are moderate.

*The Six-banded Tatou or Armadillo. Encoubert and Cir-
quirisa Buff.* (Das. Sexcinctus et Octodecimcinctus, L.)
Buff. XXIII. et Supp. III.* LVII.

With six or seven bands; the compartments
smooth, large, and angular; with a moderate
tail, annulated only at its base; five toes
throughout; the posterior buckler indented like
a saw; the non-scaly parts furnished with hairs,
longer and thicker than in the other species.

*The Nine-banded Tatou or Armadillo. Tatou Peba, Margr.
Black Tatou, d'Azara. Cachicame, Buff. Das. Novem-
cinctus, Das. Octocinctus, et Das. Septemcinctus, L.) Buff.
X.* XXXVII. *III.* LVII.

With nine intermediate bands; the tail long,
and annulated on nearly the whole of its length;
the compartments of the buckler small and
round; four toes only before; the helmet generally blackish. It is the most common in
Guiana or Brazil. It has sometimes eight
bands, but seldom seven or six; its body is
about fifteen inches long, and the tail is nearly
the same.

*The Twelve-banded Tatou or Armadillo. Cabasson, Buff.
Tatouay, d'Azara. (Das. Unicinctus, L.) Buff. X.* XL.

With twelve intermediate bands; the tail long

---

* The Tatou Peba, or Encouberto of Margrave, is the Novem-
cinctus. The Weasel-headed Tatou of Grew, Cirquinson of Buff.,
Das. Octodecimcinctus, is the same; but Grew has considered as
moveable the ranges of the helmet of the head. Even reckoning
them, there are in all but sixteen, and his figure shews no more.

and tuberculous; the compartments of the bands and the bucklers square, wider than long; five toes on all the feet, four of which on the fore-feet have enormous nails, trenchant on their external edge. It becomes very large.

The Giant Tatou, or Armadillo, Geoff. Great Tatou, d'Azara. Dasypus Gigas, Cuv. Deuxieme Cabasson, Buff. X. xiv.

With twelve or thirteen intermediate bands; the tail long, and covered with tily scales; the compartments square, wider than long. It is the largest of the Armadillos; sometimes more than three feet long, without the tail.

The Orycteropes (Orycteropus, *Geoff.* *)

Have long been confounded with the Ant-eaters; because they subsist on the same kind of food, have the head similarly formed, and the tongue long and extensible; but they are distinguished from the Ant-eater by having cheek-teeth, and flat nails, constructed for digging, and not trenchant; the structure of their teeth is different from that of all other quadrupeds; they are solid cylinders, traversed like the pores of a cane, according to their length, with an infinite number of little canals; their stomach is simple, muscular toward the pyloris, and the cæcum is small and obtuse.

There is but one species known.

* Orycteropus. The feet fitted for digging.

The Orycterope of the Cape (Myrmecophaga Capensis, Pall.), Buff. Supp. VI. xxxi.

Which the Dutch of that colony name the Ground Hog. It is an animal about the size of a Badger; low on the legs, with scanty hair, brownish gray; with the tail shorter than the body, and equally scanty of hairs; it has four toes before, and five behind. It inhabits burrows, which it digs with extreme ease. Its flesh is eaten.

The other Common Edentata have no cheek-teeth, and, consequently, are totally without teeth: there are two genera of them.

The ANT-EATERS (MYRMECOPHAGA, *L.*)

Are villose animals, with a long muzzle, terminated by a small toothless mouth; whence is protruded a filiform tongue, capable of considerable extension, and which the animals thrust into the Ant heaps and the nests of Termites, whence they draw these insects by means of the viscous saliva with which the tongue is furnished; their nails on the fore-feet, strong and trenchant, and which vary in number, according to the species, serve to tear the nests of the Termites, and furnish the animal with a good defence. In a state of repose, these nails remain constantly half bent inward, like a callosity of the foot; hence the animal only brings the side of the foot to the ground. The stomach of the Ant-eaters is simple, and muscular toward the pyloris; the canal is moderate, and without cæcum *.

They live in the hot and temperate parts of the New World, and bring but one at a birth, which they carry on the back.

The Tamanoïr (Myrmecophaga Jubata), Buff. X. xxix. *and Supp. III.* lv.

Is upwards of four feet long, with four nails before, and five behind; the tail is furnished with long hairs, directed vertically above and below; the fur is gray-brown, with an oblique black band bordered with white on each shoulder. It is the largest of the Ant-eaters. It is said that it can defend itself against the Jaguar. It inhabits low places; does not mount trees, and walks slowly.

The Tamandua (Myrmecophaga Tamandua, Cuv. Myrm. Tetradactyla et Tridactyla, L.), Schreb. LXVI.

With the form and feet of the preceding, but not more than half its size; the tail has the hairs scanty, is prehensile and denuded at the end, and enables the animal to hang by it to the branches of trees. There are some yellowish-gray, with an oblique band on the shoulder,

---

* Daubenton has made known, in the M. Didactylis, two very small appendices which may, in strictness, be taken for cæcums. I have satisfied myself that they do not exist in the Tamandua.

visible only in certain angles of light; some yellow, with a black band; some banded, with the crupper and belly black, and some entirely blackish. We do not know as yet whether these differences are specific.

The Two-toed Ant-eater (Myrm. Didactyla, Lin.), Buff. X. xxx.

As big as a Rat, with woolly fur, yellow, with a red dorsal line; tail prehensile, naked at the end; two nails only before, one of which is very large, and four behind *.

The PANGOLINS (MANIS, Lin.), commonly called Scaly Ant-eaters,

Are destitute of teeth; have the tongue very extensible, and subsists on Ants and Termites, properly speaking; but their body, their limbs, and tail, are covered with thick trenchant scales, disposed like tiles, and which they raise in rolling themselves up into a ball when they defend themselves from an enemy. All their feet have five toes. Their stomach is slightly divided in the middle. They have no cæcum. They are found only in the ancient continent.

---

* The *Myrmecophaga Tridactyla*, L. Seba. pl. F., is only a Tamandua ill-figured. The M. Striata, Shaw, Buff. Supp. III. pl. LVI. is a Coati disfigured by the stuffer.

*The Short-tailed Manis (M. Pentadactyla, Lin. M. Bra-
chyura, Erxl.), Buff. X.* xxxiv.

Three or four feet long; with the tail less than
the body. Of the East Indies. It is the
Phattagen of Elien, *Lib. XVI. Cap.* vi.

*The Long-tailed Manis. Phatagin of Buff. (M. Tetra-
dactyla, Lin. M. Macroura, Erxl.) Buff. X.* xxxiv.

From two to three feet long; with the tail
twice the length of the body; the scales armed
with points. Of Senegal, Guinea, *&c.* *

The third tribe of Edentata includes the animals
which M. Geoffroy has named MONOTREMES, because
they have but one external opening for the semen,
urine, and other excrements. Their organs of generation present some extraordinary anomalies; although without a ventral pouch, they have on the
pubis the same supernumerary bone as the carnivorous pouched animals; the different canals terminate in the urethra, which opens into a *cloach,* or
common termination, at the base of the penis, which
is not furnished with any pipe or conduit for the
semen. The only matrix consists of two trumpet-
shaped canals, which open separately into the
urethra, which conduct to the cloach. As it has

---

* We have verified the habitat of the Long-tailed Manis from
the statement of M. Adanson, and other travellers.

been hitherto impossible to find their mammæ, we are still ignorant whether these animals are viviparous or oviparous. The singularities of their skeleton are not less remarkable, especially on account of a sort of clavicle to the shoulders, placed before the common clavicle, and analogous to a similar bone in birds. Finally, besides five toes to all the feet, the males have a spur on the hind-feet, analogous to that of some gallinaceous birds. These animals have no external ears, and their eyes are very small.

The Monotremes are found only in New Holland, where they have been discovered since the settlement of the English there. There are two genera known.

The ECHIDNES, (ECHIDNA, *Cuv.* TACHYGLOSSUS, *Illig),*
otherwise Spiny Ant-eaters.

Their elongated muzzle, terminated by a small mouth, contain a tongue extensible, like that of the Ant-eaters and Pangolins. They live, therefore, on Ants, in the manner of these two genera. They have no teeth, but their palate is furnished with several ranges of small spines, directed backward. Their short feet have each five long toes, strong and constructed for digging, and the whole upper part of their body is covered with spines, like that of the Hedgehog. It appears that in the moment of danger they are also equally enabled to roll themselves into a ball. Their tail is very short; their stomach is ample, and almost globular, and their cæcum moderate. The penis is terminated by four tubercles. There are two species known.

The Spiny Echidne (Echidna Hystrix, Ornithorinchus Histrix, Home. Myrmecophaga Aculeata, Shaw.)

Entirely covered with thick spines.

The Bristly Echidne (Echidna Setosa), Ornithorinchus Setosus, Home.

Covered with hair, in which the spines are half hidden.

The ORNITHORINCUS (ORNITHORYNCUS,) *Blumenbach.* PLATYPUS, *Shaw.*

Their elongated and, at the same time, singularly enlarged and flatted muzzle, gives the greatest external resemblance to the beak of a Duck; the more so, as their edges are furnished with similar small transverse laminæ. There are teeth only at the bottom of the mouth, to the number of two on each side, in each jaw, without roots, with flat coronals, and composed, like those of the Orycteropus, of small vertical tubes. The fore-feet have a membrane which not only unites the toes, but considerably exceeds the nails; in the hind-feet the membrane terminates at the root of the nails, two characters which, with the flat tail, constitute the Ornithorinci aquatic animals. Their tongue is, in some degree, double; one in the beak studded with villosities, and a second, at the base of the first, thicker, and carrying forward two small fleshy points. The stomach is small, oblong, and has the pyloris near the cardia. The cæcum is small. In the intestines are seen several saillant parallel laminæ. The penis has only two tubercles. The Ornithorincus inhabits the rivers and marshes of New Holland, near Port Jackson. Only two species are known; the one with reddish, thin, and smooth fur, (*Ornithoryncus Paradoxus*, Blum.) The other with blackish-brown, flat, and frizzled hair. Probably these are only varieties of age, *Voy. de Peron, I. pl.* xxxiv.

SUPPLEMENT TO THE ORDER EDENTATA.

The order Edentata, or toothless animals, (that is, either partially or altogether so,) includes but a very few genera, and these again but a very limited number of species ; they are, however, decidedly among the most curious of the mammiferous tribes.

The characters of the order, although before mentioned in the text, may, for convenience, be thus briefly recapitulated.

Incisive teeth wanting in most of the species *, some with canines and cheek-teeth, others with cheek-teeth alone, and a few absolutely toothless ; toes varying in number, generally armed with powerful nails ; orbits and temporal fossæ united.

The anomalies and singularities of the component species of this order are so various, that it seems necessary to proceed at once to a consideration of the genera, without any further previous collective observations on the whole.

The genus Bradypus is not less remarkable for the singularity of organization and external form in the animals which it comprehends, than for the habits of these animals themselves, which have procured them the peculiar designation of *Sloths*.

Linnæus places the Bradypus in his order of *Bruta*, with the Edentata, properly so called, and with the Morse, the Elephant, and the Rhinoceros. Erxleben classed them between the Lemurs and the Didelphis, on the one side, and

---

* One, at least, of the Armadillos it is now ascertained has these teeth.

the Edentata on the other. Boddaert removed them from the Lemurs, and ranged them with the Bats. The Baron first, in his *Tableau Elémentaire d'Hist. Nat.*, formed them into a separate genus of Edentata ; and afterwards, in the Tables attached to his Lessons of Comparative Anatomy, he established the family of *Tardigrada*, containing these animals alone, and constituting the link between the Edentata and the Pachydermata. Blumenbach placed them at the head of the *Mammalia Fissipedes Edentata*, which were, according to him, the Ant-eaters, Pangolins, and Tatous. M. Lacépède has composed of them his seventh order of Mammalia, constituted of those animals which are digitigrade, and which have only the *dentes laniarii et molares.* M. Dumeril, in his *Zool. Analyt.*, composes his ninth family of them, called *Tardigrada*. Illiger places them in a particular order, which he puts after the Ruminantia and before the Edentata. In fine, our author, in the Animal Kingdom has placed them as we have seen, with much propriety at the head of the Edentata. Illiger divides his order Tardigrada into three genera, which M. Desmarest considers as sub-genera, but which we, with our author, can only consider as species. The first is the *Aï (Bradypus Tridactylus;)* the second is the *Unau (Brad. Didact.)* the *Cholœpus* of Illiger ; and the third, the *Prochilus* of Illiger, and *Ursine Sloth* of Shaw, which we have already, we trust satisfactorily, proved to have no pretensions to be classed with the other two, but decidedly to belong to the Ursi. M. de Blainville (*Nouv. Dict. Syst. des Anim.*) *Bull. de la Soc. Phil.*, 1816, having observed many characters in common between the Bradypi and Quadrumana, places them in this order as anomalous quadrumana organized for climbing. Of this distribution it is impossible to approve, as the genus Bradypus is destitute of the essential characteristic of the Quadrumana. We might as well arrange the Squirrels and many of the Marsupiata with the Simiæ and Lemurs.

The Bradypodes have a tolerably elongated body, covered with dry and stiff hairs, like hay, and which circumstance gives them an external appearance of bulk which is, however, altogether deceitful, as they are exceedingly lean. Their extremities are also very slender, and terminated by nails very strong and extremely arched.

These animals have but two sorts of teeth, canines and molars. The place of the incisors is altogether vacant. The canines are altogether four in number; one on each side in each jaw. They are in general more elevated than the molars, and their form is pyramidal. The molars, which are conical in the young individuals, are cylindrical, and have a hollow coronal in adults. They have a case of enamel, which encloses a bony substance, formed of transverse laminæ, which laminæ, though piled upon each other, still exhibit distinct marks of separation. There are four of these molars on each side in the upper jaw, and three only in the lower. The head is slightly rounded, and the muzzle short. The eyes are tolerably distant from each other, and directed forwards; the nostrils a little separated, and placed at the extremity of the muzzle.

The anterior extremities are longer than the posterior, especially in the Aï, where they are nearly double the length, while in the Unau they are little more than one-sixth of the length of the fore legs. The toes are covered by the skin as far as the root of the claws. These claws are strong, very long, compressed, arched, and hollowed channel-wise within. They are parallel to each other, and fastened closely to their basis, as well as the phalanges of the toes, the number of which varies according to the species. There are always three upon the hind feet, and sometimes three, and occasionally but two, on the fore feet.

They have two pectoral mammæ.

In one species there is no tail. In another its place is supplied by a slightly projecting tubercle.

THREE TOED SLOTH. *VAR.*

BRADYPUS TRIDACTYLUS. VAR.

London. Published by G.B.Whittaker June 1825.

The hair is abundant. That on the fore-arm, particu-, larly in the Unau, is directed forwards towards the arm, as in Man and the Orang-Outang. Such is the external conformation of the species of this genus. Their internal organization, which has been attentively studied by Daubenton and by our author, is not less extraordinary. The pelvis is remarkably wide, and the cotyloïd cavities are placed so far behind, that they have no power of approximating the thighs one to another. The sacrum, instead of being joined by a single point only to the ossa innominata, is absolutely soldered, as it were, to the tuberosity of the ischion. This character is not observable in any other animal excepting the Phascolomys, a species which is also equally conspicuous for the slowness of its movements. The bones of the tarsus are disposed in such a manner, that the foot turns upon the leg like a vane upon its axis, and can only rest its external edge on the ground. The phalanges of the toes, both of the fore and hind-feet, are closely articulated together, and are almost without motion ; and some of them are literally pasted to those that are next to them. The first phalanx, in particular, is joined to the metacarpian or metatarsian bone, in such a manner, that the toes appear to have one phalanx less than they ought to have. In a state of repose, their enormous claws are folded inwards, and the upper or external part of them rests upon the ground. Their levator muscles are situated in the upper part of the phalanges. There are clavicles in one species, but not in another. The number of the vertebræ vary. A remarkable character in the Aï is, that there are nine cervical vertebræ, while in the other mammalia there are but seven.

These animals have a stomach divided into many sacs or lobes. But these lobes have not upon their surface the folds or sheets, or the papillæ which are observed in the Ruminantia. Moreover, although their nutriment is purely

vegetable, they do not ruminate. Their intestines are very
short, and they have no cæcum, which circumstance is a
very singular phenomenon in herbivorous animals. There
is one common cloaca or passage for the expulsion of urine
and fæces. These animals are natives of South America. The excessive slowness of locomotion in the Bradypodes has caused
these animals to be remarked by most travellers who have
surveyed the countries in which they inhabit. This slowness
is the result of their peculiar conformation. Their long
and weak limbs, their thighs so considerably separated, their
toes lumped as it were together, and appearing externally
only by their immense claws, which answer little other
purpose than that of hooks to fasten on the branches of
trees, or to pass from one to another ; all these defects of
organization deprive these animals of the faculty of moving
with swiftness, and of grasping extraneous objects. The
want of naked parts precludes all delicacy in the sense of
touch. Their ears, almost divested of all external conch,
cannot communicate impressions to the sensorium with that
vivacity that such organs would do in a state of greater
development. Finally, their motions on the ground are
constrained and impeded by the excessive length of the anterior extremities, which obliges them to rest rather upon
the cubitus than the carpus and metacarpus.

This last character, as the Baron remarks, they have, in
common with the Ateles, the Orang-Outang, and the Loris.
These animals, too, especially the first and last, are slow in
their motions. The Orang certainly possesses sufficient
activity in climbing ; but on the ground his movements
are heavy and awkward. A remarkable trait of affinity
between the Sloths and the Loris, to which some allusion has been made before in the present work, was discovered by Mr. Carlisle. The arteries of the legs and arms
in both these genera form a multitude of ramifications,

THE THREE-TOED SLOTH OR AI—_common var._

BRADIPUS TRIDACTYLUS.

T.Landseer del.t et. fec.
Mus.Brit.

London Published by G.B.Whittaker.Dec.r 1825.

which finally unite, and in which the blood circulates much more slowly than if it was propelled into a single arterial trunk.

The nutriment of these animals consists of the leaves of trees, on the branches of which they ascend with considerable difficulty. They strip these branches totally of their leaves, after which, to escape the trouble of descending, they suffer themselves to fall upon the ground. This, however, they can never resolve to do, until they have supported an abstinence of very long duration. Their long, tufted, and stiff hair, and the strength and solidity of their ribs, prevent any accident, which might otherwise occur, from these frequent falls.

The female has but one young one at a time, which fastens itself upon her back.

The first species of Bradypus is the *Aï*, (*Bradypus Tridactylus*, Linn.) The face and head of this animal are rounded, furnished with stiff hairs of a yellowish colour, but having the eyes encircled with brown; the fur is in general composed of stiff and dry hairs of a particular character, some being brown, and others white. This causes the skin to be varied by spots of these two different tints, variously dispersed, according as one or the other colour may predominate. The middle line of the back, indeed, is more inclining to the brown. Between the two shoulders there is a place of an oval form, the hairs of which are short and silky, of a lively orange colour, with a longitudinal streak of a beautiful black in the middle.

The neck is yellowish, like the forehead; the hairs on the top of the head part from the sinciput, diverging on all sides; these are extremely fine, rounded, and then flattened for about three-fourths of their length. There is, besides, upon the skin, a sort of down, extremely fine and soft, of a brown colour at the basis of the brown hairs, and white at the basis of the white.

U 2

The claws are nearly of an equal length, though the middle one is rather the longest; they are very strong both on the hind and fore-feet, but especially on the latter. There is a small external ear concealed under the hair.

The Aï is said to be about as big as a Cat, but it varies considerably in size; its head is a little more than three inches in length; its body about fourteen; its tail eleven lines; the arm eleven inches; and the leg six only; the claws of the fore-feet, when measured underneath, and in a right line, or, mathematically speaking, taking the *chord* between their base and extremity, are thus: the middle one two inches six lines, and the two external ones two inches three lines; on the hind-feet, the external claw measures one inch six lines; the middle, two inches and one line, and the internal, one inch nine lines.

The Aï has not, at all times, the mark on the shoulders of which we spoke above, and many individuals are to be found without it. Sonnini even asserts that the *Aï à dos brûlé* (a name which he gives to that one which has the orange spot and the black line) is more rare at Cayenne than the other variety, which he considers as that which properly constitutes the species Aï. One variety of this species has the fur generally gray, the throat being covered with a short and brown hair; it has a yellowish band over the forehead, and which also extends across the cheeks; the hairs of the head do not diverge from a common centre, but are directed right and left, on each side of the neck; it has also the aforementioned spot between the shoulders.

There is another with a brown throat, but without the spot on the shoulders; it belongs to Brazil.

It is not yet ascertained whether these different varieties may not constitute so many distinct species. Against the multiplication of species we have too often declared ourselves, to incline to the affirmative of such a question with-

THE THREE-TOED SLOTH. _1ã?: Two feet one inch in length._

out the amplest evidence. However, be this as it may, every one of these animals exhibits, in the highest degree, that mode of organization which renders them the slowest of all animals, and even the slowest in their own genus.

In the Aï a very great degree of muscular power is united with an extraordinary portion of vitality. It seizes, with its enormous claws, the branches of trees, and hooks itself upon them with such tenacity, that it is exceedingly difficult to make it let go its hold; there it will remain suspended, with the body turned upside down, and describing the figure of an arch. If any one is desirous of taking it, the shortest way is to cut down the branch to which it is attached. Thus it may be carried away, and it will never change its position; but it is an acquisition of no great value, as a pupil most invincibly stupid, and a prodigiously dull companion. Its flesh and fur are good for nothing; it seems incapable of every kind of sentiment; it is never agitated by fear; it shews neither aversion nor inclination for a domestic state; it testifies neither joy, nor gratitude, nor astonishment, nor uneasiness: all its sensations seem obtuse, and it presents an image, which we can scarcely term *living*, of the most perfect apathy. Its plaintive cry inspires melancholy, as well as its aspect; it is a feeble sound, which strikes upon the ear like the accents of sorrow, and were we not acquainted with the immoveable dulness of the animal that utters it, we might fancy it a lament over his weak and helpless condition. The savages of America have characterized this sound very well by the vowels *a, i,* of which they have formed the name of the animal itself.

Old travellers have related a pleasant passage connected with the love-affairs of these interesting animals; they say, that at the approach of the female Aï, the male, by way of preliminary endearment, falls fast asleep several times. This is a sort of amorous eagerness not unworthy of a being that is equally sluggish in its sensations and its motions. The female has but two mammæ, situated on the breast. She usually produces but one young one, which is covered with hair at its birth, and which the mother drags languidly about on her back.

The Aï is an inhabitant of the southern regions of the New Continent, from Brazil as far as Mexico. The application of the term *Sloth* to these animals, and to the other species of Bradypus, is not, strictly speaking, perfectly correct; their extraordinary slowness of motion is not the effect of indolence or sloth; it is a part of the organization of the animal, an essential of its nature, and it is no more in his power to accelerate his movements than it is permitted to the Hare to creep, or the stag to crawl; it is in vain to urge, to stimulate, or to strike him; nothing in the world can quicken him. Leaning upon one side, he raises one of his fore-legs, makes it describe a long arch, and then lets it fall again with the most extreme indifference; afterwards, as if fatigued by such an amazing effort, the animal rests on the side where the leg was advanced with so much difficulty, and in a few moments puts the other in motion in a similar manner. The hinder part of the body follows with equal slowness. It has been calculated that the Aï would employ an entire day to make fifty steps; from this it follows, that, supposing it to proceed without interruption, it would take nearly a month to travel a single mile.

Although, from the nature of its aliments, this animal is obliged to ascend trees, yet it appears to climb with as much difficulty and pain, as it walks with on level ground. It takes nearly two days before it reaches the branches of a tree; it feeds upon the leaves, the buds, and the fruits; it never quits the tree until it is completely stripped; it gnaws branch after branch, and when it no longer finds any thing to browse on, it will yet remain there for many days, enduring extreme hunger, before it can prevail on itself to descend, or more properly speaking, to fall down. When, however, the demands of nature become too imperious to be resisted, it rolls itself up into a ball, drops down plump upon the ground, and drags itself heavily along to another tree, in search of fresh nutriment. This long abstinence, which lasts, it is said, for fifteen days, is not more the effect of the slothful disposition attributed to the Aï, than of its vacillating and constrained walk; it is organized for this singular degree of privation. Had nature bestowed upon it a greater appetite, she would also have afforded it a greater portion of activity for its satisfaction. It is reported that the Aï never drinks. Its dry and flat hair forms a sort of thick mantle, which shelters it from the rain.

The second species, of which Illiger makes his genus *Chælopus*, is the *Unau* of Buffon, (*Bradypus Didactylus* of Linnæus.) The Unau is larger than the Aï; its head is more elongated; the face more oblique; and the forehead less prominent and defined; its fur is a composition of long rough hairs, both brown and white, the result of which mixture is a brown-grayish tint, paler under the throat and belly than above, and more especially so than the upper part of the neck, where the tint is deeper than on any other part of the body; the hairs of the fore-arm are directed backwards; there is no down at the basis of the hair, as in the Aï; the longest hairs are those of the occiput and neck, which form a sort of peruke, or rather mane, behind the head; those on the thighs are also very long.

The canine teeth are stronger and more apparent in this species than in the preceding; they are also less subject to wear; the bones of the hands and feet are less frequently soldered together, as it were, than in the Aï; accordingly, the Unau can perform more facile and varied motions than its congener. A Unau, whose body is two feet in height, has generally a head about five inches long. The external claw of the fore-foot is one inch nine lines, and the internal two inches.

Although this animal is very heavy, and its walk vacillating, it is yet not so slow in its motions as the Aï; its pace, however, is seldom quicker than that of a Tortoise. It is fond of suspending itself by the four feet to the branches of a tree, with its back downwards, and describing the arch of a circle; it even sleeps in this posture. Its forefeet answer the purpose of seizing whatever it wants to eat, and carrying it to its mouth; but this motion is imperfect and painful, for its toes and claws being, like those of the Aï, inseparable, extend and contract all together, and thus perform but the office of a single toe. The Unau often suspends itself by three of its feet, and eats with the fourth; it subsists on the leaves of various trees; its cry is feeble and plaintive; it has scarcely any power of smell, and sees but badly, especially during the day; it has no violent appetite of any kind; it can remain a long time without eating. All its senses are obtuse, and the hardest blows and deepest wounds make little or no impression on its invincible apathy.

The Unau, like the Aï, is an inhabitant of the southern regions of America; it is frequently to be met with in the forests of Brazil and Guiana; its flesh is coarse, and never sought after, except by persons of no great delicacy of palate, such as the negroes, and the savage aborigines of the country. The female brings forth but one young one, which fastens on her back, and exhibits no greater portion of life and vivacity than its mother.

To this species must, doubtless, be referred the *Kouri*, or Little Unau of Buffon. This quadruped has, like the Unau, two toes on the fore-feet, and three on the hinder; but it is only a dozen inches in length from the extremity of the nose to the origin of the tail, which is nothing but a simple tubercle; its hair is of a musk-brown, shaded with

THE TWO-TOED SLOTH OR UNAU.

BRADYPUS DIDACTYLUS.

T. Landseer delt. et. fec.
Mus. Brit.

London Published by C. B. Whittaker, Dec.r 7th 1825.

gray and yellow; this hair is shorter and duller in colour than in the Great Unau; under the belly, it is of a clear musk-colour, shaded with ashen, and this colour grows still clearer under the neck, as far as the shoulders, where it forms a slight streak of pale fawn-colour. The largest claws of this Little Unau are not longer than nine lines.

Such is, in substance, the account given by Buffon of an individual of this species, which was sent to him from French Guiana, under the name Kouri, without any information concerning its natural habits.

M. Desmarest enumerates a third species, which he calls the *Bradypus with collar*, (*Bradypus Torquatus* of Illiger.) This Bradypus is smaller than the Unau; its body is about seventeen inches in length, and its head three inches and a half; its face is naked and black; the hairs of the forehead, temples, chin, throat, and chest, are red, short, and frizzled; those on the top of the head, which are longer than the rest, are yellowish; it has, round its neck, a large collar of long black hairs; all the rest of the body is of a dirty yellow. Like the Aï and the Unau, its hairs are long and dry; but they are less flatted. At their base is a down very soft and fine, of an extremely deep brown, near the collar, but which diminishes in intensity of colour, from that point as far as the crupper, where it is entirely white. This animal has a very small external ear, concealed under the hair. In this species, as in the preceding, the palm of the hand, sole of the foot, and heel, are completely naked.

That there are considerable differences in the superficies of different specimens of Sloths, as they successively occur to observation, is certain; whether these differences constitute specific characters, or merely those of variety, is not so clear; nor is that point, in fact, to be satisfactorily ascertained without a much more extended observation on these curious creatures. In the mean time, multiplication of figures is rather a desideratum, as being calculated to assist in the progress of future investigation. With this view, therefore, we have engraved a figure from our collection of drawings by Howitt, from a specimen which appears formerly, if not at present, to have belonged to Mr. Swainson. It was about two feet in length, and was therefore intermediate between the two other specimens engraved. The colour was uniformly of a dirty yellowish white, with a tinge of brown on the back and about the neck; but from the top of the head and neck proceeded a pretty considerable patch of long black hairs; the face was swarthy.

Major Smith has also a drawing from the same specimen; and from a note of his upon the drawing, it appears that M. Temminck had seen the specimen whence it was taken, and had applied the epithet *Cristatus* to it as a specific name. There is decidedly a much nearer alliance between this and the specimen next mentioned than between that and the third sloth with three toes we have figured, which, though so widely different in size and colour, is referred to the same species.

To this is added a figure of a specimen which has long been kept in spirits in the British Museum, which presents considerable differences in the arrangement and number of spots from the Common Aï as ordinarily described, to which, at least, as a variety, it is referred. This is closely covered, as appears by the figure, with small black spots on a dirty white ground.

Another specimen in the British Museum, of the Three-toed Sloth, has also been engraved, which presents differences from the last, as to colour, equal to those of dimensions. The specimen is uniformly of a light ochrey brown, approaching to fawn-colour, rather darker on the back, and lighter on the underparts and round the face. This may be the Aï à dos brulé of Sonnini.

This, and the spotted specimen last mentioned and figured, both appear, by the state of the teeth, to be adult if not old

individuals. They perfectly accord in all their relative proportions; but when it is considered that their differences
of colour are so great (an effect, by the way, not likely to
ensue with animals of the same species, whose habitat is
very circumscribed,) and still more, that while the body
and head of one measures two feet all but one inch, the
other does not exceed ten inches, and that both are adults,
it seems difficult to conclude that they are mere varieties of
a single species.

We have now laid before our readers every thing important in the description and natural history of these very
singular animals; if we have indulged in greater minuteness of detail concerning them than we generally permit
ourselves to do, their extraordinary conformation and
habits must be our apology; and, in fact, such minuteness
was, in some measure, requisite, to enable the reader to
understand the drift of a few general reflections, with
which we shall venture to close the subject.

We have seen, from the description now given, that the
Unau and the Aï, though congeners, are as decidedly distinct in species as it is possible for two animals to be. To
say nothing of external distinctions, there is a different
conformation of the viscera in many essential points; but
what constitutes the most striking mark of distinction
between these animals is the circumstance of the Unau possessing forty-six ribs, while the Aï has but twenty-eight;
this, as Buffon remarks, in an animal whose body is so
remarkably short, is, or rather appears to our limited
faculties to be, an error or excess of nature. Even of the
largest animals, and those whose body is the longest in
proportion to their breadth, there is none in whose skeleton
such a number of ribs are to be found. The Elephant has
but forty; the Horse, thirty-six; the Badger, thirty; the
Dog, twenty-six; and Man, twenty-four. This difference
of internal structure, undoubtedly, removes the Unau and

the Aï to a greater distance from each other than many species are whose internal forms differ very considerably; but it is certainly the external conformation that properly constitutes the animal; external forms are of little importance in themselves; and, in truth, must be considered to bear the relation of effect to cause, when compared with internal. The exterior, as Buffon well remarks, may be considered as nothing but surface or drapery; and it often happens that a very different drapery will cover forms whose interior resemblance is perfect; but, on the other hand, the least difference of internal structure will produce the most marked difference in the external form, and change not unfrequently altogether the habits, faculties, and attributes of the whole animal; but this is a subject well worthy in itself of a separate dissertation, and would require, to be properly unfolded, a consecutive examination of all the different parts of organized beings.

To return to the Unau and Aï; it is impossible, in a number of instances in the animal world, to fathom the design of nature, in the peculiar conformations which she has produced, or to assign any cause for the apparent imperfection and imbecility of some of her works; but it would be the height of presumption, in consequence of our own ignorance, to question the existence of any design or any cause. From a superficial consideration of the structure of these animals, one might be apt to conclude that their lives must be perfectly miserable; but such a conclusion would be equally rash and remote from truth. We are apt to suppose beings to be miserable themselves, because we should be miserable if we exchanged situations with them; but herein we deceive ourselves. "To be weak is miserable," was a maxim put by Milton into the mouth of a being whose authority is not likely to pass for gospel; and, indeed, it is not altogether unworthy of the grand deceiver. The fact is, that weakness does by no means necessarily involve misery as its constant accompaniment, and the absence of acute feeling is as often a source of comfort and tranquillity as the presence of strength. The Sloths are deprived, it is true, of all weapons offensive. They have neither strength to resist, nor speed to fly, nor cunning to contrive any means of escape. Their senses are too obtuse to apprise them of impending danger; they cannot climb like some animals, or burrow like others. Prisoners, as Buffon says, in the immensity of space, and confined almost to the tree under which they were born, they are exposed, helpless, to every attack of danger, and every accident. They certainly remind us, in some measure, of those imperfect outlines of organized life, those *lusus* that nature, with an apparent caprice, sometimes produces, and which the defects of their structure soon cause to be blotted out from the list of living beings. Were it not for the circumstance of the Sloths inhabiting deserted regions far from the sojourn of man, and where the more powerful animals have not greatly multiplied, it is more than probable, if we can suppose such consequence in any case to ensue, that their race would long ago have become extinct. From their conformation they appear decidedly to be the lowest of the Mammalia, and to form almost a connecting link between that order and the Reptile tribes.

Still we must regard these species as perfect in their kind, properly organized for the existence to which they are destined; and we must also admit a final and a wise cause for the anomalies they present to our view. We cannot, indeed, agree with Buffon, that species of animals have been created or organized, to use his powerful language, for misery. We suspect, on the contrary, that there is little or no misery in the animal world, or, at least, among animals in a state of nature; among us, indeed, there is a sufficiency of evil; but it is, for the most part, the work of our own hands. To those animals which the Deity in his wisdom has thought proper to subject to us, and for any abuse of which trust a severe account will, questionless, be exacted, to these indeed Man has contrived to communicate no small portion of the misery which he has himself created. Yet even these are happier than Man; they have no regret for the past, nor apprehension of the future; the present is all to them, and that present is not always miserable. As for the wilder tribes, the perfect possession of liberty, and the care of procuring subsistence, is enough to constitute their happiness, and in this respect the majority of them are far from being legitimate subjects of compassion.

As for the species on which we have been writing, we must conclude, that notwithstanding all the drawbacks under which they labour, they cannot be unhappy; their strong and hard bodies, on which almost every blow falls in vain; the obtuseness of all their sensations, which must render them insensible to pain; their extraordinary capacity of enduring privation of nourishment; and, above all, their wonderful degree of vitality*, which is never accompanied with acute feeling, all must contribute to shelter these animals from any thing like misery. Their pleasures, it is true, can be few or none; but then the amount of the pain in the balance sheet must be trifling indeed. We shall find this to be pretty generally the case throughout the universe. Where the Creator has vouchsafed to any being an extraordinary capacity of enjoyment, it appears to be necessarily accompanied by a proportionate susceptibility of pain; while the pent-house of insensibility that intercepts the rays of the sun of joy, is also an adequate

---

* The tenacity of life in the Aï and Unau approximates them to the Reptile tribes: this is so great, that on one of these animals being opened and dissected, he did not die immediately, but the palpitation of the heart continued for a considerable time after the operation.

protection against the storms of sorrow. When the great account with every being shall be closed at last, we shall find the balance even, and that the mighty calculator is not less just than wise, less merciful than omnipotent.

The genus DASYPUS of Linnæus, the species of which we most commonly know by their Spanish American name of Armadillo, is very singular in the nature of its coat of mail, with which all the species are invested. This and the analogous covering of the Pangolins, or Scaly Ant-eaters, like the spines found in a few genera already alluded to, seem to present anomalies in the general nature of the covering of animals, as much as distinct descriptions of such covering.

The nature of the integuments with which the vertebrated division of the animal kingdom in general is invested, is observed to have a general relationship with the respiration of the different orders; so much so, that the one seems dependant entirely on the other. Feathers are universally accompanied with a very complete system of respiration; hair is the common covering of animals whose breathing is less perfect; and scales are proper to a third large division, whose pulmonary system is much less complete than that of the other two. The first and third of these divisions present perhaps no exception to this rule] We know of no Bird, or Reptile, or Fish, covered with hair, but we find Mammalia invested with scales, or plates, and spines. The generality of this observation is such, that we may be assured there is a constant relationship between the nature of the integuments of animals and the state of the atmosphere to which their existence is adapted ; but nature, constant in nothing more than her inconstancies, furnishes an exception to this rule in the genus now before us and in those alluded to.

It is indeed hardly possible to find any pretty general

rule in nature void of its exceptions, and the genus we are now considering, placed among the edentatous animals on account of its want of incisive teeth, presents in one, at least, of its species, an exception to its arrangement in this order.

We propose availing ourselves rather largely of the information afforded by M. F. Cuvier on the Encoubert, one of the species of this genus, and furnished with incisive teeth, in describing which he refers to other species; and limiting our other observations rather to generic than specific descriptions.

The body of these animals is covered with large scales or plates, forming altogether a complete armour, at least on the upper parts, and consisting of four or five different parts or divisions. The head may be said to have a helmet, the shoulders have a buckler, composed of several transverse series of plates. Transverse bands, varying in the different species from three to twelve, which are moveable, cover the body; the crupper has its buckler similar to that on the shoulders, and the tail is protected by numerous rings.

The fore-feet have in some five, in others four toes; the posterior feet have five. The muzzle is more or less pointed. The hairs of the, body are few, springing from between the plates; the under parts, however, which are without armour, have rather more hairs.

The plates which compose the bands and the bucklers are articulated only at the commissure of the bands, and are united by symphyses. In a living state, the whole armour is capable of yielding considerably to the motions of the body.

The skin of the under parts of the body is covered with warts or tubercles, which are indurated towards the feet.

The tail is in general straight, and incapable of flexure; the ears vary in length in the different species; the eyes are small; the legs are thick and strong, and only long enough to raise the body from the ground ; the nails are very powerful, and organized for digging.

We possessed very scanty information on these animals, till D'Azara published his Essay on the Quadrupeds of Paraguay, which includes eight species. He tells us that most of them dig burrows in the earth, which they commonly direct under an angle of 45°; but that they turn so as to make it difficult to ascertain their length, which is presumed, however, to be from six to eight feet.

Some of the species have nocturnal habits, and are very timid, flying to their burrow the moment they hear a noise. These are very much quicker in their motions than might be supposed, from the hindrances incident to their heavy covering. Other species quit their retreat equally by day and night, and these are said not to be so rapid in their motions as the others.

When the Armadillos are pursued, and find flight ineffectual, they withdraw the head under the edge of the buckler of the shoulders; their legs, except the feet, are naturally hidden by the borders of the bucklers and the bands; they then contract the body toward a spherical form, so far as the stretching of the membrane which unites the different moveable pieces of the armour will permit.

Till the time of D'Azara, the idea was general that these animals fed exclusively on vegetable matter ; that naturalist, however, considers them as insectivorous, and even carnivorous ; the character of their excrement in a natural state indicates it. The directions of their burrows evince that they pursue the Ant-heaps, and these insects quickly disappear where they inhabit. Nobody doubts in Paraguay that the largest species described by the Spanish naturalist feeds on the carcasses of animals; and the graves of the dead which are necessarily buried at a distance from the usual places of sepulture, and in countries where the great

Armadillo is found, are protected by strong double boards to prevent the animal from penetrating and devouring the body. It appears also that they eat young birds, eggs, snakes, lizards, &c.

The new character given to the Armadillos, as to the nature of their aliment, is confirmed, and perhaps explained, by M. F. Cuvier's observations on the teeth of such of the species as he has been enabled to inspect. The fact of one division of them being found with incisive teeth, and another with cheek-teeth of a different character to those of the third, induce a great probability, however, that the carnivorous habits of this genus are confined to some particular species; and that the different sub-genera, when fully understood and arranged, will be found possessed of different habits, in accordance with different modes of procuring their subsistence.

The dentition of the genus varies, then, to a considerable extent, in the different species, and assumes all the following different formulæ:—Incisors $\frac{0}{0}$, or canines $\frac{8}{8}$; cheek-teeth $\frac{7}{7}$, or $\frac{8}{8}$, or $\frac{8}{8}$, or $\frac{9}{10}\frac{9}{10}$, or $\frac{17}{17}\frac{17}{17}$.

All the species are South American, principally of the province of Paraguay. Some inhabit the forests, others are found in the open countries. Their flesh is eaten even by the Spaniards.

It is generally understood that the Armadillos bring forth but once in a year, and differ in the different species as to the number at a birth. D'Azara ascertained that one species brought from seven to twelve young: the female, however, is said never to have more than four mammæ, which has induced an idea that, however numerous the litter, they bring up only that number.

The old mode of distinguishing the species by the number of the bands, is clearly objectionable, inasmuch as D'Azara has established that not only the number of these bands varies, in the different individuals of the same species, but further that there are individuals of different species which have the same number of bands.

The number of species and their synonyma is by no means satisfactorily explained. The eight mentioned by D'Azara from actual observation may be admitted as distinct, but their identity or distinctness from such as are already described must, in a great measure, be subject to conjecture.

After these general observations on the genus, we shall select such of the species as may be best known for further description; and as the order of their succession in this division of the subject is immaterial, we shall take those first on which there exists the most information.

To invent an artificial system of zoology, though founded on the best basis, that is, the natural analogies and deviations of animals, without many anomalies, without some absolute contradiction, is utterly hopeless. It is, perhaps, indeed, not too much to say, that such an Utopian system is hardly to be desired, as it would be destructive of the very objects of the science. If a certain number of analogies be not allowed to warrant the association into one order, though the genera and species which compose such order may all differ amongst themselves in various particulars, we are driven to the baneful alternative of multiplying divisions, and burthening the memory with terms of art as numerous as the individual aberrations of nature.

Thus, until lately, the absolute want of incisive teeth in both jaws was considered essential to the admission of any genus or species into the order Edentata. Canine teeth might or might not exist, but the order was, in fact, named from the total absence of incisors: but M. F. Cuvier has lately described the *Encoubert*, (for whose synonima we refer to the table,) which is possessed of incisive teeth, and must therefore be considered as an anomaly in this order.

2 X 2

In his description of this species, that able zoologist has published some valuable allusions to others of its congeners. We shall therefore proceed with his account, as not merely descriptive of the particular species named, but as illustrative of the whole genus.

These animals are still very imperfectly known, though d'Azara has given us much information about them. There are a few indifferent figures of some of the species, imperfect descriptions of others, and scarcely any anatomical observations on their organs; hence it is impossible to refer the different members of this family, the various species of Armadillos, to the different subgenera into which they ought to be divided. We are enabled, however, to mark the characters of some of these sub-genera, and the Encoubert may be considered as the type of one.

It has been long known, and was confirmed by d'Azara, that some species of the Armadillo had five toes on all the feet, while others had only four on the anterior feet, and although these differences might have no great effect on the qualities of the species, they were of that character which ought, in conformity with received practice, to distinguish different subgenera.

But the character which more particularly distinguishes this species, alluded to in our introduction of it, that is, the presence of incisive teeth, is more important than the variations in the number of toes ; for dentition, whether considered as cause or effect, is ever attended with corresponding habits in the animal. The Encoubert, then, in this particular, recedes considerably, not only from its genus, but from its order.

The group of Armadillos, without incisors, will again be found to require subdivision, not merely from the character of the extremities, but also by the articulation of the lower jaw, the form of the teeth, and the relationship of those in one jaw to those in the other. The Encoubert, as to this articulation, belongs to the first of these subdivisions, though it differs from it in the incisive teeth.

The articulation of the lower jaw in the second of these subdivisions, similar to that of the rodentia, is by a longitudinal condyle; the cheek-teeth of these, of a laminous form, and not cylindrical, and far from being alternate in their reciprocal action, are in this manner: the teeth in the lower jaw are in communication, by their external surface, with the internal surface of those in the opposite jaw, and as the lower jaw has only a longitudinal movement, and the teeth placed successively, touch only at their edges, their mode of trituration may be compared to the action of two plates placed side by side, operating like two saws.

Organic characters of this importance bespeak animals of different dispositions and habits, which ought not to be called by one common name. M. F. Cuvier therefore proposes to preserve the name Tatous for those of this genus which have insisive teeth, and to call those Tatusees (a common name for all the species mentioned by some travellers) which are without; and he proposes the name Priodonte as the sub-generic for such as have their teeth in a laminous form.

The covering of this animal consists, on the upper part of the body, of several large plates of different shapes, with a few single hairs from between them; on the lower part there is nothing but a smooth naked skin. The covering of the head, which may be called the cap or helmet, is three inches and a half in length, by two and a half in breadth; it commences about half an inch from the extremity of the muzzle, and extends to the occiput, and is composed of irregularly formed plates; under the eye may be seen a line of small plates, which joins the lateral angle of the helmet; these diminish in size regularly from the first to the last:

under the plates surrounding the lower part of each eye, there are about twelve hairs, much thicker than the rest, which appear to be whiskers: adjoining the helmet and over the neck, is a moveable band, composed of seven long rectangular plates, and of one at each end of the band of a triangular form; on the shoulders, the plates form anteriorly regular fixed bands, and posteriorly two other bands, like the first; all the intermediate plates are irregularly placed.

On the back there are seven moveable elongated bands, similar to that band which terminates the armour on the shoulders; but the plate which terminates each end of each of these bands is longer than the rest, and narrower at its commencement, that is, at the base of the preceding plate, so that there is a small naked space at the bottom, between each band.

The armour on the crupper is composed of a band, free at its extremities, but fixed in the middle, and then follow nine fixed bands. All these bands are formed of plates similar to those of the moveable bands, except only that the plates diminish in length from the first to the last band, so that those nearest to the tail become almost square.

The tail, at its base, has a moveable ring of small round plates, which appear almost rudimentary; then follow four other rings of perfect plates, and the rest, that is about three inches, is covered by semi-elliptic plates disposed in checkers.

The whole armour is of a pale white colour, and the plates of which it is composed, except these of the head and tail, which are smooth, are formed of tubercles thus arranged; in the middle is one of elongated oval form, and round the edge of the plate there is a suite of three, four, or five (according to the size of the plate) round tubercles; the furrows or impressions which separate these tubercles are deeper between the great central tubercle and the late-

C. Hamilton Smith Esq.r del.t

GIANT ARMADILLO.

DASYPUS GIGANTEUS.

I. Bradley Sc.

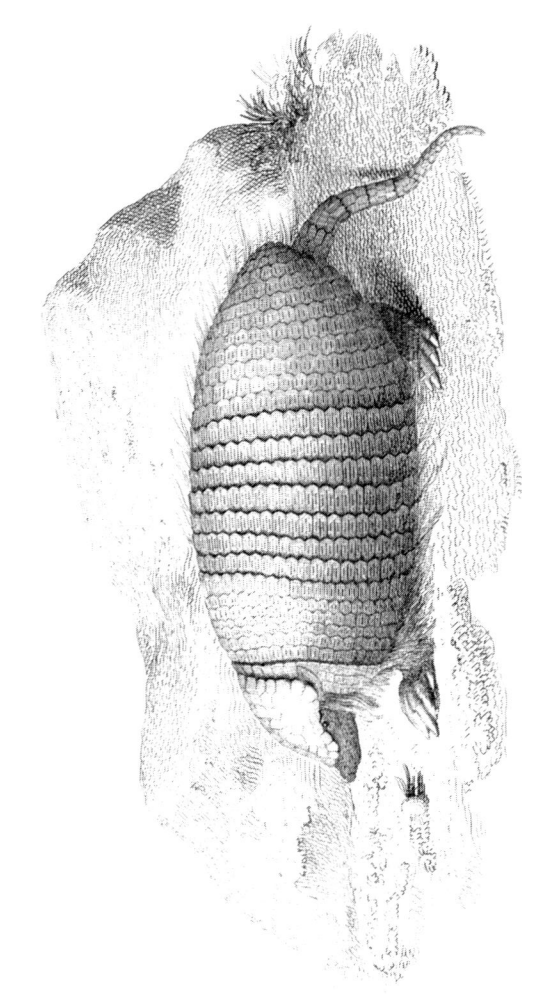

THE HAIRY ARMADILLO OR FOURTH TATOU OF D'AZARA.

ral than they are between the latter, so that these last-mentioned furrows are effaced by the friction of the plates on the ground, whence the surface of the plates changes, according to the age of the animal. From the posterior edge of each plate of the moveable bands spring three or four white hairs. The skin of the lower parts is covered with small tubercles, which become plates analogous to those of the upper parts of the animal on all the feet; the sole of all the feet is naked, and entirely smooth, without those tubercles generally observed on the naked sole of the feet of most plantigrade quadrupeds; the skin is of a violet-brown colour.

The organs of motion in the Encoubert are essentially constituted for digging; the shoulder and the arm of the anterior limbs are covered by the scapular buckler, so that they cannot be seen; the fore-arm is thick and short; the carpus is terminated by five unequal toes, armed with strong compressed nails, round above, flat below, and with trenchant edges; these toes, which are short and thick, are united, as far as the last phalanx, by a strong light membrane; the nail on the third toe is the longest and strongest.

The thigh of the hind legs is also hidden by the buckler over the crupper; the leg and the tarsus are thick and short, terminated by five short thick toes; the nails of which are not so long as those on the fore-feet, but of the same thickness, convex above, and flat underneath; all these toes are also united, as far as the last phalanx, by a membrane. The tail is strong, round, thick, at its insertion, and terminating in a point.

Smell appears to be the only sense which is acute in these animals; the rest are obtuse, but the perfection of this affords a compensation for the deficiency of the others; the nostrils open at the end of the muzzle.

The eye is very small, placed in the middle of a naked space, and protected by thick eye-lids, bordered with a few small lashes. The internal lid is very thick, but not extensive, and the eye-pupil is round.

The external ear is short, terminated in a point, and of a very simple structure.

The tongue is long, narrow, pointed, soft, and covered with soft papillæ, some round, others flat, and two conical.

The *Hairy Armadillo*, is so named by D'Azara, not as being singular in its genus in this respect, but merely as being more hairy than the rest. It may answer the purpose of that brevity which we are obliged to study, without, we hope, destroying the utility of the work, to refer for the specific descriptions of the remaining species to the short statements in the Table; a note of the points in which the species differ from each other may, after the details already given, suffice in the stead of a particular description of each.

The Encoubert and the Pichiy of D'Azara have hairs, but not so long or flexible as those which are found on the sides, and proceed half way down the tail of this species; these hairs are brown, about an inch and a half in length; there are some also on the upper parts of the body, but they are less numerous than those on the sides, and they are shorter, arising apparently from the friction of them; there are some, also, on the forehead, which are short, and the largest of all appear on the outside of the fore-legs. All the armour is dark brown, except that of the feet, which is of a reddish-brown, with a mixture of pale orange colour.

These Armadillos are not found in Paraguay, or north of the River Plate, and only in the open country. D'Azara, in one excursion between thirty-five and thirty-six degrees of south latitude, found them by thousands, and there was scarcely a man of the party of a hundred which was with him who did not take every day one or two. This species is much more easily to be caught, as it quits its burrow at all times, as well by day as night.

It is said that this species is carnivorous, and that when it finds the dead body of a Horse, it burrows a way through the lower sides of the carcass, and feeds on the interior of it, particularly the putrefied parts, leaving the bones and skin untouched.

In the Encyclopédic Méthodique, M. Desmarest seems to doubt the existence of this species as distinct from the Encoubert, and there is a contradiction in the history of its habits from the statement in the Nouveau Dictionnaire d'Histoire Naturelle, as in the former work the species is said not to dig burrows. We have engraved a figure which seems referable to it.

The *Pichiy* of D'Azara is more like the Hairy Armadillo than any other species, in its thickness, the size of the body, and head, the number of toes, and the general appearance; but it is rather less than that species, and has less and shorter hair.

The number of the moveable bands of this species varies according to the age and sex of the individual. In a young male, and in an adult female, D'Azara found seven; but there were but six in an old male. The edges of the plates are indented; there is a single long hair to each plate of the back, but those on the body are in small clusters. The colour of the whole animal is dark, with whitish interstices. The skin under the body, the hair, and the feet, are like the same parts in the Hairy Armadillo. The female has but two mammæ.

The habits of this animal are described as similar to those of the last, and it quits its burrow by day as well as night, in the manner of that species. It is said, also, to be good food.

This species is found south of Buenos Ayres, from thirty-six degrees of south latitude to Patagonia.

The *Giant Armadillo* is principally distinguishable on account of its large dimensions, when compared with the other species. From the muzzle to the tail it measures upwards of three feet six inches, and the tail is eighteen inches more. The specific characters are stated shortly in the Table.

This species is found principally in the most northern parts of Paraguay, where it is called the Great Black Wood Armadillo, as it seldom quits the forests. It is this species, in particular, which is referred to, as disinterring and devouring the bodies of the dead.

Of the *Mule Armadillo*, another new species, described by D'Azara, we shall merely add here that it is said not to dig burrows in the earth; but even this statement is insufficiently substantiated. The female brings forth from eight to twelve, toward the month of October.

The *Three-banded Armadillo*, *Tatou*, *Apar* of Buffon, is remarkable among its congeners for the faculty of rolling itself up more completely than the other species. It can, in so doing, totally conceal the head, the tail, and the fore-feet, which none of the others can completely effect. This species inhabits the vicinity of Buenos Ayres, a little south of thirty-six degrees south latitude.

The whole series of these very singular animals offer a notable example of the exclusive locality of genera. We have observed that they all belong to South America, for Seba was certainly incorrect in referring one of the species to Africa; nor do we find in any parts of the old world, or, indeed, in the great northern division of the new, any races of quadrupeds which may be considered as analogous with them, or as in any manner meriting even a comparison. Hypothetical conjecture may amuse itself to a very considerable extent in tracing out a fancied alliance, or connecting link, between very many different genera; indeed, generally, it is hardly possible to select a genus without finding some one or more species in it which approximates some other; intermediate genera placed by their describers first with one genus and then with another, to both of which they may have considerable general analogies, are eventually separated from both, and found to be properly distinct, though intermediate. Thus we have already noticed canine Cats, feline Weasels, Hyæna Dogs, Hyæna Civets, &c. &c.

But the genus Dasypus, all the species of Armadillos, stand perfectly insulated; they exhibit all the characters of a creation perfectly distinct, and, except as to the general characters of mammiferous quadrupeds, perfectly *sui generis*. There is no break in the whole circle of them, no deviation, or leaning towards any other organized form; the boldest conjecture will hardly venture to guess at any other than a separate creation for these creatures, and a distinct allocation in South America.

The confined natural habitat of this genus is also the more remarkable, when it is considered that from the facility with which it seems to endure removal even to our latitudes, there is much reason to conclude that its nature would become reconciled to a transportation to parts very distant from where it is now exclusively indigenous; and, therefore, that its present confined habitat is not altogether the result of its physical necessities.

When the type of a new creation is discovered, it seems generally to follow, as our knowledge of it is extended, that such creation is not of the confined character we at first imagined. Thus it is now ascertained, that the pouched animals, although artificially placed merely as a division of the flesh-eaters, includes not merely several genera, but genera in fact of very opposite characters, and that but for the mischief attendant on multiplying divisions, at least without something like a complete know-ledge of the things divided, the Marsupiata should form a class, including its orders, analogous to those in the Mammalia, with a single matrix; so it seems likely to turn out, though on a much less extensive scale, that a more complete knowledge of the Armadillos will induce the necessity of further subdivisions, and that so insulated a type of creation will not exist in a single form alone.

Three of such supposed divisions have, as we have already seen, been alluded to, rather than established, by M. F. Cuvier; but till we know more of the genus, the complete application of these divisions must be suspended. In the table, therefore, of this genus, we shall enumerate all the species as forming but one group, only endeavour-ing to point out, though as shortly as possible, the dif-ferent specific peculiarities, some of which may eventually be elevated to generic distinctions.

The genus ORYCTEROPUS includes at present but a single species, or, in other words, an animal is found in Southern Africa, which, when all the peculiarities of its structure are considered, cannot with propriety, in the opinion of systematic zoologists, be referred to either of the pre-esta-blished genera. Such insulated instances in artificial sys-tems are by no means uncommon; but when we consider that the operations of nature are generally of a comprehen-sive character, we may fairly be suspicious that we are de-parting from a natural system when we establish genera which include but a single species.

The *Orycteropus*, or, as we would rather call it after its English name, the *Cape Ant-eater*, when first described by Pallas, was referred to the Ant-eaters; but Kolbe and Buffon have named it, with reference to a very different genus, the Ground Hog. All its analogies agree with the Ant-eater except its system of dentition, on which natu-

THE CAPE ANT-EATER.

ORYCTEROPUS CAPENSIS. *Cuv.*

Published by G.B.Whittaker. Sept.1826.

ralists have grounded its generic separation. The genuine Ant-eaters, it must be considered, are purely edentatous— they have no teeth whatever; the animal in question, though destitute of incisors and of canine teeth, is furnished with six cheek-teeth on each side in each jaw. The structure of these cheek-teeth is perfectly *sui generis;* they differ from those of all other quadrupeds; each tooth stands insulated, at a small distance from the next, and is without a root or a very distinct crown.

The first tooth is very small; the second rather larger, formed of two united cylinders; the third and fourth are of the same form, but larger; the fifth is the largest; and the sixth is about the size of the third.

The body of this animal is thick, heavy, and pig-like, standing low on its legs. It is covered with a short pale gray fur, rather longer on the back and flanks; rather reddish on the flanks and belly, with the feet dark brown.

The head is very much elongated : the muzzle rather pointed, but obtuse at the end; the ears are very large and pointed; the eyes are moderate in size, and situated nearer to the ears than to the end of the muzzle; the tongue is narrow, and flattish, extensible, and exuding a viscous secretion on its surface.

The fore-feet have four toes, but the hind feet, which are plantigrade, have five; the nails are very strong and rounded, those on the hind feet much larger than those before, all of them approximating the character of solid hoofs, and eminently qualified for digging.

The tail is round, very long, strong and thick, near at its insertion and tapering gradually to a point.

The skin is very thick; and the hairs, being harsh and scanty, give it another pachydermatous character, and the arrangement of the bones of the tarsi and metatarsi is also analogous to the animals of that order. We, therefore, observe in this animal another instance of a connecting link in the great chain of animal existence.

The Cape Ant-eater, whatever may be its generic analogies or anomalies, is aptly so named from the principal food on which it lives. It is said to thrust its extensible tongue into the Ant-hills, and this organ being furnished with a glutinous secretion, is thereby the better enabled to secure by adhesion a greater number of these insects, which are thus conveyed into the mouth of the animal. A similar mode of feeding was attributed to the Ant-eaters of America, till D'Azara stated that they merely scratched the Ant-hill, which induced the insects to come to the surface in great numbers, when the animal passed his tongue horizontally along the surface among them, and thus secured his prey. This we suspect also to be the mode of operation of the present species.

Hardvark, says Mr. Burchell, is the colonial name of this animal and signifies Earth-Hog; and indeed it may, from its appearance and size, be more justly compared to the Hog than to the other Ant-eaters ; but in its mode of life, however, it exactly resembles the latter. With its fore-feet, which are admirably formed for that use, it digs a deep hole, wherein it lies concealed the whole of the day, never venturing out but at night, when it repairs to feed at the Ant-hills, which abound in many, parts of the country. Scratching a hole on one side of them, it disturbs the little community, on which, the insects running about in confusion, are easily drawn into the animal's mouth by the long slender tongue with which nature has provided it for this purpose. Without tusks, or any efficient teeth, this animal is quite defenceless, and depends for its safety solely on concealment; in which it so completely succeeds, that no animal is less seldom seen, and from its power of burrowing with incredible rapidity away from those who endeavour to dig it out of its retreat, few are more difficult to be obtained. Its flesh is wholesome and well tasted.

It measures from three feet six inches to four feet from the nose to the anus, and the tail is nearly two feet long.

The separation of the Orycteropus, whether as genus or sub-genus, from that of MYRMECOPHAGA, to which we now proceed, seems proper, when it is considered that these latter are purely edentatous—are absolutely toothless. The head and muzzle in the Myrmecophaga have also little of the pig-like character of the Cape Ant-eater ; it is generally greatly elongated, excessively small, with a very narrow little mouth at the termination. The ears also are small and round ; the tongue very long, cylindrical, protractile to a considerable distance ; and the tail in some species subprehensile.

D'Azara, the best modern original describer of South American Mammalia, has furnished some additional information on the natural history of the *Great American*, or maned *Ant-Eater, Myrmecophaga Jubata, Lin.*, aptly called the Gnouroumi, or Little-mouth. The Spaniards of Paraguay call it the Bear Ant-eater, and the Portuguese, Tamandua.

This singular animal is an assemblage of anomalies : its head, formed like a trumpet, does not equal in its thickest part the size of the neck ; the tail has some analogy with that of fishes, being, when stript of the fur, extraordinarily thick at its base, and compressed laterally ; the arms are immeasurably strong in reference to the body, very much compressed on the sides, and apparently without any play in the elbow ; moreover they are nearly as large below as above, and are altogether much more so than the hind limbs ; the fore feet have scarcely any resemblance to feet, and the animal does not bring them to the ground in the ordinary way, but more as if they were the hoof of a horse, treading only on a sort of pulp, or hard excrescence, and upon the exterior toe, which, contrary to the general rule, is the largest; the figure carefully represents this peculiarity; the others do not appear like toes, and the animal can open them no farther than the point, where the nails become perpendicular to the line of the fore-arm; the hind feet do not appear calculated for walking, and are very ill formed—the sole is swelled, and the inner toe is the shortest and weakest; the mouth is nothing more than a small horizontal cleft at the end of the conical, or trumpet-formed, head; its length from the tip of the nose to the insertion of the tail, averages about four feet six inches;. the ears are small, round, and large at the base; the eyes are small, imbedded, and without lashes or lids; the nostrils are large, of the figure of the letter C; the tongue is fleshy, very flexible, sharp, not altogether round, but similar to the tongue of some birds, and the animal càn protrude it sixteen or eighteen inches; the nails are generally bent, very powerful, and trenchant in the internal part; a callous pulp, about two inches in height, and almost entirely united to the fourth toe, but altogether destitute of a nail, may be said to represent a fifth toe— the animal treads on this in walking; in the posterior part of the palm of the fore feet is a callosity, against which he supports the point of the largest nail in seizing and holding, which last he persists in with incredible force and effect: the sole of these feet is callous; the hind feet have five short toes a little inclined inward, the three middle are of equal length, and the internal toe is shorter than the external; the nails of these feet are bluntish, but slightly bent, and useless for the purpose of seizing. The bones of the tail are flatted laterally, and diminish to a point at the termination; the whole tail is amply furnished with very long hairs; the animal generally carries it horizontally with the long hairs trailing on the ground, but it

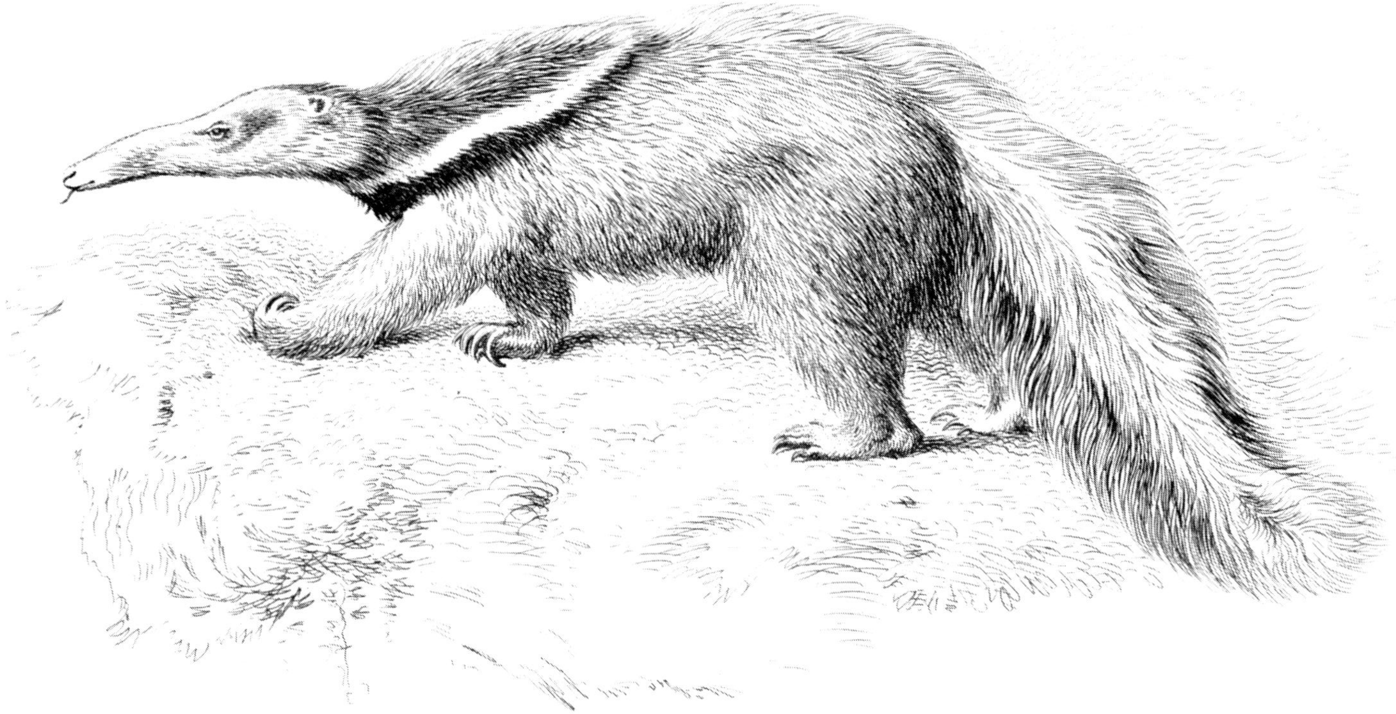

THE GREAT ANT-EATER.

shakes and elevates it when angry. It has two pectoral
teats. The fur is thick, harsh, short on the head and ears,
but rather longer on the shoulders. Between the ears commences a band or mane of erect hairs, which runs down
the spine and tail. An oblique, black, slightly-crescented
stripe, bordered with white on each side, passes across each
shoulder from the side of the neck, inclining toward the
upper part of the back. Under the third ray is a slight
mixture of dark and white, lighter on the sides of the
body, which forms, indeed, the ground colour of the whole
animal. On the toes of the fore feet is a black spot, above
this another, perfectly white, which surrounds the foot, and
still higher is another of a deep black.

The Maned Ant-eater inhabits the low and swampy
country from Paraguay to the river Plate; it is also found
occasionally in the woods, but it never climbs the trees.
It moves with the muzzle close to the ground, with a slow
and heavy step; and although it gallops when hunted, its
utmost speed is not equal to one-half that of a man.
When met with, it may be driven before with as much facility as an ass; but, if over-pressed, it seats itself to receive
its aggressor with its large strong nails and powerful arms,
which constitute, in fact, its only means of defence. It
has been said that the Jaguar dare not attack this animal,
and that the Maned-Ant-Eater will embrace and squeeze
him in his arms, drawing at the same time his great nails
close together, till he has deprived the Jaguar of life,
and that sometimes both these animals are thus killed by
each other at once. It is certain that the animal defends itself in this manner; but it is perfectly incredible, says D'Azara, that it can prevail against the Jaguar,
an animal which by a single bite, or a single blow of the
paw, could lay the Ant-Eater dead at its feet long before
it could have any opportunity of embracing its adversary,
an action, however effectual, which the animal cannot per-

form with more celerity than any other; and further, it must be remembered that this animal is incapable of leaping or springing, and can, in fact, only seize and hold any thing that is presented before it. " I have killed several," says M. D'Azara, " by striking them on the head with a thick stick with the same security as if I had struck the trunk of a tree."

It is, however, excessively strong, but of very lethargic habits. When about to sleep, it lies on the side, places the head between the arms, joins the fore feet to those behind, and spreads its tail over the upper side so as to cover the whole body. It is solitary in its habits, and the female brings but one at a birth, which is constantly in company with its mother, even after it is able to walk, and till it is nearly a year old.

In a state of liberty, this animal subsists perhaps altogether on ants. To procure its food it scratches the ant-hills with its long nails, when the insects immediately come forth from their retreat in great numbers. The Tamandua then draws its long, thin, narrow tongue among them, and withdraws it to the mouth laden with ants. The animal repeats this action with so much celerity, that in the space of a single second it protrudes and withdraws the tongue twice, which is, in fact, never thrust below the surface.

It appears incredible that such diminutive insects can suffice to support an animal of such size and strength. The wonder, however, is said to diminish much to those who have the opportunity of seeing the immense multitude of these insects contained under a single ant-hill, which, in these countries, are so numerous as frequently almost to touch each other.

The Maned Ant-eaters have, occasionally been domesticated, and in this state have been fed on soft bread, morsels of meat, and flour steeped in water; food, we may·observe, all of an artificial preparation, and not such, therefore, as they could procure for themselves in a state of nature.

By this account, the result apparently of observations on many living specimens, we may observe, that the old assertion in relation to this species, that it climbs trees, that it inserts its long tongue into the ant-hills, that it inhabits only the very hottest climates, that it is as quick in its motions as a man, and that it can destroy the Jaguar by pressure in its arms,—are not founded in fact.

There is some confusion in the synonyma of the species of Ant-Eaters. The Great Ant-Eater already mentioned may, at least so long as the epithet continues applicable, retain it; but it is better distinguished by the Linnean specific Jubata, or the Maned Ant-Eater. This, however, is in fact, according to D'Azara, the Tamandua of the Portuguese; a name which appears in Europe to have been erroneously applied to the second species, to which we shall now shortly refer.

This species is the *Tamandua* of Buffon, which name we have just seen from D'Azara is referable to the last. The name Cagouare is an abbreviation, signifying inhabitant of the woods, and of stinking and infectious places, and is therefore very applicable as distinguishing this species, being in fact a great climber of trees, from the M. Jubata, which is incapable of that action. The Spaniards call it the little Bear Ant-Eater, in reference to the M. Jubata. Comparative epithets, we have before observed, frequently become inapplicable by the progress of discovery.

This species measures two feet five or six inches in the length of the body, and the tail is one foot five or six inches more. The colour varies from pale gray to deep black, and it is sometimes found like the Maned Ant-Eater with an oblique stripe of another colour on each shoulder. The fore feet have four toes; the posterior, five; the tail is nearly round, covered with fur near the base, but denuded as it approaches the end. The head is elongated, and forms with the neck a slightly-arched cone ; the opening of the mouth is very small ; the eyes are small ; the ears round ; the body elongated and cylindrical ; the legs very strong; and the fur is silky and shining, and between two and three inches in length in the longest parts.

It is very subject to deviate from the common type in the colour of the fur : the established varieties will be found noticed in the table.

This species differs in its habits from the last, principally in being a climber, in doing which it is assisted by its prehensile tail, like the division of American monkeys distinguished by the same peculiarity. It emits a strong scent of musk, which may be smelt at a considerable distance, particularly when the animal is irritated. It seems, moreover, to eat honey, and probably the bees, which do not sting in its native habitat, and live in the clefts of trees. In sleeping the animal lies on the belly, the muzzle is placed under the chest, and the head is hidden under the neck ; the anterior limbs and the tail are placed along the side. The details of its organization accord with those of the Maned Ant-eater, except, as already stated, that this has the body proportionally thicker, and the tail round without long hairs, and the lower third of it altogether naked.

We have again to refer to one of our drawings of an inedited variety, as we presume, of this species. The fur of this is of two colours : the whole of the head and muzzle, the legs and feet, and the tail, except for about one-eighth of its length from its insertion, are of a pale straw colour ; the same colour is also continued over the top of the shoulders and down the spine more than half its length, where it terminates in a point ; the rest of the fur, including a narrowish stripe over the shoulder, is of a very dark brown, approaching to black, so that the animal has

THE URSINE ANT - EATER

THE TAMANDUA. *ANNULATED VAR?*

MYRMECOPHAGA TETRADACTYLA. L?

London, Published by G.B.Whittaker, Feb.ʸ 1825.

the appearance of being uniformly straw colour, with a mantle or covering over the belly, sides, crupper, upper part of the thighs, and upper end of the tail, of the dark colour, held on the body by means of shoulder-straps of the same colour. It seems nearest to the brown belly variety of the Table. The name Ursine Ant-eater on the plate might be retained if the specimen should seem to demand a specific separation, which, when the known varieties of the Cagouaré of D'Azara are considered, it hardly seems to do.

This drawing was made some years ago by Howitt, from a stuffed specimen exhibited in London ; if we refer to the known peculiarities of the feet of these animals, it is easy to suppose that the specimen was placed on the toes of the hind feet by mistake of the stuffer, and that the animal was in fact a plantigrade. If this be a variety of Cagouaré, the end of the tail it may be concluded would be denuded as in that species. Again, the position of the fore feet was in all probability forced and unnatural, as both the large species before noticed walk in a very awkward manner on the sides of the paws, with all the claws turned inward. We have little confidence, therefore, in the drawing, except so far as it represents the superficies of the specimen, the peculiarity of which is probably referable to accidental or permanent variety.

Krusenstern, in his voyage round the world, figures a new species, or a variety of the Cagouaré. The fur is uniformly brown, deeper at the end of the muzzle and extremity of the paws ; the cheeks are bright, with a long triangular brown spot round the eye ; the tail is yellow, rather shorter than the body, with eleven annuli of a dark brown colour.

The figure we have engraved, under the title of the Tamandua annulated variety ? seems likely to be the same as that indicated by the circumnavigator, differing principally in the absence of the dark spot round the eye. This was drawn also from a stuffed specimen, and is subject to the same observations as to the position of all the feet as that last mentioned.

It is clear from the several specimens already described, either that the Tamandua is considerably subject to vary beyond what has been already noticed, or that there are several cognate species.

The *Little Ant-eater* is not more than seven or eight inches long, with a prehensile tail rather longer than the body. It has but two nails on the fore feet, one of which is very large; the hind feet have four; the fur is woolly, of a uniform brownish-yellow; darker or approaching a red tint along the dorsal line.

This little species resides continually in the trees, where it attacks ants and insects, which retire under the bark. It suspends itself to the branches by means of its prehensile tail, and of the paws, the naked parts of which are constructed for seizing and holding with the same relative force as in the Maned Ant-eater. Its loco-motion is slow and silent. The female brings but one young at a time, which is concealed in a bed of leaves in the cleft of some tree.

The Genus Manis, by the internal organization of its limited species, approximates the Ant-eaters, but the external characters of the two are widely different. They have the body elongated, the muzzle pointed, the tail thick at its base, more or less long, and tapering to a point at the end. All the upper parts are covered with strong, horny scales: these are triangular, trenchant, and imbricated, which gives the animals the appearance at first sight of reptiles.

Their elongated muzzle is terminated by a little mouth which is perfectly destitute of teeth of all sorts; the tongue

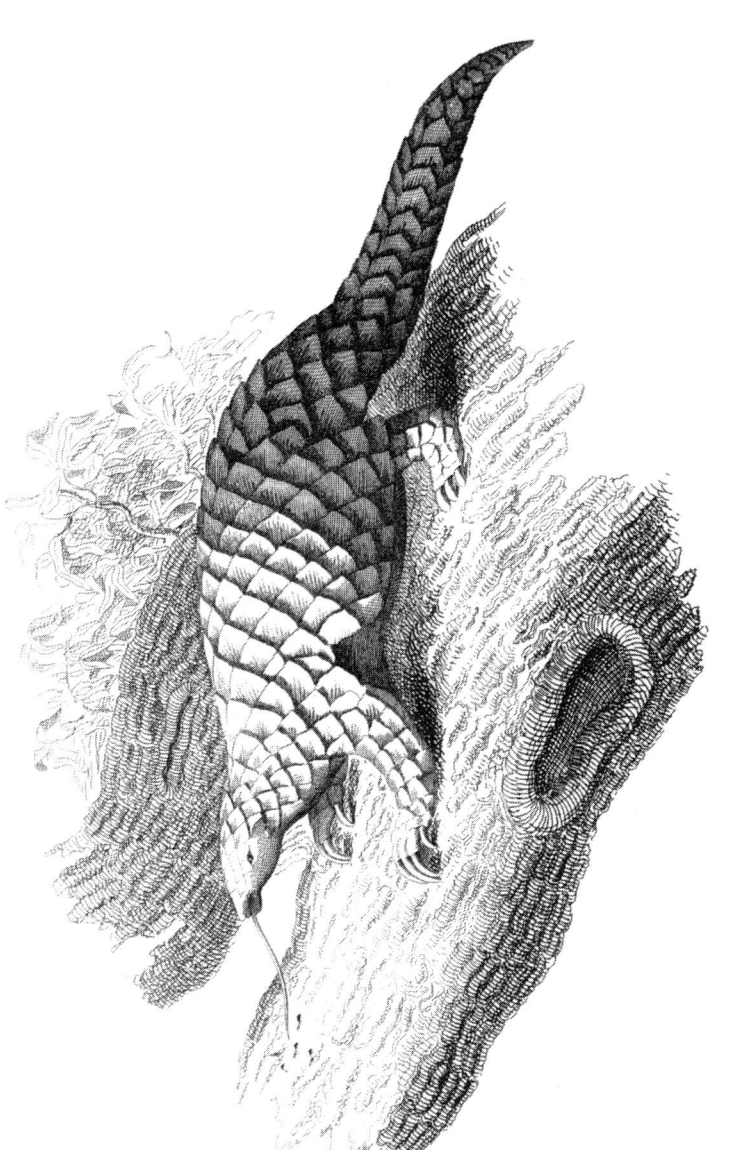

T.Bradley. sc.

SHORT TAILED MANIS.

M. BRACHYURA

London Published by C B Whittaker. May 1824.

is long, round, and capable of extension, but not to the same degree as in the Ant-eaters ; the neck seems confounded with the head and body ; they have no external ears ; all the feet have five toes furnished with long and strong nails, and they have no cæcum.

The Pangolins (an African name having reference to the capability of the animal of rolling itself up), while they are assimilated to the Ant-eaters on the one hand, are not unlike the Armadillo on the other; their creation has the appearance of being borrowed from these two types.

The *short-tailed Manis (M. Pentadactyla,* Lin.) has the tail shorter than the body, and extremely thick at its base. The scales of the back form eleven longitudinal ranges ; the under part of the head and body, and the extremity of the paws, are naked. The length of body from nose to tail is about one foot ten inches, the tail measures about one foot five inches.

It was for a long time understood, that this species, as well as the following, was common both to Africa and India; but our author has ascertained that the habitat of the two is distinct; the present species belongs to India alone, and the other to Africa.

Cased in armour, the Pangolin has nothing to fear from the most ferocious of the mammiferous tribes ; it is said that the Tiger and larger Cats can do it no harm with all their hostile efforts ; their attempts to crush are equally vain with those of biting or tearing.

The Pangolin does not assume a sperical shape like the Hedgehog when rolled up. Its body is round, but its thick and long tail (which is nearly equal in volume to the rest of the animal) winds around it.

The tail is not so long in the young as in the adult; the scales also of the former are smaller, thinner, and of a paler colour ; when adult the colour of the scales becomes of a deeper brown; in old age they become hard enough to turn a musket-ball.

This species is found in Bengal and in the Indian Islands.

The *Long-tailed Manis,* (*M. Tetradactyla,* Lin.,) or the African species, is smaller than the preceding, and much thinner. It is particularly characterized by the length of the tail, which is at least double that of the body ; the tail is also less bent upward, and is more flat horizontally ; the scales of the body, which are smaller than in the Indian species, are armed at their extremity with three very sharp points, which are not found in the other species. The belly and chest in this species are covered with rough brown hairs. The nail of the hinder thumb is very small.

The species is very gentle in disposition ; it lives on insects which it catches by the aid of its long tongue, in the manner of the Ant-eaters ; its legs being short, and the toes furnished with very long nails, the animal is necessarily a bad runner, and cannot escape the hunters except by hiding itself in the clefts of rocks, or in its burrows. In these the female brings forth, and nurses her young, secure from almost all hostile attacks.

The mammiferous character of this genus separates it from the Lizards, to which, in all the other particulars of its organization, it exhibits a decided approximation.

In the Museum at Philadelphia there is an edentatous animal which has been described with minuteness in the annals of the New York Lyceum of Natural History, by Dr. Harlan, whose account has been transcribed into the Zoological Journal The Doctor names his genus CHLAMY-PHORUS, and he distinguished his type by the specific addition *truncatus.*

The specific characters of this animal may be thus described. The body, on the upper part, is covered with a sort of leathern-like shell, and is abruptly truncated behind ; the shell consists of rhomboid scales, disposed in tranverse lines, soldered together ; the under part is covered with white silken hair. The upper part of the head is covered with a continuation of the dorsal scale. The toes are five on all the feet ; the anterior feet are furnished with excessively long laterally-compressed nails, with the external edges and the points sharp. The tail is rigid, bent under the abdomen. The total length is not much above five inches, the girth posterior to the shoulders four inches, and of the tail little more than an inch.

From this abbreviated statement, it may be collected, that the body altogether is of a cylindrical shape, and in the figure, the line from the tip of the nose to the rectangular angle at the tail, deviates but little from a straight line ; underneath also, with the exception of the insertion of the limbs, the line is almost parallel with that above, but inclines regularly from the throat to the tip of the nose, which terminates in a sharp point.

The teeth are incisors, $\frac{2}{0}$; canine-teeth, $\frac{0\,0}{0\,0}$; cheek-teeth, $\frac{8\,8}{8\,8}$; the crowns of the two first approach to a point, and thus much resemble canine-teeth ; the six remaining are all nearly flat on the crown, their structure is simple; a cylinder of enamel surrounds a central pillar of bone, there being no division into body and root; the lower half is hollow, the cavity representing an elongated cone. The cavity of the cranium is spacious ; two long processes project obliquely forward, upward, and outward, from the os frontis anterior to the cavity of the cranium, and directly above the malar bone, giving to the front of the skull an aspect totally unique ; these prominences are hollow, communicating with the frontal sinuses, and must contribute in a great measure to enlarge the organ of smell. The shell which covers the upper part of the body and neck is composed of about twenty-four transverse series of plates. The ear is patulous, situated behind the eye, which is small and black, and like the ear, nearly hidden by long silky hair. The extremity of the snout is furnished with an enlarged cartilage, as in the Hog. The anterior portion of the lower jaw is shaped like that of the Elephant, much elongated ; but the general form and proportions resemble very much the lower jaw of the Sheep.

The animal before us, says Dr. Harlan, unites in its external configuration, traits peculiar to the genera *Dasypus*, *Talpa*, and *Bradypus* ; yet a very superficial observation will unfold characters generically distinct from either. It will be observed, that though this singular being is clothed with a coat, or rather cloak of mail, in a slight degree resembling the Armadillo, yet it differs remarkably in its texture, form, situation, arrangement, and mode of attachment to the body. In the Armadillo the body is covered with a hard scaly shell, and consists, 1st, In a plate upon the forehead; 2d, A vast shield situated upon the shoulders, and formed of small rectangular compartments disposed in transverse bands ; 3d, In bands of similar plates, but moveable, and varying in number from three to twelve, more or less according to the species; 4th, In a shield upon the rump very similar to that on the shoulders ; 5th, In rings more or less numerous on the tail ; five toes behind ; before, sometimes five, at others four ; hairs sparse. The whole shell is covered by a thin transparent epidermis, which is joined to the skin of the body, which gives to the shell a shining aspect, as if it were varnished; the extremities are entirely covered with strong scales. The Armadillo burrows in the earth; is sufficiently quick in its motions; is capable of rolling its body into the form of a ball; and is omnivorous. The external ear is sometimes large, and always very apparent.

From this statement, it appears, that there exists only the most distant analogy in the external covering of the *Dasypus*, with that of the new genus before us; other analogies will be found in the comparison of the skulls.

The lower portions of our animal, as well as that beneath the scales, will bear a pretty close comparison with the same parts of the Mole. The hair is finer and longer than in the Mole, and at a distance resembles long staple cotton in appearance. The eye is small; the neck, breast, and shoulders are very powerful; the posterior extremities are short and weak; the anterior short and strong, and furnished with large claws, as in the Moles; but in the form of the head, the structure and form of the claws, and the external ear, which is apparent when the hair is separated, our animal is totally dissimilar to the Mole. The claws bear some analogy to the sloth, but are articulated to the last phalanx, as in the Mole. Thus far, like the Mole, this animal is eminently constructed for subterranean progression; and here in all probability any strict analogy with that animal ceases.

The palm of the anterior extremities is directed rather inwards, whereas in the Mole it is directed outwards, and the nails are destitute of the cutting edge so remarkable in the former.

The skull and teeth, which differ in shape and in the edentatous character from that of the Mole, are analogous to that of the Armadillo, particularly in the edentatous character and its consequences. The teeth, in structure, however, are more nearly allied to those of the sloth, as they consist of a simple cylinder of bone surrounded with enamel, except the crowns, which are destitute of enamel in the centre; the roots are hollow. In these particulars of the teeth, together with the short process descending from the zygomatic arch, as well as in the form of the fore claws, there is considerable similitude between this animal and the Sloths, but in all other points of organization they are widely separated.

As far as the nature of the subject will admit, continues

Dr. Harlan, in conclusion, I have gone through with the detail of the organization of this most singular quadruped. During the investigation I have had frequent occasion to admire those laws of co-existence which regulate the structure of organized beings; nature, true to herself, in this as in all other instances, has pursued an undeviating course. We have been presented in the subject before us with a new form : an animal combining in its external configuration a mechanical arrangement of parts which characterizes respectively the Armadillo, the Sloth, and the Mole, constituting in themselves individually and separately, of all other quadrupeds, those which offer the most remarkable anatomical characters. Pursuing the investigation step by step with the skeletons of the above-mentioned animals before me, it was not till I had completely finished every point of observation, that I perceived in the skull alone of the new animal, a reunion more or less complete of all those remarkable traits that an external view of the animal had offered for contemplation, which, taken collectively, furnishes us with an example of organic structure, if not unparalleled, at least, not surpassed in the history of animals.

The most peculiar and unique characters consist—1st, In the general contour of the animal ; 2nd, In the form, texture, and disposition of its scaly cloak, which would very much confine the power of flexion and extension of the body, and nearly altogether impede lateral motion ; the greatest freedom of motion would consist in the extension of the head on the body ; 3rd, In the position of the organs of generation ; 4th, In the form, structure, position, and use of the tail ; 5th, In the peculiar and complicated structure of the feet and claws ; 6th, In the structure of the organ of hearing ; 7th, In the bony protuberance of the os frontis ; 8th, In the disposition of the teeth ; 9th, In the form of the lower jaw, which separates the animal in this respect from the order *Edentata,* and approximates it to the *Ruminantia* and *Pachydermata.*

The third tribe of the Edentata of our author is distinguished by exclusive, or anomalous characters, still more strongly marked than the two former, though of a nature very distinct from them in kind. This tribe includes, at present, but two genera, the species of which are very different in external appearances, though they accord in all those peculiarities noticed by the Baron in the text, and which renders a repetition of them, as common to the tribe of Monotremes, unnecessary.

Dr. Shaw was the first naturalist who introduced these singular creatures to notice, and Sir Everard Home was the first physiologist and comparative anatomist who described their internal structure. The systematic zoologists were much puzzled in allotting them a place in their systems, more especially as the questions of their parturition and lactation were, in some degree, matter of doubt. M. Geoffroy proposed making a distinct order of them; Duméril placed them after the Edentata; Tiedman appended them as an addition to the class Mammalia; Lamark proposed a distinct class for them; Illiger associated them with one of the Tortoises (*Testudo Squammata Bontii*), and named the group Reptantia; Blainville treated them as anomalies among the Didephous animals, and would, with Lamark, separate them into a class, and our author, as we have seen, refers them as a third tribe, or group, to the order Edentata.

Dr. Shaw, in 1792, described in the Naturalist's Miscellany, the *Aculeated Ant-eater;* but as as he had not then examined the internal structure of the animal, his conclusions drawn from superficial appearances only, that the animal in question, while it had the spines of the Porcupine, and the mouth and tongue of the Ant-eater, exhibited a connecting link between these two genera, were, in fact, groundless, or, at least, had only a superficial application. The Doctor merely repeats his former account of the animal in the General Zoology, in 1800; but Sir Everard Home, in 1802, published the result of his inspection of it in the Philosophical Transactions. This gentleman then discovered the analogies between this species and the Ornithorynchus, and associated the two together as a single genus, which he considered intermediate between the classes Mammalia, Birds, and Reptiles. Having, therefore, removed the animal from among the Ant-eaters, where Dr. Shaw had placed it, he named the species *Ornithorynchus Hystrix*, or the Porcupine Ornithorynchus.

The French naturalists, however, were not satisfied with treating the animal in question as a congener with the Ornithorynchus, and the Baron in his *Tableau Elémentare des Animaux*, established a distinct genus for it, under the name ECHIDNA, having reference to its spiny covering.

This animal, says Dr. Shaw, so far as may be judged from the specimens hitherto imported, is about a foot in length; the whole upper parts of the body and tail are thickly coated with strong and very sharp spines, of a considerable length, and perfectly resembling those of a Porcupine, except that they are thicker in proportion to their length; and that instead of being encircled or annulated with several alternate rings of black and white, as in that animal, they are mostly of a yellowish white, with black tips, the colour running down to some little distance on the quill, and being separated from the white part by a circle of dull orange; others have but a very slight appearance of black toward the tips. The head, legs, and whole under part of the body, are of a deep brown or sable, and are thickly coated with strong close-set bristly hair. The tail is extremely short, slightly flattened at the tip, and coated at the upper part of the base with spines at least equal in length to those of the back, and pointing perpendicularly upwards. The snout is long, and tubular, and perfectly resembling, in structure, that of the M. Jubata, or Great Ant-eater, having only a very small rictus, or opening at the tip, from whence is protruded a long lumbriciform tongue, as in the Ant-eater. The nostrils are small, and seated at the extremity of the snout. The eyes are very small and black, with a pale blue iris. The legs are very short, and thick, and are each furnished with five-rounded broad toes ; on the fore-feet are five very long and blunt claws, of a black colour, the two exterior being the shortest; the second toe on the inner side of the hind-feet has much the largest and strongest claw ; the third, fourth, and fifth diminish successively, and the first is the smallest.

The Echidna has been found principally in Van Dieman's Land, and some of the neighbouring islands; it lives on insects, which, like the Ant-eater, it secures by means of its long and sticky tongue. It burrows in the earth, and appears, like the Hedgehog, to have the faculty of assuming a spherical shape, and thus opposing its spines to any hostile attack. We are, however, as yet but little informed on the subject of the habits of the animal, its gestation, mode of parturition, number of young, &c.

The specific characters of the second species, or bristly Echidna, being all that is known of the animal, will be found stated briefly in the Table.

The form of the paws calculated for digging in the Echidna, and for swimming in the Ornithorynchus ; the spiny covering of the former, which is not found in the latter ; the duck-bill shaped muzzle of the one, and trumpet-formed tube of the other, and the difference in the size and shape of the tail, are the principal characters which

have induced the generic separation of the ORNITHORYN-
CHUS from the Echidna; which, as already stated, are
united by Sir Everard Home, their original commentator.

Dr. Shaw, the first describer of this animal, named it
the Duck-billed Platypus, but Sir Joseph Banks having
shortly after sent a specimen to Blumenbach, that eminent
physiologist, preferred the name Ornithorynchus for the
newly-discovered creature; the merited celebrity of the German writer prevailed, and the genus has retained the name
of his choosing almost universally.

Of all the Mammalia yet known, says Dr. Shaw, speaking purely of its superficies, this seems the most extraordinary in its conformation, exhibiting the perfect resemblance
of the beak of a duck engrafted on the head of a quadruped. So accurate is the similitude that, at first view, it
naturally excites the idea of some deceptive preparation by
artificial means, the very epidermis proportion, sereatous
manner of opening, and other particulars of the beak of a
Shoveler, or other broad-billed species of Duck, presenting
themselves to the view; nor is it without the most minute
and rigid examination that we can persuade ourselves of
its being the real beak or snout of a quadruped.

The body is depressed, and has some resemblance to that
of an Otter in miniature. It is covered with a very thick,
soft, and beaver-like fur, and is of a moderately dark
brown above, and of a subferruginous white beneath; the
head is flattish, and rather small than large; the mouth, or
snout, as before observed, so exactly resembles that of some
broad-billed species of Duck, that it might be mistaken
for such; round the base is a flat, circular membrane,
somewhat deeper or wider below than above, viz., below,
near the fifth of an inch, and above about an eighth.
The tail is flat, furry like the body, rather short and obtuse, with an almost biped termination; it is broader at

RED PLATYPUS.

ORNITHORHYNCHUS RUFUS.

London. Published by G.B. Whittaker June 1825

the base, and gradually lessens to the tip, and is about three inches in length; its colour is similar to that of the body.

The length of the whole animal, from the tip of the beak to that of the tail, is thirteen inches ; of the beak, an inch and an half. The legs are very short, terminating in a broad web, which on the fore feet extends to a considerable distance beyond the claws, but on the hind feet reaches no farther than the roots of the claws. On the fore feet are five claws, straight, strong, and sharp-pointed, the two exterior ones are somewhat shorter than the three middle ones. On the hind feet are six claws, longer and more inclining to a curve than those on the fore feet ; the exterior toe and claw are considerably shorter than the four middle ones ; the interior, or sixth, is seated much higher up than the rest, and resembles a strong sharp spur. The fore feet are naked both above and below, but the hind feet are hairy above, and naked below. The nostrils are small and round, and situated about a quarter of an inch from the tip of the bill, and are about the eighth of an inch distant from each other. The ears, or auditory foramina, are placed about an inch beyond the eyes, they appear like a pair of oval holes of the eighth of an inch in diameter, there being no external ear. On the upper part of the head, on each side, a little beyond the beak, are situated two smallish oval white spots, in the lower part of each of which are imbedded the eyes, or at least the parts allotted to the animal for some kind of vision, for from the thickness of the fur, and the smallness of the organs, they seem to have been but obscurely calculated for distinct vision, and are probably like those of moles, and some other animals of that tribe, or, perhaps, even sub-cutaneous, the whole apparent diameter of the cavity in which they are placed not exceeding the tenth of an inch.

In the place of teeth, the cartilaginous edges of the beak,

toward the lower part, are furnished with vertical fibres, flat at top, not planted in alveoli, but simply attached to the gum; the tongue is short, and furnished with papillæ, and two aculeated horny points. The organs of generation are not outwardly apparent, and in common with the rest of the tribe, the animal has but one opening. There are no visible mammæ; but whether the animal be absolutely without them, or whether they have hitherto escaped detection, or are only developed when they are wanted, is not quite ascertained. If they are altogether wanting, we can only suppose that the young are contained in the uterus long enough to exist without lactation when they are born.

Some details of the internal organization may be desired of an animal so new and curious. The cavity of the cranium is spacious; the maxillary bone singularly elongated before, and is furnished with a long spatulous apophyse for supporting the horny beak; there are altogether forty-nine vertebræ—seven cervical, seventeen dorsal, two lumbar, two sacral, and twenty-one caudal, the eight first of which are especially provided with very elongated transverse apophyses; the first piece of the sternum has also on each side a sort of transverse apophyse. There are seventeen ribs, eleven of which are false; the clavicles are very thin; the omoplate is shaped like a hedging-bill, and has a thick part appended, which joins the sternum; the basin is provided, as in all the marsupiata, with two long triangular bones articulated at their base to the pubis, stretching forward and diverging; there is an epiglottis to the larynx; the heart, like that of the Mammalia, has two auricles and two ventricles; the lungs are large, elongated, and free, with three lobes on the right side, according to Sir Everard Home, and four according to Cuvier; the diaphragm is large; the stomach is very small; the cæcum is small; the spleen is larger than the stomach, rectangular, and formed of two elongated lobes; the liver is large, composed of four lobes and a lobule; the gall vesicle is large and elongated; the hepatic vessels are very short; the reins globular; the bladder is very large, thin, and pyriformed; the testicles are enclosed in the belly near the reins; the urethra does not take the ordinary direction, but terminates in the anus, as in the birds; the glans penis is double, and terminated by some hollow spines, through which, it appears, the semen passes; the orifice of the vagina is placed also in the anus; at the bottom of the vagina is the opening of the urethra; and in the same situation are found the two trumpet-formed openings, which may be considered as separate matrices.

The Ornithorynchi have hitherto been found only in the rivers in the vicinity of Port Jackson, especially the river Nepean, on the eastern coast of New Holland. Those found in 1815, in Campbell River, and the River Macquarie, beyond the Blue Mountains, are larger than those before known: though they do not appear to differ specifically.

These animals are expert swimmers, and seldom quit the water; on shore they crawl rather than walk, occasioned by the shortness of the limbs and comparative length of the body. Nothing certain is known as to their food, but the singular resemblance of their beak to that of ducks, induces a strong probability that, like these birds, they live on worms and aquatic insects.

The above description applies to the Red or Common Ornithorynchus. Peron and Leseur have pointed out and figured a brown species which differs from the other only in having the fur of a blackish-brown flatted and frizzled, instead of inclining to a red tint, and being quite smooth. The existence, however, of this, as a distinct species, is doubted, but there is a specimen in the British Museum which seems referable to it, which, in addition to the particularities noticed, has the tail broader at the extremity, and the beak more square than in the other; the spur of the male also is straight and obtuse, while in the red species it is longer and more bent. This spur, according to the observation of M. Blainville on the red species, exists in the male only; it is situated on the inner side of the metatarsus; it does not belong to a sixth toe as has been imagined, but is simply attached to the skin; it has a groove toward the point, and within it, at its base, is a vesicle filled with a particular liquor, which introduced into the wound made by the nail, in the opinion of Blainville, is destined to poison the part.

As every thing human must partake of the imperfections of humanity, nothing is more easy than to cavil and find fault; and while we admit the difficulty of placing the Monotremes in some more distinct and insulated situation among the creatures of this earth, without many and weighty objections, we cannot but confess a dissatisfaction at finding them where they are. Indeed, the whole of the New Holland Quadrupeds have, as we have already noticed, very powerful claims to a distinct allocation in artificial systems; the difficulty is, to know where to stop with due regard to the object and utility of zoology as a science. The zoological novelties of America puzzled their first describers, but they almost all soon found their analogies in the creations of the Old World; we have little analogy to assist us in referring the Australasian animals to genera and divisions previously known.

The continent of New Holland, for it can hardly be considered as an island, is even less known to us than the long forbidden regions of central Africa; the novelties in nature presented by the shores only of that immense tract of land afford an earnest of what may be expected from a knowledge of its interior,—the Marsupiata may be, in all probability will be, increased both in genera and species, and the Monotremes will, by parity of presumption, be found to include more forms than those of the two genera at present known. Should this become the case to any extent, a reformation of these animals will probably become matter of propriety in every point of view, and the novice who will venture into the study of organized, or even of animated creation, must submit to charge his memory with a still greater extension of terms of art, as well as of objects of nature.

If the Monotremes be considered as analogous to, or as connected with, the Marsupial animals, their dislocation seems unnatural, and it is to these, as Mammalia, that they most approximate; and if the varied dentition of the several genera of Marsupial animals be not considered, a sufficient objection to their association with the order in which we find them, the edentatous peculiarities of the Monotremes seems to form no bar against placing these animals with them.

In one point of view, however, the Monotremes are conveniently placed as the third tribe of Edentata; if we refer to the other two, we cannot but observe a decided leaning, or inclination, in the species, to deviate from the class to which they belong, to vertebrated animals lower in the general scale than the Mammalia—that is, to the reptiles; the helpless imbecility, mere animal vegetation, the mental as well as bodily apathy of the Sloth, bespeak, as it were, the gradual departure of that mammiferous genus, by a gentle inclination, to an inferior type; in short, the Sloth has many analogies with the reptiles; in the Armadillos, and still more in the Pangolin, the analogy is so far increased, that the latter in particular.assume the very dress and covering of the reptiles. Hence it seems natural to place by the side of these, animals which evince so decided a departure from their class, as the group of Monotremes which are equally if not more remarkable in this particular. The exterior similarity of the Echidna to the Porcupine in person, and the Ant-eater in habits, is by no means so really important as the physical anomalies of that genus, in common with Ornithorynchus, and their striking analogy both with birds and reptiles. All the component genera of the order Edentata may be considered as mammiferous animals exhibiting more or less analogies with the oviparous classes.

THE

SIXTH ORDER

OF THE

MAMMALIA.

THE PACHYDERMATA.

THE Edentata terminate the series of unguiculated animals, and we have seen that there are some of these which have the nails so large, and so far covering the extremity of the toes, as to approach to a certain degree the hoofed animals. They have, nevertheless, the faculty of bending the toes round an object, and of seizing with more or less effect. The entire absence of this faculty characterizes the hoofed animals; using their feet merely as supporters, they never have clavicles; their fore-arm remains always in a state of pronation, and they are reduced necessarily to feed on vegetables; their forms and modes of living present fewer varieties than those of the unguiculated Mammalia; only two orders of them can be established, those which ruminate and those which do not; but the latter, which we designate under the common name of PACHYDERMATA, admit of some divisions into families.

The first will be that of the Pachydermata, with a proboscis, and with tusks, or the PROBOSCI-DIANA*; which have five toes on all the feet; very complete in the skeleton, but so encrusted in the callous skin which surrounds the foot, that they do not appear externally, except by the nails attached to the edge of this sort of hoof. They have no canines, or incisors, properly speaking, but two tusks are implanted in the incisive bone, which spring from the mouth, and frequently attain to an enormous size. The necessary magnitude of the alveoli of these tusks renders the upper jaw so high, and so far bends back the bone of the nose, that the nostrils are found in the skeleton toward the top of the face; but in the living animal they are elongated into a cylindrical proboscis, composed of many thousand small muscles, variously interlaced together, moveable in every direction, and endowed with exquisite sensibility, and terminated with an appendix formed like a finger. This proboscis gives the Elephant nearly as much address as the per-fection of its hand can give to the Monkey. The Elephant employs it to seize every thing he wishes to carry to the mouth, and also to suck up his water, which, by bending the trunk round, he then conveys into the mouth; it supplies the place also of a long neck, which could not have carried his thick head and weighty tusks. Within the head are

---

* The proboscidiana have many analogies with the Rodentia; first, their large incisors; second, their cheek-teeth, often formed of parallel laminæ; third, the form of many of their bones, &c.

large vacancies, which render it lighter; the lower jaw has no incisors; the intestines are very voluminous; the stomach is simple; the cæcum enormous; the mammæ, two only in number, placed on the breast. The young suck with the mouth, and not with the trunk.

But one genus of the Proboscidiana is known in living nature, which is that of

The Elephant (Elephas, L.),

which comprehends the largest of terrestrial animals. The astonishing services they derive from their trunk, at once the active powerful organ both of touch and smell, contrasts with their clumsy aspect and heavy proportions; these, accompanied with rather an imposing physiognomy, have contributed to an exaggeration of the intelligence of these animals. After studying them a long time, we have not been able to observe that their intelligence equals that of the Dog, or of many other carnivorous animals. Naturally gentle, the Elephants live in troops, under the guidance of the old males; they feed exclusively on vegetables.

Their distinctive character consists in the cheek-teeth; the body of which is composed of a number of vertical laminæ, formed each of a bony substance, enveloped in enamel, and tied together by a third substance, called cortical, similar, in a word, to that we have noticed in the Cabiaias, and in many other of the Rodentia. These cheek-teeth succeed each

other, not vertically, in the manner that the adult-
teeth succeed the milk-teeth, but from behind toward the front, so that as fast as one tooth becomes
worn, it is at the same time pushed forward by its
successor from behind; hence the Elephant has
sometimes one and sometimes two cheek-teeth on
each side, four, or eight altogether, according to
circumstances. The first of these teeth have but
few laminæ, and their successors alaways have
more. It is said that some Elephants will in this
manner change their teeth as many as eight times.
They change their tusks only once.

The existing Elephants, covered with a rough
skin, nearly destitute of hair, inhabit only the
torrid zone of the old continent: only two species
are hitherto known.

The Indian Elephant (Elephas Indicus, Cuv.) Buff. xi. 1.
and Supp. iii. 54.

Has an oblong head; concave forehead, with the
tops of the cheek-teeth presenting undulating
transverse ridges, which are the separations of
the laminæ which compose them, worn by trituration. This species has the ears smaller than
the other, and has four nails on the hind feet.
It inhabits from the Indies to the eastern sea,
and in the large islands to the south of India.
They have been used from time immemorial as
beasts of draught and of burthen; but they have
never hitherto been propagated in a domestic
state, although the assertion of their pretended

modesty and repugnance to· copulate in the presence of persons is without foundation. The females have only short tusks, and many of the males are like them in this respect.

The African Elephant (E. Africanus, Cuv.)   Perrault,
Mém. pour l'Hist. des Ann.

With a round head; convex forehead; large ears, and cheek-teeth, with lozenge-shaped sides on their crowns. It seems to have only three nails to the hind-feet. This species inhabits from Senegal to the Cape. It is not known whether they are to be found on the eastern side of Africa, or whether they are displaced in these parts by the other species. The females have tusks as large as the male, and these are generally larger than in the Indian species. The African Elephant is not hitherto tamed; but it appears that the Carthaginians employed them for the same purposes as the Indians do theirs.

In almost all parts of the two continents are found under ground the bones of a species of Elephant allied to that of India; but having the ridges of the cheek-teeth narrower and straighter, and the alveoli of the tusks much larger in proportion, and the lower jaw more obtuse. A specimen, recently drawn from the ice on the coast of Siberia by Mr. Adams, appears to have been covered with thick fur of two kinds; so that it is possible this species may have lived in cold climates. It has long disappeared. (See Cuvier's Researches on Fossil Bones, Vol. III. ɪɪ.) The second genus of the Proboscidiana, or

THE MASTODONS (MASTODON, Cuv.)

Has been entirely destroyed, and has left no species living. It had the feet, tusks, trunk, and many other details of conformation common with the Elephant; but it differed in the cheek-teeth, whose top was furnished from the issue from the gum with thick conical points. These, in proportion to their detrition, presented disks of a size proportioned to the wearing of the points *. These teeth, which succeeded each other from rear to front, like those of the Elephant, were furnished with pairs of points, numerous in proportion to the age of the animal.

The Great Mastodon (Mastodon Giganteum, Cuv.)

In which the section of the points were lozenge-shaped, is the best known species. It equalled the Elephant, but its proportions were more heavy. Its remains are found, wonderfully well preserved, in vast abundance, throughout all the whole of South America. They are infinitely more rare in the old continent.

---

* This confirmation, common to the Mastodons, the Hippopotamus, the Swine, &c., has induced a mistaken notion that the first of these were carnivorous.

The Mastodon with narrow teeth (Mastodon Angustidens, Cuv. loc. cit.)

The cheek-teeth of this species, narrower than those of the last, have the points of these teeth, when worn down, forming a disk, in the shape of a trefoil, which has caused them to be confounded by some authors with the teeth of the Hippopotamus. It was one third less than the great Mastodon, and much lower on its legs. Its remains are found throughout nearly all Europe, and in the greater part of South America. In some places, its teeth tinted by the iron, become in heating them of a fine blue, and give what are called Eastern turquoise *.

Our second family will be that of the ordinary Pachydermata, which have four, three, or two toes to their feet.

Those which have a single pair of toes have the foot in some sort cleft, and approximate the Rodentia in many particulars, in the skeleton and in the complication of the stomach. But two genera are commonly made.

The Hippopotamus, (Hippopotamus, L.)

Which have on all the feet four toes, nearly equal, terminated by little hoofs: six cheek-teeth in each jaw on each side, the three first of which are conical, and the three posterior are furnished with

* Some species have been found less extended. See Cuv. loc. cit.

two pairs of points, which assume by detrition a trefoil shape; four incisives in each jaw; in the upper jaw, short, conical, and bent; in the lower, long, cylindrical, pointed, and bent forward ; one canine tooth in each jaw on both sides, the upper one straight, the lower thick, bent, the two rubbing against each other.

These animals have a very massive body, without fur; the legs very short ; the belly nearly touching the ground; the head enormous, terminated by a large thick muzzle, which encloses the accommodation for their thick anterior teeth ; the tail short ; the eyes and the ears small. The stomach is divided into many sacs. They live in rivers on roots and other vegetable substances, and display a great deal of ferocity and also of stupidity.

Only one species is known, found only in the rivers of the South of Africa. It came formerly by the Nile to the south of Egypt; but it has long disappeared from that country *.

THE SWINE (SUS, L.)

Which have on all their feet two middle toes, large formed, with strong hoofs, and two exterior toes much shorter, which scarcely touch the ground in walking. The incisors vary in number, but the lower

* The fossil bones of the Hippopotamus are very common in Tuscany, and it is not yet ascertained whether they come from the existing species or of some lost species ; but some bones of a small species of Hippopotamus, now lost, have been found in France. See Cuv. loc. cit.

ones are always bedded in front; the canines spring from the mouth and bend upwards; the muzzle is terminated by a truncated sort of button, fitted for grubbing the earth; the stomach is little divided.

The Swines, properly speaking, have twenty-four or twenty-six cheek-teeth, the posterior of which have tuberculous crowns, the anterior are more or less compressed, and six incisors in each jaw.

The Sanglier, or Wild Hog (Sus Scropha,) Buff. V. xiv. *and* xvii.

Which is the root of our domestic Hog, and of its varieties, has the tusks in a prismatic form, bent outward, and a little upward; the body is thick, the ears erect, the fur erect and black; the young are striped black and white; it does great damage to the cultivated fields in the vicinity of forests, in burrowing in search of roots.

The Domestic Hog varies in size, in the height of its legs, in the direction of its ears, and in colour, sometimes white, at others black, red, or varied. Every one knows its utility by the facility with which it is fed—the flavour of its flesh—the facility of salting it; and, finally, by its fecundity, which surpasses by far that of other animals of the same size, sometimes fourteen being brought at one litter; the female is gravid four months, and produces twice a year; the Hog will continue to grow five or six years ; is prolific at the age of one year, and will sometimes live for twenty. Although these animals are very brutal, they are nevertheless social in disposition, can defend themselves against the wolf by forming a circle, and presenting their snouts to the enemy in every direction. Voracious and noisy, they do not even spare their own young. It is spread throughout the world, and none but the Jews and Mahometans refuse to eat its flesh.

The Masked Sanglier (S. Larvatus, F. Cuv.) Sus Africanus, Schreb. cccxxvii. *The Sanglier of Madagascar, Daub.* mdccclxxxv. *S. Daniels, African Scenery, pl.* xxi.

Has the tusks of the Common Hog, but on each side of its muzzle, near the tusks, is a large tubercle, nearly similar to the mammæ of a woman, supported by a long prominence, which gives the animal a very singular appearance. It inhabits Madagascar and the South of Africa.

The Babiroussa, or Stag Hog (S. Babirussa,) Buff. Supp. III. xii.

Higher and lighter on the legs than the others, with the tusks long and thin, inclining vertically, those in the upper jaw bending behind spirally. It inhabits some of the islands of the Indian Archipelago.

We may separate from the Swine

PHACO-CHŒRES, (*F. Cuv* *.)

Which have the cheek-teeth composed of cylinders joined together by a cuticle nearly similar to the transverse laminæ of the teeth of the Elephant, and succeeding each other from rear to front. The cranium is remarkably large; the tusks rounded, directed laterally and upwards, of a frightful magnitude, and on each of the cheeks hangs a thick fleshy lobe, which assists in deforming their figure: they have only two incisors above, and six below.

The specimens brought from Cape Verd (*S. Africanus*, Gm.) have the incisors in general very complete; those which come from the Cape of Good Hope (*S. Æthiopicus*, Gm.) scarcely ever show them, only there are sometimes found vestiges of them under the gum; this difference may result from age, these teeth being used in the latter species; probably also they indicate diversity of species, more especially as the head of that from the Cape is rather larger and shorter than in the other.

We should still less leave in the genus of the Swine,

The PECARY, (DICOTYLES, *Cuv.* †)

Which have very nearly the same cheek-teeth and incisors as the Swine properly speaking, but their canines directed like those of ordinary animals, do

* *Phaco chærus*, swine with warts.

† *Dicotyle*, Double Naval, on account of the opening of its back.

not pass out of the mouth, and they have no external toe to the hind-feet. They have no tail, and upon the lumbus is a glandular opening, from which issues a fetid excretion. The metatarsian and metacarpian bones of their two great toes are united together, like those of the ruminantia, with which their stomach, divided into many pouches, gives them also a direct relation. One singularity is, that the aorta is often found very enlarged, but without having the enlargement fixed, as if they were subject to a sort of aneurism.

Only two species are known, both of South America, which have only been distinguished by M. D'Azara. Linnæus confounds them under the name of *Sus Tajassa*.

The Pecary, with collar, or *Patira*, (*Dic. torquatus, Cuv.*)

With fur annulated gray and brown, with a whitish collar, passing obliquely from the angle of the lower jaw on the shoulder. One half less than our Wild Boar.

The Tagnicati, *Taitetou*, *Tajassou*, &c. (*Dic. labiatus, Cuv.*)

Larger; brown; with white lips.

Here may be placed a genus now unknown in living nature, which we have discovered and named.

ANOPLOTHERIUM, (*Cuv.*)

This exhibits the most singular relationship between the several tribes of the Pachydermata, and attaches itself in some respects to the order of Ruminantia. Six incisors in each jaw, four canines nearly similar to the incisors, and not surpassing them, and twenty-eight cheek-teeth, forming a continued series, without a void interval, which is only to be observed in man. The sixteen posterior cheek-teeth are similar to those of the Rhinoceros, the Damans, and the Palæotheriums, that is to say, square at top, and with two or three crescents underneath. The feet terminated by two large toes, as in the ruminantia, have this difference, that the metatarsian and metacarpian bones are always separated. The arrangement of the Tarsi is the same as in the Camel.

The bones of this genus have been found hitherto only in the plaster quarries of the environs of Paris. We have there recognised five species of them, one about the size of a small ass, with the low form and long tail of the Otter. (*A. commune*, Cuv.) The fore-feet have on the internal edge a small accessory toe; one of the size and lightness of the Gazel (*A. medium*); one of the size and nearly of the proportions of the Hare, with two small accessory toes on the side of the hind-feet, *&c.* (See Cuv. Researches in Fossil Osteology, tom. III.)

The Common Pachydermata, which have not the feet cleft, comprehend moreover three genera very similar to each other in the cheek-teeth, and having seven on each side of the upper jaw, with square

2 A 2

crowns, with several salient lines, and seven in the lower jaw, with the crown with a double crescent, the last of all with a treble crescent; but their incisors differ.

The RHINOCEROS, (RHINOCEROS, *L.*)

Vary even among themselves in this particular. These are large animals, with each foot divided into three toes, and with the bones of the nose very thick, and united into a sort of vault, carry a solid horn, as if it were composed of agglutinated hairs. Their disposition is stupid and ferocious; they like humid places; live on herbs and the branches of trees; have a simple stomach, long intestines, and a very large cæcum.

The Rhinoceros of India, (Rh. Indicus. Cuv.) Buff. XI. VII.

Has, besides its twenty-eight cheek-teeth, two strong incisive teeth in each jaw, and two still smaller on the outside the upper. It has but one horn, and its skin is remarkable by the deep folds it forms behind and across the shoulders, in front and across the thigh. It inhabits the East Indies, especially beyond the Ganges.

The Rhinoceros of Sumatra, (Rh. Sumatrensis, Cuv.) Bell. Philosophical Trans. 1793.

With the same four large incisors as the last,

has scarcely any folds on the skin, and carries a second horn behind the common horn.

The Rhinoceros of Africa, (*Rh. Africanus, Cuv.*) *Buff. Supp. VI.* vi.

Carries two horns, like the last, and has no folds of the skin, nor any incisive teeth, the cheek-teeth occupying nearly the whole length of the jaw.

There have been found under ground in Siberia, and in different parts of Germany, the bones of a two-horned Rhinoceros, whose cranium, much larger than that of the existing species, is distinguished also by a vertical bony partition, which supported the bones of the nose. It is a lost species, and a carcass nearly entire, which has been drawn from the ice on the borders of Vilhoui, in Siberia, has shown that it was covered with a thick fur. It could therefore live in the north, like the fossil Elephant

Other bones of a Rhinoceros, which appeared to be much more nearly related to that of Africa, have been recently disinterred in Tuscany and Lombardy. (See Cuv. Researches in Fossil Osteology, vol. II. and vol. I. articl Corrections and Additions.)

The DAMANS, (HYRAX, *Herman.*)

Have been for a long time placed among the Rodentia, on account of their very small size; but by a close examination of them, it is found that even to the horn, these are in some sort Rhinoceroses in miniature, at least they have exactly the same molars; but their upper jaw has two strong bent incisors, and during nonage two very small canines; the lower has four incisors without canines. There are four toes to the fore-feet, and three to those behind, all with a sort of very small hoof, thin and round, except the internal toe behind, which is armed with a bent oblique nail. These animals have the muzzle, and the ears short, are covered with fur, and have only a tubercle in the place of a tail. Their stomach is divided into two pouches; besides a large cæcum and many dilatations to the colon, toward the middle of which there are two appendices, analogous to the two cæca of birds.

One species is known of the size of a Rabbit, of grayish colour, pretty common in the rocks of all Africa, where it frequently becomes a prey to rapacious birds, and which seems also to inhabit some parts of Asia. At least we do not find any certain difference between the *Hyrax Capensis* and the *Syriacus*. (Buff. Sup. II. XLII et XLIII, et VII. LXXIX*.

The PALÆOTHERIUM, *(Cuv.)*

Is also a lost genus, with the same cheek-teeth as

---

* I doubt very much the authenticity of the *Hyrax Hudsonius,* Shreb. CCXL. c. It has only been seen in a museum.

the two last, six incisors, and two canines in each jaw, like the Tapir, and three visible toes on each foot ; they had also, like the Tapir, a short fleshy trunk, for the muscles of which the bones of the nose were shortened, and left under them a strong slope. We have discovered the bones of this genus mixed up with those of the Anoplotherium, in the plaster quarries of the environs of Paris, and they exist in many other parts of France.

Eleven or twelve species are already known. At Paris alone we find five, one of which is the size of a Horse, two that of the Tapir, two that of a small sheep. Near Orleans are found the bones of a species, which nearly equalled the Rhinoceros. These animals appear to have frequented the vicinity of lakes and marshes ; inasmuch as the stones which conceal their bones contain also fresh-water shells. (See Cuv. Researches in Fossil Osteology, III.)

To these three genera should succeed that of

The TAPIRS (TAPIR, *L.*)

In which the twenty-seven cheek-teeth present before the effect of trituration two transverse rectilinear hills ; in front there are in each jaw six incisors and two canines, separated from the cheek-teeth by a void space. The nose is in the form of a little fleshy trunk ; the anterior feet have four toes, the posterior three.

Only one species is known —

(Tapir Americanus, L.)

As big as an ass, with a brown skin nearly naked, scull moderate, neck fleshy, forming a sort of crest on the nape. It is common in the humid places and the sides of rivers of the warm parts of South America. Its flesh is eaten. The young are spotted white like fawns.

The third family of Pachydermata or hoofed non-ruminating animals, will comprehend

The SOLIPEDES,

Or quadrupeds which have only one apparent toe and a single hoof on each foot, although they have under the integument on each side of their meta-carpe and metatarse two protuberances, which represent as many lateral toes.

Only one species is known, which is that of

The HORSE (EQUUS, *L.*)

With six incisors in each jaw, which during non-age have a sort of dimple on the crowns, and six cheek-teeth in each side in each jaw, marked with laminæ of enamel which penetrate the tooth with four crescents, and in the upper with a small disk at the internal edge. The males have moreover two small canine teeth in the upper jaw, and sometimes also in the under, but these are almost always wanting in the females. Between the canines and the first cheek-tooth is a void space, which corresponds with the angle of the lips where the bit is placed, and by means of which alone man is enabled to overcome this vigorous quadruped. Their stomach is simple and moderate, but their intestines are very long, and the cæcum enormous. The mammæ are between the thighs.

The Horse, (Equus Caballus, Lin.) Buff. IV. 1.

This noble companion of man in the chase, in war, in the works of agriculture, the arts, and of commerce, is the most important and the best treated of all the animals which are submitted to us. It appears that the Horse exists in a wild state only in those places where horses, formerly domesticated, have been left in freedom, as in Tartary and America. In such places they live in troops, conducted and defended each by an old male. The young males, driven away as soon as they are adult, follow these troops at a distance until they are fit for the mares ; the domesticated foal sucks six or seven months ; the sexes are separated at two years. They are broken in at three years old, but they should not be ridden till they are four years old, when also they can propagate without injury. The Mare is gravid eleven months. The age of the Horse is known especially by the incisive teeth. The milk-teeth fall out about fifteen days after birth ; at two years and a half the middle teeth are replaced ; at three and a half the two following ; at four and a half the two outermost, called the corner teeth. All these teeth have the crowns at first indented, and lose their indentation through detrition by degrees. At seven years and a half or eight years, all these are effaced, and the horse is no longer marked.

The lower canine teeth come at three years and a half; the upper at four; they continue pointed till six; at ten they begin to peel off.

The life of the Horse seldom exceeds thirty years.

All the world knows to what an extent this animal varies in colour and size. Its principal races even have sensible differences in the form of the head ; the proportions and aptitude for different sorts of employment.

The lightest and most fleet are the Arabian Horses, which have assisted in improving the Spanish race, and contribute with that to form the English race. The heaviest and strongest come from the coasts of the North Sea. The smallest from the north of Sweden and Corsica. Wild Horses have the head large, the fur crisp, and inelegant proportions.

The Dziggetai, (Equus Hemionus, Pall.) Schreb.

Is a species which, as to its proportions, is

intermediate between the Horse and the Ass, (it is properly the wild Mule of the ancients,) and which lives in troops in the sandy deserts of the centre of Asia. It is Isabella, or light coloured bay, with the mane and dorsal line black ; its tail is terminated by a black tuft.

The Ass, (Equus Asinus, Lin.) Buff. IV. xi.

Is known by its long ears, and the tuft at the end of the tail ; by the black stripe it has on the shoulders, and which is the first indication of the bands which distinguish the two following species, originally of the great desert of central Asia. It is there still to be found in a wild state in countless troops, which migrate from north to south, according to the season ; hence they do not do well in countries too much to the north. Every one knows its patience, sobriety, robust constitution, and the services it renders to the poor in the country.

Its rough voice, or braying, results from two small peculiar cavities at the bottom of the larynx.

The Zebra, (Equus Zebra, Lin.) Buff. XII. i.

Nearly of the form of the Ass, with transverse stripes all over, white and black, with perfect regularity. It is originally from all the southern parts of Africa. We have seen a female

Zebra produce successively with an Ass and with an Horse.

The Couagga, (Equus Quaccha, Gm.) Buff. Sup. VII. vii.

Resembles the Horse rather than the Zebra, but comes from the same country. Its fur on the neck and shoulders is brown, striped across with whitish ; the crupper is reddish-gray, the tail is less whitish. Its name expresses its voice, which resembles the barking of a dog.

SUPPLEMENT TO THE ORDER
PACHYDERMATA.

THE notice of the Baron in the text, and the specific characters which distinguish the two species, render any
further or lengthened description of the ELEPHANT as
a genus unnecessary ; and as we propose selecting these
animals as a type from which to treat of the mental faculties and instincts of the mammiferous class, we shall appropriate a larger space to our supplementary observations on
them, in reference to these particulars, than our limits will
generally allow ; a space, however, very inadequate to the
importance, extent, and difficulty of the subject.

It is but lately that the differences between the Asiatic
and the African Elephants have been noticed by naturalists.
The Baron Cuvier was the first who compared them, and
determined their specific characters.

The Elephant is the largest of existing quadrupeds: its
proboscis is an organ of seizing and of touch, of feeling
and of respiration, but not of smell. With this it can hold
a pole or branch, and strike with tremendous violence, and
with this it conveys both food and water to the mouth. The
tusks are changed but once during the life of the animal,
but the molars change as often as detrition makes it necessary ; they do so, however, not in the ordinary manner,
by the new teeth pushing the old up, but by a lateral succession from rear to front. The nasal apertures are not
prolonged beyond the bones of the nose, and do not pass
through the proboscis, and the lower lip has a very little
motion. The shortness of its neck not permitting the animal
to lower the mouth to the ground to pasture, it collects the
grass and leaves of trees with its proboscis. When it eats,
the muscles of the cheeks seem, by a sort of spontaneous

action, to push the food between the teeth for trituration. Sight and hearing are acute, though the eyes arc small, compared with the enormous head ; smell however appears to be the most perfect of the senses of these animals : before death they usually discharge a considerable quantity of aqueous liquid through the proboscis.

The *Asiatic Elephant* is distinguished from its African congener, principally by the character of the teeth already noticed in the text; the head moreover is oblong, the forehead concave, and the ears do not descend lower than the neck. This species is found in the whole of Southern India, and in the neighbouring islands. Though so extensively employed by man, it can hardly be considered a domestic animal, as it is not bred in captivity; but when a fresh supply is wanted for general purposes, they are hunted or rather sought for in their sequestered retreat, and after being captured, are quickly reduced to servitude. Taking and taming wild elephants is an affair of great moment in India, a description of which, however amusing, we feel constrained to forego.

A strong Elephant can carry 2000 pounds weight, and can travel without difficulty fifty miles in a day; in long marches, however, they become very tender-footed, as may be seen by their gait, and by their feeling with the proboscis on the ground where they are about to tread for a footfall without stones or sharp rocks, otherwise they are very nimble for their bulk, walk up and down footways into ravines where camels cannot pass, and where horses find difficulty.

The period of gestation is twenty months; the new-born Elephant is about three feet long, and all its senses are perfect : it sucks with the mouth and not with the proboscis, turning the latter back in that operation. Lactation continues nearly two years, and between fifteen and twenty years old they may be said to be adult : though they have a great affection for their young, it is understood that these suck indifferently all the females in the herd to which they belong. They are gregarious, in herds of about 100, and inhabit the humid forests and vicinity of rivers, in which they swim with great ease, sometimes having no part above the surface of the water but the end of the proboscis through which they respire. When they quit the water they are fond of collecting the soil and dust with the proboscis, and covering their body with it.

Though gregarious in their habits, solitary wild Elephants are sometimes met with, but these are always observed to be males, and are in general extremely furious, attacking everything they meet, and doing the greatest damage. It seems probable that these have been driven by stronger rivals from the herd.

It is the opinion in India that they live three centuries, and several now in the service of the East India Company were old when they came into possession of the Europeans upwards of eighty years ago. These old animals, however, dislike to rise from the ground, and are at first unwilling to move forward, piping an angry note of dissent.

It appears probable, though it is not determined, that there may be more than one variety of the Asiatic Elephant distinguished by the size of the tusks : those of the females are in general less than those of the males. These tusks weigh sometimes as much as 150 pounds the pair.

The fine specimen of the Asiatic species in Mr. Cross's collection at Exeter 'Change, which he has had many years, measures between ten and eleven feet in height, and weighs at least, by computation, between four and five tons. This huge mountain of flesh consumes daily three trusses of hay, and about two hundred weight of carrots and other fresh vegetables, together with from sixty to eighty gallons of water; this animal is observed, when turning the proboscis up to the mouth in feeding, to incline it always on the left side, which may be a specific and not an individual character, though certainly unimportant *.

* Since the above very short allusion to Mr. Cross's elephant was written, the death of that magnificent animal has excited a great deal of public attention. The editor conceiving the circumstances, considered purely in a zoological point of view to demand a notice here, applied to Mr. Cross for authentic information, and full particulars of the affair, who, with his accustomed alacrity, was quite ready to give them, stating however, at the same time, that he had already furnished a minute account to Mr. Hone, which had been published in the *Every-Day Book*. Mr. Hone, in his introductory remarks, states that, except as to language, the narrative is in fact that of Mr. Cross, and although it may contain a few details rather more minutely than the editor would have stated them, it seemed to him quite unnecessary to repeat the same tale in other words, merely to give it an air of originality, and he has therefore subjoined Mr. Cross's account, as published by Mr. Hone.

" The first owner of the lordly animal, now no more, was Mr. Harris, proprietor of Covent-garden theatre. He purchased it in July, 1810, for nine hundred guineas on its arrival in England, aboard the Astel, Captain Hay, and the elephant ' came out' as a public performer the same year, in the procession of a grand pantomime, called ' Harlequin Padmanaba.' Mrs. Henry Johnstone was his graceful rider, and he was ' played up to' by the celebrated columbine, Mrs. Parker, whose husband had a joint interest with Mr. Harris in the new performer. During his ' engagement' at this theatre, Mr. Polito ' signed articles' with Messrs. Harris and Parker for his further ' appearance in public' at the Royal Menagerie, Exeter 'Change. On the death of Mr. Polito, in 1814, Mr. Cross, who for twenty years had been superintendent of the concern, became its purchaser, and the elephant, thus transferred, remained with Mr. Cross till the termination of his life. From his ' last farewell' to the public at Covent-garden theatre, he was stationary at the menagerie, from whence he was never removed, and, consequently, he was never exhibited at any other place.

" On the elephant's first arrival from India he had two keepers; these accompanied him to Exeter 'Change, and to their control he implicitly submitted, until the death of one of them, within the first year after Mr. Cross's proprietorship, when the animal's in-

Griffiths sc.

C. Hamilton Smith, Esq.r del.t
Berlin Library

THE ELEPHANT OF AFRICA. Adult Female
ELEPHAS AFRICANÙS.

The *African Elephant* is distinguished by a round or cylindrical head, with the face more protruded than in the

creasing bulk and strength rendered it necessary to enlarge his den, or rather to construct a new one. The bars of the old one were not thicker than a man's arm. With Mr. Harrison, the carpenter, who built his new den, and with whom he had formed a previous intimacy, he was remarkably docile, and accommodated himself to his wishes in every respect. He was occasionally troublesome to his builder from love of play, but the prick of a gimblet was an intimation he obeyed, till a desire for fresh frolic prompted him to further interference, and then a renewal of the hint, or some trifling eatable from the carpenter's pocket, abated the interruption. In this way they went on together till the work was completed, and while the elephant retained his senses, he was happy in every opportunity that afforded him the society of his friend Harrison. The den thus erected will be particularized presently: it was that wherein he remained till his death.

" About six years ago this elephant indicated an excitement which is natural to the species, and which prevails every year for a short season. At the period now spoken of, his keeper having gone into his den to exhibit him, the animal refused obedience; on striking him with a slight cane, as usual, the elephant violently threw him down: another keeper seeing the danger, tossed a pitchfork to his comrade, which the animal threw aside like a straw. A person then ran to alarm Mr. Cross, who hurried down stairs, and catching up a shovel, struck the animal violently on the head, and suddenly seizing the prostrated man, dragged him from the den, and saved his life.

"This was the first appearance of those annual paroxysms, wherein the elephant, whether wild or confined, becomes infuriated. At such a period it is customary in India to liberate the elephants and let them run to the forests, whence, on the conclusion of the fit, they usually return to their wonted subjection. Such an experiment being impossible with Mr. Cross, he resorted to pharmacy, and, in the course of fifty-two hours, succeeded in deceiving his patient into the taking of twenty-four pounds of salts, twenty-four pounds of treacle, six ounces of calomel, an ounce and a half of tartar emetic, and six drams of powder of gamboge. To this he added a bottle of croton oil, the most potent cathartic perhaps in existence: of this a full dram was administered, which alone is sufficient for at least sixty full doses to the human being; yet

Asiatic species, a convex forehead and enormous ears, which descend as far as the legs. They are indeed so large, that at

though united with the preceding enormous quantity of other medicine, it operated no apparent effect. At this juncture Mr. Nyleve, a native East Indian, and a man of talent, suggested to Mr. Cross the administration of animal oil, as a medicine of efficacy. Six pounds of marrow from beef bones were accordingly placed within his reach, as if it had been left by accident; the lickerish beast, who would probably have refused it had it been tendered him in his food, swallowed the bait. The result justified Mr. Nyleve's prediction. To my inquiry whether the marrow had not accelerated an operation which would have succeeded the previous administration, Mr. Cross answered, that he believed the beef marrow was the really active medicine, because, after an interval of three weeks, he gave the same quantity wholly unaccompanied, and the same aperient effect followed. He never, however, could repeat the experiment; for the elephant in successive years wholly refused the marrow, however attempted to be disguised, or with whatever it was mixed.

" In subsequent years, during these periods of excitement, the paroxysms successively increased in duration; but there was no increase of violence until the present year, when the symptoms became more alarming, and medicine produced no diminution of the animal's heightened rage. On Sunday, (the 26th of February,) a quarter of a pound of calomel was given to him in gruel. Three grains of this is a dose for a man: and though the entire quantity given to the elephant was more than equal to six hundred of those doses, it failed of producing in him any other effect than extreme suspicion of any food that was tendered to him, if it at all varied in appearance from what he was accustomed to at other times. On Monday morning some warm ale was offered him in a bucket, for the purpose of assisting the operation of the calomel, but he would not touch it till Cartmell, his keeper, drank a portion of the liquor himself, when he readily took it. The fluid did not appear to accelerate the wished-for object, and in fact, the calomel wholly failed to operate. Though in a state of constant irritation, he remained tolerably quiet throughout Monday and Tuesday, until Wednesday, the first of March, when additional medicine became necessary, and Mrs. Cross conceived the thought of giving it to him through some person whom the elephant had not seen; and whom therefore he might regard as a casual visitor and not sus-

The African Elephant.

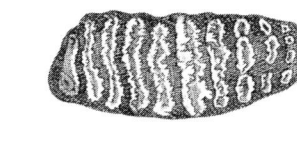

Cheek teeth of
1 The Asiatic Elephant.
2 The African Elephant.

London, Published by C.B. Whittaker, March 1.1827.

The Asiatic Elephant

the Cape they make sledges of them to draw the heavy tools
to and from the field, and even convey the sick. The pecu-

pect. To a certain extent the feint succeeded. She sent some
buns to him by a strange lad, in one of which a quantity of calomel
had been introduced. He ate each bun from the boy's hand till
that with the calomel was presented ; instead of conveying it to
his mouth, he instantly dropped the bun, and crushed it with his
foot. In this way he was accustomed to treat everything of food
that he disliked.

 " It was always considered that the elephant's den was of sufficient
strength and magnitude to accommodate, and be proof against any
attack he was able to direct against it, even in his most violent
displeasure. In the course of the four preceding years the front
had sustained many hundred of his powerful lunges, without any
part having been substantially injured, or the smallest portion displaced, or rendered rickety in the slightest degree; but on this
morning (Wednesday,) about ten o'clock, he made a tremendous
rush at the front, wholly unexcited by provocation, and broke the
tenon, or square end at the top of the hinge story-post, to which
the gates are hung, from its socket or mortise in the massive cross-
beam above; and, consequently, the strong iron-clamped gates,
which had hitherto resisted his many furious attacks upon them,
lost their security. Mr. Cross was then absent from the menagerie,
and, in the urgency of the moment, his friend Mr. Tyler, a gentleman of great coolness and faculty of arrangement, gave orders for
a strong massy piece of timber to be placed in front of his den, as
a temporary fixture against the broken story-post; and offered
everything he could think of to pamper, and, if possible, to allay
the animal's fury. On Mr. Cross's arrival he rightly judged, that
another such lunge would prostrate the gates ; and, as it was
known that Mr. Harrison, the carpenter of the den, who formerly
possessed great influence over him, had now lost all power of con-
trolling him, it was morally certain, that if any other persons attempted to repair the mischief in an effectual way, their lives would
be forfeited. Mr. Cross, under these circumstances of imminent
danger, instantly determined to destroy the elephant with all possible despatch, as the only measure he could possibly adopt for
his own safety and the safety of the public. Having formed his
resolution, he went without a moment's delay to Mr. Gifford, chemist in the Strand, and requested to be supplied with a potent
poison, destitute if possible of taste or smell. Mr. Gifford, sen-

liarity of the cheek-teeth before noticed also separates it, and there is reason to think that three toes only of the

sible of the serious consequences to Mr. Cross in a pecuniary point of view, entreated him to reflect still further, and not to commit an act of which he might hereafter repent. Mr. Cross assured him that whatever irritation he might manifest, proceeded from his own feelings of regard towards the elephant, heightened by a sense of the loss that would ensue upon his purpose being effected; adding, that he had a firm conviction that unless the animal's death was immediately accomplished, loss of human life must ensue. Mr. Gifford replied, that he had never seen or complied more reluctantly with his wish on any occasion, and he gave him four ounces of arsenic. Mr. Cross declares that on his way back the conflict of his feelings was so great at that moment, that he imagines no person contemplating murder could endure greater agony. The arsenic was mixed with oats, and a quantity of sugar being added by way of inducement, it was offered to the elephant as his ordinary meal by his keeper. The sagacious animal wholly refused to touch it.

" His eyes now glared like lenses of glass reflecting a red and burning light. In order to soothe him, some oranges, to which fruit he had great liking, were repeatedly proffered; but though these were in a pure state, he took them, one after the other, as they were presented to him, and dropping each on the floor of his den instantly squelched it with his foot, and having thus disposed of a few he refused to take another. This utter rejection of food, with amazing increase of fury, heightened Mr. Cross's alarm. He again went out, and in great agitation procured half an ounce of corrosive sublimate to be mixed in a quantity of conserve of roses, securely tied in a bladder, to prevent, if possible, any scent from the poison, and with some hope that if the animal detected any effluvia through the air-tight skin it would be the odour of roses and sugar, which were substances peculiarly grateful to him. The elephant was accustomed to swallow several things lying about within reach of his proboscis, which, if tendered to him, he would have refused; and this habit suggesting the possibility that he might so dispose of this, which, it was quite certain, if presented would have been rejected, the ball was placed so that he might find it; but the instant he perceived it he seemed to detect the purpose; he hastily seized it, and as hastily letting it fall, violently smashed it with his foot.

hind feet have nails. The tusks are said to be of equal size both in the male and female of this species, and the eyes

" The peril was becoming greater every minute. The elephant's weight was upwards of five tons, and from such an animal's excessive rage, in a place of insecure confinement, the most terrible consequences were to be feared. Mr. Cross therefore intrusted his friend, Mr. Tyler, to direct and assist the endeavours of the keepers for the control of the infuriated beast. He then despatched a messenger to his brother-in-law, Mr. Herring, in the New Road, Paddington, a man of determined resolution, and an excellent shot, stating the danger, and requesting him to come to the menagerie. As he arrived without arms, they went together to Mr. Stevens, gunsmith in High Holborn, for rifles. On their way to him they called at Surgeons-hall, Lincoln's Inn Fields, where they hoped to see the skeleton of an elephant, in order to form a judgment of the places through which the shots would be likeliest to reach the vital parts. In this they were disappointed, the college of surgeons not having the skeleton of the animal in its collection; but Mr. Clift, who politely received them, communicated what information he possessed on the subject. Mr. Stevens lent him three rifles, and at his house Mr. Cross left Mr. Herring to get the pieces ready, after instructing him to co-operate with Mr. Tyler, in attempting the destruction of the animal, if it should be absolutely necessary before he returned himself. From thence Mr. Cross hastened to Great Marlborough-street, for the advice of Mr. Joshua Brookes, the eminent anatomist. He found that gentleman in his theatre, delivering a public lecture. Sense of danger deprived Mr. Cross of the attentions due to time and place under ordinary circumstances, and he immediately addressed Mr. Brookes; ' Sir, a word with you, if you please, immediately : I have not an instant to lose.' Mr. Brookes concluded his lecture directly, and knowing Mr. Cross would not have intruded upon him except from extreme urgency, withdrew with him, and gave him such instructions as the case seemed to require. Mr. Cross, accompanied by one of Mr. Brookes's pupils, hastened homeward. They were met near the menagerie by Mr. Tyler, who entreated Mr. Cross to run to Somerset-house and obtain military assistance from that place, for that they had been compelled to use the rifles in their own defence, and had put a number of shot in him without being able to get him down. Mr. Brookes's pupil accompanied Mr. Tyler, to assist are situate lower, nearer the mouth, and more forward in the African Elephant than in its Asiatic congener.

him, if possible, while Mr. Cross rapidly proceeded to Somerset-house, where he found a sentry on duty, who did not dare to quit his post, and referred him to the guard-room, where there were only two other privates and a corporal, who, at first, declared his utter inability to lend him either men or arms ; but on the earnest entreaties of Mr. Cross for aid, and his repeated representations, that he would be responsible in purse and person, and compensate any consequences that could be incurred by a dereliction from the formalities of military duty on so pressing an occasion, the corporal relented, and with one of the privates hastened to the menagerie.

" Mr. Cross now met Herring, of the public office, Bow-street, to whom he communicated the situation of affairs at Exeter 'Change, and requested his assistance in obtaining arms.   Herring suggested an application to Bow-street for that purpose.   It appears that from accident they were not procurable there, and deeming it possible that they might be got at Sir W. Congreve's office, Mr. Cross ran thither, where he was also disappointed.   Mr. Brooks, glass-man of the Strand, informed Mr. Cross there were small arms in the neighbourhood of Somerset-house ; these, on returning to that place, were discovered to be old howitzers, and therefore useless. From thence he went on board the police-ship stationed on the Thames, near Waterloo-bridge, expecting to find swivels, and was again disappointed; being informed, however, that swivels were fired during civic processions from Hawes's soap-manufactory, on the Surrey side of the river, near Blackfriars-bridge, he rowed over and obtained a swivel, with a few balls, and the head of a poker, and the assistance of one of Mr. Hawes's men.   The use for either, however, ceased to exist; for they arrived at the menagerie within a few minutes after the conclusion of such a scene as had never been exhibited in that place, nor, probably, in any other in this country.   The elephant was dead.

To describe the proceedings of Exeter 'Change, from the time of Mr. Cross's leaving it, it is necessary to recur to the period of Mr. Herring's appearance thither, on his return from Mr. Stevens's, in Holborn, with the three rifles, and one of Mr. Stevens's assistants.   He found that the violence of the elephant had increased every minute from the period of his departure with Mr. Cross, and

Major Smith has a drawing of this species, copied by him from one in the Berlin library, which belonged to the col-

that at great personal hazard Mr. Tyler, with Cartmell and Newsam, and the other keepers, had prevented him from breaking down the front of the den.

" The keepers faced him with long pikes or spears, to deter him as much as possible from efforts to liberate himself from the confinement which at ordinary periods he had submitted to without restraint. When he lunged furiously at the bars, they assailed him with great bravery, and their threats and menaces prevented the frequency of his attacks. In this state of affairs Mr. Herring concurred with·Mr. Tyler, that to wait longer for Mr. Cross would endanger the existence of every person present ; and having communicated the fact to Mrs. Cross, who had the highest regard for the animal from his ordinary docility, she was convinced, by their representations, that his death must be accomplished immediately, and therefore assented to it.

" For the information of persons not acquainted with the menagerie, it is necessary to state that it occupies the entire range of the floor above Exeter 'Change, the lower part of which edifice withinside is occupied by shops belonging to Mr. Clarke. This part of the building, on the business of the day being concluded, is closed every night by the strong folding-gates at each end, which, when open, allow a free passage to the public through the 'Change. It will be perceived, therefore, that the flooring above is Mr. Cross's menagerie, or, at least, that very important part of it which is allotted to his matchless collection of quadrupeds. A large arrangement of other animals is in other apartments, on a higher story. Nero, not Wombwell's Nero, which was baited by that showman at Warwick, but a lion not only in every respect finer than his namesake, and, in short, the noblest of his noble species in England, occupies a den in the menagerie over the western door of the 'Change. Other lions and animals are properly secured in their places of exhibition on each side of the room, and the east end is wholly occupied by the den of the elephant; its floor being supported by a foundation of brick and timber, more than adequate to the amazing weight of the animal. The requisite strength and construction of this flooring necessarily raise it nearly two feet from the flooring of the other part of the menagerie, which, though amazingly stable and capable of bearing any other beast in perfect safety, would have immediately given way beneath the tread of lection of Prince Maurice of Nassau, and which, by his permission, we have engraved ; since which, M. Frederic

the elephant; and had he forced his den he must have fallen through.

" As soon, therefore, as his sudden death was resolved on, Mr. Tyler went down to Mr. Clarke, and acquainting him with the danger arising out of the immediate necessity, suggested the instant removal of every person from the 'Change below, and the closing of the 'Change gates. Mr. Clarke, and all belonging to his establishment, saw the propriety of their speedy departure, and in a few minutes the gates were barred and locked. By the adoption of these precautions, if the elephant had broken down the floor no lives would have been lost, although much valuable property would have been destroyed ; and, in the event contemplated, the animal himself would have been confined within the basement. Still, however, a slight exertion of his enormous strength could have forced the gates. If he had made his entry into the Strand, it is impossible to conjecture the mischief that might have ensued in that crowded thoroughfare, from his infuriated passion.

" On Mr. Tyler's return-up stairs from Mr. Clarke, it was evident, from the elephant's extreme rage, that not a moment was to be lost. Three rifles therefore were immediately loaded, and Mr. Herring, accompanied by Mr. Stevens's assistant, entered the menagerie, each with a rifle, and took their stations for the purpose of firing. Mr. Tyler pointed out to the keepers the window places, and such recesses as they might fly to if the elephant broke through, and enjoining each man to select a particular spot as his own exclusive retreat, concluded by showing the danger of any two of them running to the same place for shelter. The keepers, with their pikes, placed themselves in the rear of Mr. Herring and his assistant, who stood immediately opposite the den, at about the distance of twelve feet in the front. Mr. Herring requested Cartmell to call in his usual tone to the elephant when he exhibited him to visitors, on which occasions the animal was accustomed to face his friends with the hope of receiving something from their hands. Cartmell's cry of " Chunee ! Chunee ! Chuneelah !" in his exhibiting tone, produced a somewhat favourable posture for his enemies, and he instantly received two bullets aimed from the rifles towards the heart ; they entered immediately behind the shoulder-blade, at the distance of about three inches from each other. The moment the balls had perforated his

Cuvier has also done us the honour to transmit us a beautiful side-view drawing of this species, together with outlines

body, he made a fierce and heavy rush at the front, which further
weakened the gates, shivered the side bar next to the dislodged
story-post, and drove it out into the menagerie. The fury of the
animal's assault was terrific, the crash of the timbers, the hallooing
of the keepers in their retreat, the calls for " rifles ! rifles ;" and
the confusion and noise incident to the scene, rendered it indes-
cribably terrific. The assailants rallied in a few seconds, and came
pointing their spears with threats. Mr. Tyler having handed two
other rifles, they were discharged as before ; and, as before, produced a similar desperate lunge from the enraged beast at the front
of his den. Had it been effective, and he had descended on the
floor, his weight must have inevitably carried it, together with
himself, his assailants, and the greater part of the lions, and other
animals, into the 'Change below, and by possibility have buried
the entire menagerie in ruins. " Rifles ! rifles !" were again called
for, and from this awful crisis it was only in the power of Mr. Tyler
and some persons outside, to load quick enough for the discharge
of one rifle at a time. The maddened animal turned round in his
den incessantly, apparently with the design of keeping his head
from the riflemen, who after the first two discharges could only
obtain single shots at him. The shutter inside of a small grated
window, which stood in a projection into the den, at one of the
back corners, was now unshipped, and from this position Mr.
Herring fired several shots through the grating. The elephant
thus attacked in the rear as well as the front, flew round the den
with the speed of a race-horse, uttering frightful yells and screams,
and stopping at intervals to bound from the back against the front.
The force of these rushes shook the entire building, and excited
the most terrifying expectation that he would bring down the whole
mass of wood and iron-work, and project himself among his assailants.

" After the discharge of about thirty balls, he stooped and sunk
deliberately on his haunches. Mr. Herring, conceiving that a shot·
had struck him in a vital part, cried out—' He's down, boys !
he's down !' and so he was, but it was only for a moment : he
leapt up with renewed vigour, and at least eighty balls were successively discharged at him from different positions before he fell
a second time. Previous to that fall, Mr. Joshua Brookes had ar-

of front views of the heads of both this and the Asiatic Elephant. As these outlines admirably illustrate the ex-

rived with his son, and suggested to Mr. Herring to aim especially at the ear, at the eye, and at the gullet.

"The two soldiers despatched from Somerset-house by Mr. Cross, came in a short time before Mr. Brookes, and discharged about three or four rounds of ball cartridge, which was all the ammunition they had. It is a remarkable instance of the animal's subjection to his keeper, that though in this deranged state, he sometimes recognised Cartmell's usual cry of ' Chunee ! Chunee ! Chuneelah !' by sounds with which he was accustomed to answer the call, and that more than once, when Cartmell called out ' Bite, Chunee ! bite !' which was his ordinary command to the elephant to kneel, he actually knelt, and in that position received the balls in the parts particularly desired to be aimed at. Cartmell, therefore, kept himself as much as possible out of view as one of the assailants, in order that his voice might retain its wonted ascendency. He and Newsam, and their comrades, took every opportunity of thrusting at him. Cartmell, armed with a sword at the end of a pole, which he afterwards affixed to a rifle, pierced him several times.

" On the elephant's second fall, he lay with his face towards the back of the den, and with one of his feet thrust out between the bars, so that the toes touched the menagerie floor. At this time he had from a hundred and ten to a hundred and twenty balls in him ; as he lay in a posture, Cartnell thrust the sword into his body to the hilt. The sanguinary conflict had now lasted nearly an hour; yet, with astonishing alacrity, he again rose, without evincing any sign that he had sustained vital injury, though it was apparent he was much exhausted. He endeavoured to conceal his head, by keeping his rear to the front; and lest he should either make a successful effort at the gate, or, on receiving his death-wound, fall backwards against it, which would inevitably have carried the whole away, the keepers availed themselves of the juncture to rapidly lash the gates of his den with a chain and ropes so securely, that he could not force them without bringing down the entire front.

" Mr. Herring now directed his rifle constantly to the ear: one of these balls took so much effect, that the elephant suddenly rushed round from the blow, and made his last furious effort at the gates.

ternal differences of the head, which distinguish these species, we gladly insert them on a separate plate. M. Cuvier's specimen was a female about two years old, and the tusks had not appeared: the type of the Berlin drawing was a female, and evidently adult, though young, which appears principally by the convexity of the back. Full grown middle-aged Elephants (at least of the Asiatic species, and we may presume the same of the African) have the back flat or sloping down according to the breed, and very old animals have it slightly hollowed. It is found principally from the Cape of Good Hope to Senegal, and in the interior, but it is not certain whether it is to be seen on the eastern coast of Africa. Its not being tamed seems rather to be referrible to the low state of humanity in Africa, than to any difficulty on the part of the species itself.

Mr. Burchell relates the death of a native by an Elephant, which, as it displays the immense power of these animals, particularly in the proboscis, may be repeated. Carel Krieger was an indefatigable and fearless hunter, and being also a most excellent marksman, often ventured into the most dangerous situations. One day having with his party

Mr. Tyler describes this rush as the most awful of the whole. If the gates had not been firmly lashed, the animal must have come through; for, by this last effort, he again dislodged them, and they were kept upright by the chain and ropes alone. Mr. Herring from this time chiefly directed his fire at the gullet; at last he fell, but with so much deliberation, and in a position so natural to his usual habits, that he seemed to have lain down to rest himself. Mr. Herring continued to fire at him, and spears were run into his sides, but he remained unmoved, nor did he stir from the first moment of his fall. Four or five discharges from a rifle into his ear produced no effect: it was evident that he was without sense, and that he had dropped dead, into the posture wherein he always lay when alive.

"The fact that such an animal, of such prodigious size and strength, was destroyed in such a place, without an accident, from the commencement to the close of the assault, is a subject of real astonishment."

pursued an elephant which he had wounded, the irritated animal suddenly turned round, and singling out from the rest the person by whom he had been wounded, seized him with his trunk, and lifting his wretched victim high in the air, dashed him with dreadful force to the ground. His companions struck with horror, fled precipitately from the fatal scene, unable to turn their eyes to behold the rest of the tragedy; but on the following day they repaired to the spot, where they collected the few bones that could be found, and buried them near the spring. The enraged animal had not only trampled his body literally to pieces, but could not feel its vengeance satisfied till it had pounded the very flesh into the dust, so that nothing of this unfortunate man remained excepting a few of the larger bones*.

We shall here take occasion to hazard a few observations on instinct in general, and on the intellectual endowments of the mammalia in particular, and shall endeavour to illustrate the subject by the introduction of a few anecdotes of the Elephant, not before in print†, the authority for which anecdotes is not to be impeached ; most of them, indeed, are capable of verification by the testimony, not merely of those on whose veracity they are given, but also of many more eye-witnesses.

When we consider how very much of material nature around us remains to be explained and understood ; how imperfectly the various substantive objects of our senses, with all their essential qualifications and incidents are in general comprehended, we need not be surprised that im-

* Mr. Burchell's Travels in Southern Africa.

† To copy all the stories of these animals on authority good, bad, and indifferent, would swell this article to an inconvenient magnitude; to select from them would be invidious, while confining ourselves exclusively to those here stated, will sufficiently answer our purpose of displaying the intellectual qualities of the species, and analogically of the brute creation in general.

material invisible creation should be so much beyond the reach of our faculties, as we in truth find it. Metaphysical science, if indeed it merit the rank of a science, is in general a baseless superstructure; it is for the most part deduced from data altogether presumed or professedly incomprehensible : no wonder then that it is unsatisfactory and inconclusive. One set of philosophers, because they cannot comprehend what is obviously beyond the reach of human sense, because they cannot demonstrate the connexion between mind and matter, between materiality and immateriality, pertinaciously insist on their identity, and thus dispose of one great but rational mystery by the substitution of another still greater and absolutely irrational.

It is true that the very existence of a mystery is to be ascertained only by fruitless attempts to explain and un-unravel it, or by its obvious admitted impossibility by natural means; these attempts, however, when conducted with propriety, moderation, and deference to what is commonly received as sacred among us, are deserving of respect and attention, and tend eventually to the establishment of truth. Many doctrines that have given offence to the best-meaning persons, as militating against received, admitted truths have eventually, on thorough investigation, been found to support that which they were thought to destroy. Copernicus was proscribed, and Newton was deified for teaching almost the same doctrine ; and thus it not unfrequently happens, that a more enlarged conception and nobler views of the Deity are obtained by the discovery of facts which were at first sight thought incompatible with His attributes ; but the doctrine of mental materialism has been investigated, if not always temperately, at least thoroughly, and but few uninfluenced by preconceived notions, or by fancied superiority of understanding, have adopted it as more comprehensible and satisfactory than that to which it stands opposed.

Without therefore stopping to balance probabilities, or

further to discuss a question, on which novelty has been
long since exhausted, we shall confidently assume the distinctness, while we admit the connexion between mind and
matter with reference to the animal kingdom, below the
elevation of man, and as exemplified more especially in the
species before us, the Elephant.

The bodies of brutes appear to us to be acted upon by
distinct excitements, which may be thus divided into three:
1. By involuntary mental impulse, or instinct, properly
speaking. 2. By involuntary corporeal impulse, nervous
irritability or mechanism, and 3. By voluntary mental impulse on limited reason.

Instinct, εν τιξειν, is strictly an inward goading or stimulation to some particular action, in which general sense it
is applicable to the two first of these divisions; or instinct,
properly speaking, and mechanism, sometimes called mechanical instinct. Confined to instinct, properly speaking,
it may be said to be, an involuntary desire or aversion, acting on the mind without the intervention of reason, motive,
or deliberation, but tending uniformly and exclusively to
the preservation of the individual or propagation of the
race. If limited to mechanism, it may be defined as an involuntary stimulus of an innate unknown power acting on
the principles of life, both animal and vegetable, for the
preservation and propagation of organized existence : in
either of these respects, it is the very antithesis of reason,
whose acts result from volition, are perfectly free, may be
beneficial, injurious, or indifferent to the creature, may
perpetuate its race, or work its destruction.

Instinct, properly speaking, then, has more the character
of an immediate special interposition of Providence in
favour of the creature, than is ordinarily to be observed in
the common course of events, for however conscious we may
feel that the unceasing operations of the Deity must be exercised for the continuance of creation, that a momentary
suspension of His care, or as we more commonly say, of

those laws by which everything is, and continues its being, would be attended with an instantaneous subversion, if not annihilation of all things, we have not often occasion to notice any very particular special interposition for some partial purpose ; any suspension of common consequences for a particular object, except in the case of instinct, which seems to display a sort of immediate communication between the creature and the Creator, almost without the instrumentality of final causes.

Instinct, properly speaking, supplies the place of reason, and dictates to the passive mind what reason would otherwise suggest for the preservation of the creature, in whose favour it operates ; if this creature were left to its own unassisted resources, its limited inefficient faculties would not enable it to maintain its existence ; casualties, and the ordinary but unceasing operations of mutability, would soon sweep it from the surface of the earth, did not a higher power in the shape of instinct interfere for its preservation.

If this be so, we may expect *à priori* to find that the existence or extent of the instinctive faculty is proportioned to the deficiency of reason, and this we believe to be the case. It has been much questioned whether man, in whom reason, however vitiated, however perverted, sits paramount on earth, is endowed at all with instinct; nor is this question easy of solution. Association, education, and habit, obliterate in a great measure the traces of instinct in the adult, that is, assuming instinct to have any place at all ; that it has however, especially when the immaturity or deficiency of reason require, is most probable.

This question rather appertains to the zoological history of man. To illustrate the relative situation of the lower animals, it may be necessary however to advert to it for an instant here. The sole objects of instinct being, as we believe, preservation and propagation, and these objects being, generally speaking, within the reach of reason, we can only expect to see instinct displayed in the human subject: first, as connected with these ends; and, secondly, as auxiliaries to reason. It is during nonage, therefore, and during a morbid state of the intellectual organs, that we may expect to find the most decided traces of instinct. Education, prejudice, and association, will also under these circumstances have less influence, and a discrimination between their effects and those of pure instinct will consequently be more easy of detection. To content ourselves therefore with a single example, let us inquire whether the new-born babe exhibits any symptoms of the stimulation in question. The affirmative, as well as the negative of this position, have been respectively contended for. The action of sucking in the infant, it has been said, (we think indeed without foundation) is acquired. A healthy infant within a quarter of an hour after the reception of air into the lungs, will turn its head from side to side, apparently, may evidently seeking something; that this something is to suck, is certain, for whether it be the mamma of its mother, or any other object offered to the mouth, it immediately commences the operation of sucking. That a difficulty sometimes attends this desirable object of the nursery is notorious; but it frequently happens without the slightest objection; nor can it in any case result from teaching, from experience, or from anything like reason in the helpless creature in question. Reason has not assumed her seat in the vacant mind; instruction, in the simplest form, has hitherto had no access to it; association, which implies experience, observation, comparison, and deduction, is as yet an undeveloped germ, and this pitiable type of human imbecility would quickly sink into that non-entity whence it has just emerged, were it not that its Creator by His providence brings instinct to its aid, and if he do not *teach* it to suck, *impells* it by an unconscious stimulus to that mean of preservation.

But in the human mind, and probably also in the brute, a certain development of intellectual power speedily takes place ; and it soon becomes difficult to trace the actions of the growing subject to their source, whether to the voluntary result of reason, or the involuntary impetus of instinct. As the former proceeds, however, especially in man, the difficulty vanishes, and we are soon enabled to observe that the simple necessities of life are seized and applied by the action of reason ; and as this state of things advances, instinct probably recedes. There are, however, symptoms of its action even in the adult. There is strong reason to believe that the impulse to propagation, and to parental affection, is not altogether independent of passion or instinct. The fear of death also, or rather of nonentity, seems so intimately connected with our nature as to be instinctive; a natural feeling, which it is the business of reason to subdue within legitimate bounds.

But to illustrate the definition of pure unmixed instinct, let us instance two or three examples; the more common and notorious the better.

It is a curious mathematical problem at what precise angle the three planes which compose the bottom of a cell ought to meet, in order to make the greatest possible saving of material and labour. This is one of those problems belonging to the higher part of mathematics, which are called problems of maxima and minima. Maclaurin resolved it by a fluxionary calculation, which is to be found in the Transactions of the Royal Society, and determined precisely the angle required. Upon the most exact mensuration the subject could admit, he afterwards found that it is the very angle in which the three planes in the bottom of the cell of a honey-comb actually meet. "Shall we ask," says Dr. Reid, "who taught the bees the property of solids, and to resolve problems of maxima and minima ? It need not be replied, that they are as ignorant of these properties as the solids themselves, and as incapable of a problem as any inanimate body. The action in question is by a superior power, working through the instrumentality of the unconscious bee.

It is thus, in the lower divisions of the Animal Kingdom, that instinct displays itself in the most obvious manner: as we mount through the graduated scale of animal life, to those beings that stand nearest to man, instinct becomes obliterated as we advance; till at last it is to be but faintly and equivocally distinguished. To instance a case, therefore, of pure, unequivocal, unmixed instinct, it seemed necessary to select from one of the lower classes. That of the bee, however trite, is infinitely in point of ability beyond the utmost reach of the faculties proper to the animal, and it is connected with the two objects of instinct; preservation, in constructing a repository for food, and propagation, in affording a competent nursery for the young.

In the lower classes of the vertebrated division of the animal world, however, we may find various instinctive actions which exhibit more decided marks of their abstracted nature than the instinctive acts of the Mammalia. "What can we call the principle," says Mr. Addison, "which directs every different kind of bird to observe a particular plan in the structure of its nest, and directs all the same spe-cies to work after the same model? It cannot be imita tion; for though you hatch a crow under a hen, and never let it see any of the works of its own kind, the nest it makes shall be the same, to the laying of a stick, with all the other nests of the same species. It cannot be rea-son, for were animals endued with it to as great a degree as man, their buildings would be as different as ours, according to the different conveniences that they would propose to themselves."

It is obvious that the various formations of the nests of birds, like the admirable adaptation of the honey-comb, is pure instinct. It is perhaps above the attainment of man's highest reason, and not to be compared with the other acts of the creature in question. It has no appearance of a mixture with reason, and is perfectly without volition. Each species of bird can make its nest in no other manner, and is impelled by an impetus, over which it has no control, to make its nest in its own particular form, in due season. The migrations of certain species, and many other actions of birds, are not less unequivocal in their nature than nidification, though less universally applicable.

The utmost refinement of human reason, by the aid of artificial education, can only succeed, in some cases, in discovering the perfections of instinct, while the creature in whom it is exercised is so low in the intellectual scale in all other respects, as to render its possession of a reasoning faculty perfectly problematical. There cannot be a doubt, therefore, that pure instinct is in all cases the act of a higher immaterial power, whose providential care is ever active in preserving what was at first thought worthy of creation. Like a tender parent with his children He leaves indifferent matters to the indifferent choice of His weak creatures, but restrains them by an uncontrollable, yet gentle force, from that individual as well as total destruction on which through their own weakness and inefficacy they would otherwise rush.

But the great question is as to the *modus operandi*. Can we presume that the great Creator and Governor of the Universe descends in person to superintend the construction of a honeycomb, or a nest, or the migration of a swallow? Have we any data to warrant the supposition of intermediate superior agency. Does not Nature, that is, God, act by general, not by partial laws, and is it not agreeable with all analogy, that, as he endowed bodies with gravitation (a principle we are as ignorant of as that of instinct), and subjected matter to those conditions under which it exists,

2 C 2

so when He united mind to matter for a time, and made each race of His creatures subservient to its own particular laws, He deprived it of free agency in certain respects, and placed an involuntary impulse in its stead, the real objects of which impulse the creature itself is perhaps as unconscious of, as matter is of its own gravitation?

We must confess the subject not within the reach of our finite reason. Insatiable curiosity soon finds a limit to its bold excursions, and when we once quit the world of sense for that of immateriality, we soon find that all our fancied discoveries are nought but hypothesis and conjecture. The sacred volume, whose sole object is moral excellence, and not abstracted knowledge, loses the character of inspiration the moment it is perused with a view to the latter alone. All the information it affords us is so much only as is necessary to its higher objects of moral improvement and faith.

The accumulated intellectual inquiries of ages have therefore left this part of the subject perfectly unexplained, though, as in the case of almost all other impenetrable mysteries, the successive hypotheses of men of reputation have assumed the similitude of established truths, at least with that preoccupied or idle portion of society, who take things for granted on the authority of others, without the trouble or the means of personal examination. The Cartesian philosophers, who attribute all the actions of animals to mechanism, have certainly much more celebrity than merit.

> L'impression se fait mais comment se fait-elle?
> Selon eux par nécessité
> Sans passion, sans volonté:
> L'animal se sent agité
> De mouvements que le vulgaire appelle
> Tristesse, joie, amour, plaisir, douleur cruelle;
> Ou quelqu'autre de ces états.

Referring the actions of instinct to mental powers, as Dr. Darwin and Smellie have done, seems a very unsatisfactory attempt at explanation, which, while it leaves the original difficulty untouched, makes a distinction without a difference between instinct and reason ; intermediate agency, as assumed by Cudworth and a more modern writer on the subject, is purely hypothetical, and may be thought to be rather derogatory than otherwise to that Omnipotence which can surely act universally by natural final causes: the conjecture in question (and it is utterly without proof) seems to make creation a continued act, instead of one single simple piece of exertion, if such a word be venial so applied, which once having made the machine, set it in motion by a simultaneous act for the destined period of its existence.

Leaving the final causes, then, of involuntary mental impulse, or instinct properly speaking, we may proceed to a brief notice of involuntary corporeal impulse or mechanism.

Any explanation of this particular excitement to organic action, or of this division of instinct, if it can be treated as such, seems equally impossible with the last. It comprehends however a much wider field of action, pervades the vegetable as well as animal kingdom, and is found to exist either with or without sensation. The involuntary action of breathing, the mechanical spring to so many other movements; the involuntary peristaltic motion of the bowels, and such like functions of animal economy, are carried on without any command from the mind, apparently without any interference of the cerebral organs, at least in a mass. In mental instinct the mind seems to give the word to the nerve, the nerve acts upon the muscle, and thus the motion of the body is accomplished: the question is in what manner the mind is impelled ; but in involuntary corporeal movements, mind seems to be no party, or at least is an unconscious agent.

An organized body, whether animal or vegetable, exists by the means of generation alone, and is thus distinguished from an inorganized body, whose being may be said to be fortuitous. The organized body must be kept alive, must be fed from without, and must propagate ; and these processes presuppose some original active agency, which agency is the living principle, or nervous irritability, of physiologists, the instinctive principle of metaphysicians, and the involuntary corporeal impulse or mechanism, sometimes called mechanical instinct, to which we allude. To enumerate the various mechanical actions which must be going on unceasingly and unconsciously, in every organized body, would exceed both our means and limits. Nutriment taken in by the animal body at the mouth, and by the vegetable at the root, is constantly undergoing a most complicated and wonderful process : is first reduced to a fluid, purely *sui generis*, is impelled and driven to every part of the little world it is employed in sustaining; is again, by a sort of chemical analysis, of which we have no notion, separated, and while the essential parts are left by something like precipitation, the useless residuum is passed off. What power is it which impels these wonderful and complicated actions, both in the vegetable and in the animal ? What, but this living principle, this instinctive action, this involuntary corporeal impulse, or mechanism, to which we allude. To understand its nature however, we must first be divested from the trammels of this material world, with which alone we compare everything, and associate everything, and with which, in fact, however connected, it bears no analogy.

The third spring of action, in animals next below the elevation of man, or voluntary mental impulse, we shall venture, not indeed without authority, though with respect for the contrary opinion of others, to consider as identified with reason, limited indeed in its extent, and but very little capable of improvement.

Many persons are of opinion that to admit brutes to be endowed with reason, the same in kind with ours, however inferior in degree, is an insult to man: but let them endeavour to analyze this opinion; to separate its component parts; and the whole will probably be found resolvable into pride. This bane of society, this curse of all civil relations, not only pervades social life, but contaminates all our intercourse with the whole range of creation. The invalid, writhing under a complication of that frightful list of corporeal maladies which follow in the train of civilization and improvement, is with difficulty persuaded that he has a body in all respects the same, only more exposed to acute sensation, as that of his humble canine companion: he forgets also that this companion is capable of a degree of attachment and gratitude, qualifications surely of a mental character, which, while they will survive the severest shocks of adversity, triumph over every temptation, and often shame the frail friendship of proud, reasonable man. The material part of man, though most harmoniously and aptly adapted for his mental powers and mode of life, is in fact in many respects beneath the average level of animal organization. Why then should we, who are on a level with or beneath other animals in some respects, on the one hand be so unwilling to acknowledge the justice of their very humble comparative intellectual pretensions on the other?

One of the deepest thinkers of this country, the celebrated Locke, was willing to concede the reasoning faculty in a restricted form to brutes. " It seems," says he, " as evident to me that some of them do in certain instances reason, as that they have sense."

Mammalia, birds, and fishes are capable of intoxication; and in one of the following anecdotes of the species which we must not forget is more immediately before us, the elephant, it will be seen that a promise to the brute to make it drunk with arrack, induced it to work when threats had failed.

The mental faculty of these three classes, if not of all vertebrated animals, will undergo a transient but material change by the operation of spirits a fact which seems not very easily to be reconciled with the total absence of reason or mind.

They are individually under the influence of different tempers: horses will strongly evince this: some of these animals are to be impelled by severity ; others a similar government will ruin, but these yield to gentleness and mild treatment.

Brutes certainly dream : the Dog and the Cat frequently show the most unequivocal signs of it; and although dreaming among ourselves is included in that general obscurity which involves the subject of intellect, we have no reason to suppose that it is a mere instinctive action in us, and by analogy therefore not so in lower animals. If our dreams were uniformly directed to the subject of self-preservation, or of propagation, the legitimate obects of instinct, their origin as connected with the instinctive principles would be more probable; but when we know that circumstances of the most trivial or indifferent character, generally, though not always such as have preoccupied the mind, as the meeting a friend, affairs of business, a prize in the lottery, an impossible event, and such like, are frequently the subject of dreams, it seems to follow that they have no connexion with that irresistible impulse of a power from without, uniform in its objects, and perfect in its means, which marks the character of instinct.

But it is said that brutes possess a principle different from instinct, and analogous to reason, but having no affinity with it ; or, in other words, that they have a faculty different from reason, but like it. This position appears, however, to be extremely difficult, if not incapable, of proof, or even of an adequate degree of probability: such an assumed principle seems totally abstracted from the intellectual powers we possess; we cannot therefore examine, compare, and understand it. It is impossible for us to study the principle of brute actions anywhere but in our own actions, and the limits of our intellect are also the limits of the intellectual world. We can explain, distinguish, and appreciate the nature of the actions of brutes, only by means of the light which a study of ourselves affords. It is of the utmost importance to bear this truth always in mind. The comparison of our actions with theirs can be our only guide to a determination of their character. What we are conscious in ourselves to be the proximate cause of the one, we may, nay must, from analogy presume to be that of the other; we have in fact, no other means of judging. Why should we suppose two dissimilar causes for two similar effects? It is in vain that we are told that the cause of one is dissimilar to that of the other: let those who assert this, prove it if they can; in the interim let us content ourselves with abiding by the rules of sound philosophy, and not suppose an unknown cause, when one that is known will explain the phenomena. In fact, if the Deity had created a different faculty for the actions of animals, from that to which the determination of our own is owing, all our endeavours to discover it would be utterly in vain; it would escape the most piercing ken of human intelligence, and remain an everlasting enigma.

Not, therefore, to attempt any definition of a principle in which we are supposed not to participate, which is totally abstracted from us, and on which therefore we can have no definite notions, we shall proceed to lay before the reader a few well-authenticated instances of what, in common conversation, would be called the sagacity of the Elephant,— a word we believe to be strictly applicable to the cases in question, which appear to us to have no connexion or affinity with instinct, properly speaking, and at the same time sufficiently to evince the limited reason of the species in question, and, by analogy, a similar faculty progressively diminishing in the lower classes.

An elephant which a few years ago belonged to Mr. Cross, at Exeter 'Change, attained to the practice of a curious trick, which by repetition might be said to have acquired, if indeed instinct could be acquired, something of an instinctive character ; but which, the first time it occurred, at least, seems attributable to nothing short of reason. It is the usual part of the performances of an elephant at a public exhibition, to pick up a piece of coin thrown within his reach for the purpose, with the finger-like appendage at the extremity of the trunk : on one occasion a sixpence was thrown down, which happened to roll a little out of the reach of the animal, not far from the wall : being desired to pick it up, he stretched out his proboscis several times to reach it: he then stood motionless for a few seconds, evidently considering, we have no hesitation in saying evidently considering, how to act; he then stretched his proboscis in a straight line as far as he could, a little distance above the coin, and blew with great force against the wall ; the angle produced by the opposition of the wall, made the current of air act under the coin as he evidently intended and anticipated it would, and it was curious to observe the sixpence travelling by these means toward the animal, till it came within his reach, and he picked it up.

This complicated calculation of natural means at his disposal, was an intellectual effort beyond what a vast number of human beings would ever have thought of, and would be considered as a lucky thought, a clever expedient under similar circumstances in any man. It was an action perfectly indifferent, had no relation either to self-preservation or to propagation. The picking up the sixpence and all the other tricks these animals are taught, may be strained into a mental association below the level of reason, arising from the food with which the animal is in general rewarded for such tricks : that this is the stimulus to perform them is perfectly clear ; but is it an *unreasonable* stimulus, does it not at once display the reasoning faculty ? But admitting the stimulus to the action, the new and untaught mode of performing it seems referrible to no other principle within the animal than to reason ; and to seek for a higher independent principle from *without*, and to suppose that a greater power immediately or by intermediate agency dictated the indifferent, insignificant action of picking up the sixpence, reminds us, if the subject may be treated for a moment jocosely, of Martinus Scriblerus's quotations from the poets, comparing the most sublime of all Beings, to a painter, a chemist, a wrestler, an attorney, a butler, a baker, &c.

Some young camels belonging to a much respected friend of the editor, and brother of a very valuable contributor to this work, were travelling with the army, when they had occasion to cross the Jumna in a flat-bottomed boat: the novelty of the thing excited their fears to such a degree that it seemed impossible to drive or induce them to enter the boat spontaneously; upon which one of the mohauts, or elephant-keepers, called to his elephant, and desired him to drive them in: the animal immediately put on a furious appearance, trumpeted with his proboscis, shook his ears, roared, struck the ground to the right and left, and blew the dust in clouds towards them; and so effectually subdued one great fear in the refractory camels by exciting a greater, that they bolted into the boat in the greatest hurry, when the elephant re-assumed his composure, and deliberately walked back to his post.

The same elephant was appealed to by his mohaut to remove a branch from a tree which hung too low to raise the tent-pole; the animal looked at the pole as if measuring it with his eye, then at the tree and impending branch; he then turned his rump towards the trunk of the tree, stepped a couple of paces forward, took the branch in his trunk, and felt as if examining where it would split off; finding it easy at this place, he moved a little back to where it was thicker; then taking a firm hold, he gave it three or four successive swings, increasing his force, till with one very powerful effort it tore and fell on the ground. Being appealed to, to remove another branch still higher, he looked up, stretched his proboscis, and caught only a twig or two and some leaves ; he was urged again, he shook his ears and gave a piping sound of displeasure ; but the mohaut insisting after another vain attempt, he caught the bearing pole of a dooly (a kind of palanquin,) and shook it with violence, making a poor sick soldier immediately start out of it: the hint was sufficient—he would not be trifled with.

Elephants frequently are employed to assist in the conveyance of heavy artillery, not by drawing the guns, which is left to oxen, but by lifting them when there may be occasion, as in the case of a bank, or through mud, or such like impediment to their progress. In these situations it is always observed that the elephant will not attempt to raise the gun till all the draught cattle are on a full strain to pull it forward.

This action it seems very difficult to refer to any other principle than that of an understanding, however simple, of the effect of the combined horizontal and vertical force necessary under the circumstances to extricate so unwieldy a mass as a twenty-four pounder.

At the siege of Bhurtpore in the year 1805, an affair occurred between two elephants, which displays at once the character and mental capability, the passions, cunning, and resources of these curious animals. The British army, with its countless host of followers and attendants, and thousands of cattle, had been for a long time before the city, when on the approach of the hot season, and of the dry hot winds, the supply of water in the neighbourhood of the camp necessary for the supply of so many beings began to fail; the ponds or tanks had dried up, and no more water was left than the immense wells of the country would furnish. The multitude of men and cattle that were unceasingly at the wells, particularly the largest, occasioned no inconsiderable struggle for the priority in procuring the supply for which each were there to seek, and the consequent confusion on the spot was frequently very considerable. On one occasion two elephant drivers, each with his elephant, the one remarkably large and strong, and the other comparatively small and weak, were at the well together; the small elephant had been provided by his master with a bucket for the occasion, which he carried at the end of his proboscis; but the larger animal being destitute of this necessary vessel, either spontaneously or by desire of his keeper, seized the bucket, and easily wrested it away from his less powerful fellow-servant : the latter was too sensible of his inferiority, openly to resent the insult, though it is obvious that he felt it; but great squabbling and abuse ensued between the keepers. At length the weaker animal, watching the opportunity when the other was standing with his side to the well, retired backwards a few paces in a very quiet, unsuspicious manner, and then rushing forward with all his might, drove his head against the side of the other, and fairly pushed him into the well.

An inquiry might naturally be made here, whether these animals were in the case in question possessed of anything like a moral sense? We should certainly have no inclination to refer a moral sense, strictly speaking, in any case to the lower animals : its existence, independently of education and habit in man, may be problematical ; but there seems little doubt that the animals in question had *acquired* a principle, not far, if at all removed from a partial knowledge of right and wrong: being constantly fed by portions or messes, it may be easily supposed that it attained a knowledge of *meum* and *tuum*, and such a knowledge, however limited in its beginning, might, from the constant intercourse of these creatures with man, be in some degree *improved*, (of which instinct is altogether incapable,) and more largely applied. This notion however presupposes a limited degree of reason in the animal.

It may easily be imagined that great inconvenience was immediately experienced, and serious apprehensions quickly followed, that the water in the well on which the existence of so many seemed in a great measure to depend, would be spoiled, or at least injured by the unwieldy brute which was precipitated into it; and as the surface of the water was nearly twenty feet below the common level, there did not appear to be any means that could be adopted to get the animal out by main force, at least without injuring him: there were many feet of water below the elephant, who floated with ease on its surface, and experiencing considerable pleasure from his cool retreat, evinced but little inclination even to exert what means he might possess in himself of escape.

A vast number of fascines had been employed by the army in conducting the siege, and at length it occurred to the elephant-keeper that a sufficient number of these (which may be compared to bundles of wood) might be lowered into the well to make a pile, which might be raised to the top, if the animal could be instructed as to the necessary means of laying them in regular succession under his feet. Permission having been obtained from the engineer officers to use the fascines, which were at the time put away in several piles of very considerable height, the keeper had to teach this elephant the lesson, which by means of that extraordinary ascendency these men attain over the elephants, joined with the intellectual resources of the animal itself, he was soon enabled to do, and the elephant began quickly to place each fascine as it was lowered to him, successively under him, until in a little time he was enabled to stand upon them; by this time, however, the cunning brute, enjoying the cool pleasure of his situation after the heat and partial privation of water to which he had been lately exposed, (they are observed in their natural state to frequent rivers, and to swim very often,) was unwilling to work any longer, and all the threats of his keeper could not induce him to place another fascine. The man then opposed cunning to cunning, and began to caress and praise the elephant, and what he could not effect by threats he was enabled to do by the repeated promise of plenty of rack. Incited by this the animal again went to work, raised himself considerably higher, until by a partial removal of the masonry round the top of the well, he was enabled to step out: the whole affair occupied about fourteen hours. This affair involves a series of intellectual operations, which it seems very difficult to separate from reason.

Major Smith had an opportunity of studying the character of an elephant which was exhibited a few years ago in the United States. This animal had a great affection for a dog, and the spectators, to tease her, used occasionally to pull the dog's ears and make it yelp; on one occasion when this species of joke was going on at the side of a barn, within which the elephant was kept, as soon as she heard the voice of the dog in distress, she began to feel the boards that separated her from it, and giving one blow, appeared surprised that they did not fall: she then struck with greater force, made the boards fly in splinters, and looked through with such menacing gesture, that the experimentalists thought proper to make off.

The newspapers mentioned at the time another incident of this animal: the elephant was crossing a river in a passage punt, when some men, to tease her, took the dog into a boat that was towed alongside, and began to pull its ears; the elephant resenting the ill-usage, filled her proboscis with water, and then squirted it upon the men, but finding they would not desist, she set in good earnest to the task of sucking up water and discharging it into the boat. At first the men laughed at the expedient, but she persevered till they began to bale to keep from sinking; upon this manœuvre she redoubled her efforts, and would certainly have been able to swamp the boat, had the passage across been prolonged a few minutes more.

This poor elephant was maliciously shot by a malevolent back-settler who waylaid her behind a hedge. The Major examined the skeleton, and was surprised to find that the ball had gone quite through the abdomen, and broke a rib on each side. He also found that three of the vertebræ of the back had formed an anchylosis, in consequence, as he was informed, of the animal having fallen through a bridge which gave way under her.

We have already expressed an opinion that true instinct is to be found most commonly in such animals as exhibit the smallest indications of reason, and *vice versâ*, this seems to be true to the extent of rendering it very difficult to separate the acts of reason from those of instinct, in the most rational brute: thus it is no easy matter to distinguish mental instinct from the reasoning faculty in the Elephant. In the Dog, the domestic Cat, and the carnivorous order generally, the impulse to whose food seems purely of an instinctive character, the indications of this impulse are plainer, though these animals may have an equal or superior mental capacity with it.

In man, whose physical characters display his relative situation admirably, and whose omnivorous regimen is as apparent by his bodily peculiarities, as it is by the aptness with which he adopts it practically, there is no great improbability that the ardent love of the chase, of angling, and of such like cruel sports, originates in an instinctive principle, similar to that which impels the carnivora, properly speaking, to their food. Our reason revolts so strongly against the mere idea of feeding on an animal body, or even of inflicting pain, which we are capable of feeling so acutely ourselves, that there seems to be no principle short of the uncontrollable fiat of instinct, which can reconcile us to such actions. It really, therefore, is difficult to point out satisfactorily any pure instinctive act in the adult Elephant, more than in the human race.

Seeking for the mamma of its mother, which the young Elephant, in common with all lactiferous animals does instinctively, and the spontaneous use of the organs bestowed upon it by nature, as the proboscis and the tusks, are actions of a kind common to all young animals. When adult, a discrimination in the choice of its food, with reference rather to its quality than its flavour, a parental fondness for its young, a dread of bodily harm, and perhaps a consciousness of man's superiority, are instincts probably as strong as any the animal evinces; but that it is also actuated by a spontaneous motive, that in indifferent affairs it is governed, not by an arbitary impulse, but by a free, reasonable choice, seems apparent by the anecdotes of the animal already related.

But before we take leave of the subject, at least for the present, it may be necessary to add a final sentence in vindication of man's only real superiority on earth from any presumed depreciation the foregoing remarks may be thought to countenance. Reason, that paradoxical gift bestowed upon him, at once the source of almost all his misery, and yet without which his mere vegetative existence could hardly be considered a boon, must not be treated too lightly on the one hand, or proudly and exclusively appropriated on the other. Reason in the brute mind appears to us like a seed in an ungrateful soil: it makes an effort to vegetate, it commences the operation, but circumstances arrest its progress and render fructuation impossible; while the same seed, in the mind of man, springs to maturity, and bears fruit proportioned to its natural soil and to the artificial cultivation bestowed upon it.

Gradation is a most prevalent principle in the great scheme of creation: the three material kingdoms of nature, however remote their extremes, have points of close contiguity, if not of actual contact: the immaterial world appears in this respect to be very analogous to the material. The human subject, lowest in intellectual cultivation, if we regard him for a moment as he really is, abstracted from his capability of improvement, which in fact alone forms his real superiority, is very little, if at all, above the highest intellectual quadruped. We may then trace intellect, however restricted in development as it passes, diminishing in degree through the whole of the encephalous animals : these fall into that division of the animal kingdom in which the medullary substance, no longer concentrated, is divided into portions, or spread throughout the system ; and in such the co-existence of intellect with mere sensation is uncertain, until at last it becomes evident, as in the zoophytes, that nothing but sensation is left, and we may well doubt whether even sensation can exist in matter without something like an immaterial connexion. From these sensation ceases, but the living principle glides off into the vegetable kingdom, in which it is as strong as in the animal. Sensation, indeed, must be considered to be the lowest state of intellect, which seems to quit the creatures of this world in the semi-vegetable zoophytes.

The real difference between brute reason and human does not appear to us to be in kind, but in the *capability* afforded to the latter and denied to the former. The elephant that had a *reasonable* knowledge that a current of air would move a comparatively light body, and that a sudden resistance would cause the same current to turn in a contrary direction from that which was at first given it, went as far in an intellectual operation, and availed himself of as much of past experience and association as the human savage could do: but all the experience and all the cultivation in the world would never enable the same elephant to penetrate the principles of mechanical powers, to understand the doctrine of angles, to calculate the extent of resisting mediums. Not so the savage;—we know, if not experimentally, at least by analogy, that his mind is capable in some directions of an indefinite extent of improvement: reason in the brute is rudimentary and incapable of progression ; in man it attains a degree of development proportioned, as we have said, to the pains bestowed upon its culture.

And this is perfectly compatible with the accountable condition of man, and the contrary state of lower animals. They, in all probability, have not arrived at what is called a moral sense : their rudimentary reason has not attained to a comprehension of right and wrong ;· their intellectual restricted gift, therefore, brings with it no reciprocity of obligation ; they have not eaten the fruit of the tree of knowledge of good and evil ; their eyes are not opened; and happy and enviable is such a condition compared with that of those who, while they boast of their dangerous pre-eminence, forget or disregard the obligations it imposes.

Having endeavoured, in the foregoing part of this Essay, to distinguish in general the characteristic differences between intelligent and instinctive actions, and adduced some examples in illustration at least, as we submit, of the former in the elephant, whence we would infer actions resulting from the same principle in other brutes, we shall now venture, in continuation of this difficult but interesting subject, to examine what is preliminary to the performance of each : *i. e.* what are the causes, not the primary, which appear inscrutable, but the proximate causes of both, or to speak as philosophically as may be, what are the phenomena immediately previous, in the order of succession, to the phenomena of instinctive and intelligent action. Could this be completely performed, we should be enabled to ascertain the line of separation between human and brute intelligence, which undoubtedly ought to be the chief, and perhaps the sole object of all researches of this kind.

Unfortunately, we have not the means within our power of completely resolving this problem. Before we could do so, it would be necessary to possess a scientific classification, a sort of entire synopsis of all the modifications which the human understanding can experience. In other words, it would be necessary to have a list of all the operations of which it is susceptible, or of the ideas which it can obtain. We are far from possessing any thing of this description sufficiently adequate to our purpose; on the other hand, be it also remembered that we can have no direct knowledge of the intellectual *acts*, of the internal emotions of any beings but ourselves. By the faculty of consciousness, a man becomes acquainted with what passes within himself; through the medium of language, he becomes acquainted with what passes within his fellow-men. But in the case of animals, the same degree of certainty is far from being attainable. There is no common language between us, sufficiently intelligible for the purpose. It is only by induction and analogy that we are enabled to form any judgment of the intellectual *acts* which must precede their actions. It is only from their actions, from their organic movements, that we can draw our conclusions; and certainly it must be allowed that many different causes may give rise to similar organic movements.

Still, however, enough remains to enable us to draw conclusions, which have at least the strongest probability on their side; and this is all we can hope to arrive at. When the quantity of information we possess on any subject is sufficient to justify us in coming to any judgment which is probable, we are warranted in holding by that judgment, until fuller information shall overturn or confirm it.

In investigating what may be called the inducement to the actions of brutes, the first question appears to be, do animals form *simple* or *primary* ideas, in a manner similar to that by which man forms them? The mammalia, at least, have senses exactly like our own, capable of receiving similar impressions, and, consequently, capable of resulting similar ideas.

There is, however, one condition essentially necessary to the formation of all ideas. This is a preparatory organic act, which we call *attention;* without this, no idea is ever formed. After an idea is once formed, it may certainly recur without this preliminary act ; but it is never originated without it. The mere impression on the senses is insufficient; this act is necessary to put the organ in a proper state to receive an impression essential to the formation of an idea, and we may add, that it is equally necessary to every other operation of the intelligent faculty.

During our waking moments, as our senses are continually acted upon by surrounding objects, they are constantly receiving impressions of various kinds. These impressions, however, do not necessarily form ideas ; we see objects, we hear sounds, we touch bodies, but all these impressions may be utterly inconsequential in regard to our intelligence, and take place without producing a single idea ; but if the stimulus of some want or desire produce the preparatory act of which we have been speaking, in other words, if we place ourselves in a state of attention, and fix that attention on an object, by which our senses are impressed, one or many ideas are the immediate result.

Now this faculty of attention, which follows sensation, and produces ideas, is certainly possessed by animals : it is superfluous to offer any proof of this; were it otherwise, animals could not be of the smallest utility to man. It is possessed in the highest degree by the mammiferous animals, whose senses and cerebral conformation are the most perfect, when compared with our own.

It must be confessed, however, that this faculty, on which all intellect is founded, is obviously possessed, by the most perfect of these animals, in a much inferior degree to that in which it exists in man : their senses, like his, receive impressions from external objects, but to the majority of such impressions or sensations they pay no attention; they only take notice of those which are immediately relative to their habitual wants, and their ideas must consequently be few, and very little varied ; very extraordinary circumstances are necessary to make them vary their actions, or extend in any degree the circle of their ideas. This takes place among them, to the greatest degree we ever witness it, under the influence of man : under his guidance they are susceptible of an augmented education ; but when left to themselves, their improvability is very limited. All objects, except those in which their physical wants are interested, are to them as nothing ; nature presents to their view no object of wonder, of curiosity, of admiration, or of love; nothing can interest them but what ministers to the relief of their wants, the gratification of their appetites, security from danger, or enjoyment of repose ; all else is seen without attention, and without intelligence.

There is, however, a vast difference in this respect between animals of different species. The actions of the Elephant above quoted, were almost all independent of physical wants; those of the domesticated race at large are not unfrequently of the same character: in the wild genera they are much less common, though among these even there seems a difference in this respect. Look, for instance, at the attention and apparent curiosity with which a monkey examines every object before it, compared with the apathy of the swine! But it must be questioned whether all this sagacious attention be not finally referrible to his physical wants. He appears indeed to examine bodies with the eye of a philosopher ; but it is more than probable that a discovery of their esculent properties, if they have any, is all he has in view.

Animals then possess, in a certain degree, the power of attention, and consequently of forming ideas. That these ideas remain impressed on their sensoria, and frequently recur, is quite evident; in other words, that they have memory. It is equally evident that numerous and varied associations are formed between these ideas, and that animals deduce thence many judgments, which judgments like our own are true or false according to the premises on which they rest, and the accuracy with which they are deduced ; in short, that they are as unlike the deductions of instinct, and, as far as they go, as like those of human intelligence, as sufficiently to infer identity.

But if the principle in both cases be the same, it is clearly possessed by us and by the brute in proportions immeasurably different; otherwise, indeed, the question of identity could not exist.

To point out the limits of animal intelligence is hopeless; but, to speak with becoming humility, we may endeavour to approach the truth in this respect.

It is self-evident that the intelligence of every being must be limited by the number and variety of its ideas. These, as we have seen, depend upon the degree of attention. There is a kind of successive dependence in the intellectual faculties, each one being proportioned to the strength of its precursor. We allow that animals possess attention, memory, association, and judgment, or the power of deducing inferences from comparison of ideas. But if the attention be limited, so is the number of ideas, so is the memory; the associations are consequently unvaried, and the judgments few, and resulting from very simple processes of comparison. That this conclusion is warranted by the observation of the actions of animals cannot be denied, except in the case of those instinctive operations which have nothing to do with the present question.

That animals can compare two or more objects present to their senses, discern some of their relations, and execute an act of judgment thereupon, is clear; the actions of the Elephant above quoted would be sufficient to prove this fact. That they can compare a present object of perception with one that is past, the traces of which are in their memory, is also indisputable. But that they can compare two ideas of memory together is more than we can tell. None, however, of their actions will bear us out in maintaining the affirmative. An immediate appeal to the senses seems always necessary to stimulate their intelligent faculties into action. It is certain that we cannot tell to what modes and forms their perceptions are subjected, nor what are the precise relations which they are incapable of seizing; but, at least, we can make a probable guess on the subject.

If animals are incapable of comparing two ideas of memory, it of course follows that they are incapable of all the more complex processes performed by human intelligence; that they are incapable of pursuing any train of consecutive ideas; that perhaps they have no complex ideas at all; that they do not possess the powers of generalization or abstraction, which depend on language; and above all, are utterly destitute of the faculty of imagination.

Still further, even in the objects presented to their senses they are incapable of discerning many relations; they seem to have no idea of beauty or deformity, of symmetry or disproportion; their ideas of figure do not appear to be very exact, nor those of colour very distinct in general. Magnitude, solidity, motion, and extent, are qualities more within their comprehension; of numbers we cannot presume that they have any correct notion, and though they do form collective ideas, as the dog of the flock which he is to guard, &c., we have no reason to suppose that those are always of the clearest kind.

In short, we may come to this conclusion, that animals possess intelligence the same in kind as ours, but in a very limited degree; that this limitation is owing to the limitation of their ideas, and this again to their incapacity of attention, and their want of language, and other means which man possesses for enlarging the circle of his. Thus, though they possess the original faculties of man in a rudimentary state, they are incapable of much improving them, and still more of deriving from them those acquired faculties, such as abstraction, imagination, &c., which are nothing but different phenomena of the association of ideas.

M. F. Cuvier conjectures, that if we do not find in any single species of animals all those intellectual faculties which we recognise in ourselves, it is possible that an attentive examination might enable us to discover a great number of them in that assemblage of species which constitutes the Animal Kingdom. He thinks, moreover, that such faculties might serve, as well as physical qualities, to distinguish the different species. But if, as we have seen, the ideas of animals be limited to their physical necessities, if their power of attention be little, and their capacity of combination consequently narrow, all the differences in an intellectual way, which we can hope to find among them, must consist only in the superior sagacity of some in catching their prey, or evading their enemies, and the greater susceptibility of others to receive an education from man. Let it ever be remembered, that as the foundation is, so is the superstructure: intellect, mind, understanding, or whatever else we please to term it, is nothing but a superstructure of ideas raised on the bases of sensation and at-tention.

M. F. Cuvier is of opinion, moreover, in his argument directed against those who maintain that animals reflect, but rest their affirmative on their instinctive actions,—that animals have no consciousness of their own identity or capability of reflection. His words are, " Mais ce qui nous paroît hors de doute, c'est que tous les animaux sans exception sont depourvus du sens intime de la perception du *moi* et de la faculté de réfléchir ; c'est à dire de con-sidérer intellectuelment, par un retour sur eux-mêmes, leur propre modifications ; *ils ignorent qu'ils reçoivent l'impression des corps exterieurs*, qu'ils PENSENT, qu'ils agissent." It appears to us difficult, we confess, to allow an animal the faculty of thinking, and to deny him all consciousness and reflection ; to believe that a dog does not know that he is not another dog, nor a cat, nor a bull. In fact, consciousness appears to us to be inseparable from the thinking faculty, from intelligence, in the most limited form ; the very act of thinking teaches us that *we* think. " Cogito ergo sum," says Descartes, only he had no ne-cessity to put it in the form of an enthymeme, for it seems an instinctive truth. This perception of the *moi* takes place in infants, in the very dawn of reason ; infants are much in the same situation as animals, as to the degree of their in-telligence ; perhaps they are inferior in this respect: if, however, they have this perception of the *moi*, it may be concluded, from a very reasonable analogy, that animals have it too. Even idiots and madmen possess a con-sciousness of self; the latter prove their possession of it by the very act of doubting their identity, by the very act of fancying themselves to be another ; for, after all, it is the *moi* that is still predominant, and their wildest extravagancies result from an intensity of distempered egotism.

It is not meant to assert that this consciousness of existence is as clear and strong in animals as in man. But can it be separated from the thinking faculty? And if there are no direct proofs in favour of the opinion that animals possess it, if we are obliged to reason from analogy on the subject, be it remembered, that there is an equal want of proof of any kind on the opposite side of the question.

Nor can we deny reflection to animals. Do animals never deliberate? Are all their actions headlong and precipitate? Were those already recounted of the elephant of this description? Do they never hesitate before they perform an action? Do we never see them on the point of performing a certain action, and then, suddenly perceiving that it was not necessary for their purpose, forbear to perform it? No one, however superficial his observation of animals may have been, can deny all this. The elephant in the well was induced to action alternately by caresses, by threats, and by promises. It appears, therefore, that reflection is as inseparable from mind, as extension is from matter. That which is not extended, cannot be material; that which does not reflect, cannot be intellectual.

Before we dismiss for the present the subject of animal intelligence, we would for a moment indulge ourselves in another remark, which may serve to strengthen the foregoing considerations. Without at all subjecting ourselves to the charge of materialism, we may believe that certain physical organs or instruments are absolutely necessary to the operations of mind. We have no knowledge of mind operating independently of such media. It is not necessary to stop to prove that man receives his ideas through the senses, and performs operations between them with his brain. In other words, that the brain is the organ, though not the *principle* of intelligence, and that a certain organic structure is necessary to the action of that principle in man. Now if we find a similar organic structure in other animals, we must conclude that it was intended for the accommodation of a principle of the same kind. Is this the fact, or is it not? The higher orders of mammalia have senses, nervous and cerebral systems like our own. These were, then, obviously intended as the media of action to the principle we speak of. It is true that no animal is so well organized as man for intellectual operations, though often much better for physical. This is true, even with respect to external structure. So perfect an organization in this way was not necessary for the limited degree of the thinking principle. But still there is quite enough of similarity and approximation in the organs of animals to our own, to warrant the belief that they were intended to be acted on by a principle, the same in kind, though different in degree.

Although these remarks have already reached an extent to which they are by no means entitled, we must still crave the indulgence of the reader a little longer, while we hazard a few additional observations on the nature of instinct, in reference principally to what has been lately said by others on the subject, and in which we shall avail ourselves of the ingenious and profound reflections of M. F. Cuvier; who, though too great a philosopher to pretend to remove the mystery that attaches to this subject, has still presented us with an hypothesis, which, to a certain extent, may serve to explain its phenomena.

Some philosophers have supposed that instinctive actions depended on a particular form of the brain, and were in some measure mechanical. But, asks M. F. Cuvier, what is this form, and on what analogy does such a supposition rest? It presupposes a description of proofs, of which we are not in possession. We may doubtless find, in the structure of the brain of animals, certain forms, with which their intellectual faculties may be connected; but this notion, however probable, is far from having been demonstrated to be true,* and the excessive difficulty of such a demonstration will, for a long period, render it impracticable. A theory of this description cannot be allowed to rest on mere negative analogies, on suppositions which are liable every moment to be destroyed by some new fact, some hitherto-undiscovered phenomenon of existence.

There are, however, other phenomena, in which with more foundation, and resting on surer analogies, we may form an hypothetical explanation of instinctive actions. These are the phenomena of habit. The habit of performing any action consists in the repetition of the corporeal act, without effort, and without any consciousness of the intellectual act which has been its primitive cause. The intellectual power, stimulated by some want or desire, originally set in motion the corporeal organs ; but in the course of time an immediate dependence comes to be established between them and the same want or desire, so that the intervention of the mind is no longer necessary to the production of the action : in this case, the action is no longer composed of an act of the mind, and an act of the body, but merely of the last, under the influence of the exciting cause, which originally gave birth to both. Almost all our actions may assume this character of habit, of which the slightest examination will afford us abundance of proofs.

Now if nature should have originally established this

* It may be remarked here, that the observation of instinct in animals seems to give a fatal blow to Messrs. Gall and Spurzheim's system of protuberances. Innate propensities, say these gentlemen, are marked by corresponding bumps on the head. But how account, upon this system, for the wonderful instincts of many insects, mollusca, and other species that are acephalous. In the acephala, and a multitude of other insects, remarkable for the vividness of their instincts, there is no brain, properly speaking. If an animal have no head, it can have no bumps upon the head, and consequently, according to Dr. Gall, no innate propensity, no determination to any mode of action, no instinct.

kind of dependence between the wants and organs of animals, the phenomena of instinct would admit, so far, of an easy explanation. When the organ was affected by the exciting cause, the action would follow as a matter of course, without intelligence having any thing to say to it. And query, if this is not really the case? If this hypothesis has not something more to rest on than mere analogy? If it is not to a certain extent based in fact?

In reality, do we not discover the existence of such an immediate relation in ourselves primitively established between the wants and organs? We have no more need than animals of the assistance of thought, to enable us to stop, to recede, or to fly, on the sight of a new object which affrights us. The sentiment of fear is alone sufficient to suspend or excite the action of the muscles, without intelligence taking the slightest share in the operation. Various other instances of this sort may be adduced in children, and even in men, to prove this immediate relation between the wants and organs. We should be unfortunate indeed, if our self-preservation, that of our offspring, and our species, depended always on the complicated processes of reasoning and the tardy decisions of judgment.

Now the phenomena of habit, only that we know how they originated, are exactly similar to those of instinct. In reading, in the exercise of arms, in the motion of the fingers on an instrument of music, everything is organic. When we read, we recognise characters, and articulate the sounds which they represent, while our minds are entirely occupied by the sense of what we are reading. The fencer follows with his foil the foil of his adversary, and no act of thought contributes to his rapid motions; the pianist runs over the keys in all directions with both his hands, forming all the combinations which the ten fingers are capable of, while his attention is exclusively fixed upon the notes placed before his eyes, and which he is employed in reproducing on his instrument. All these exercises, and every mechanic art, are performed with greater perfection, in proportion as they are more withdrawn from the influence of thought. As long as that is necessary, the action is badly performed, and it is only by approximating to animals, that we acquire perfection in all such exercises. In short there is nothing absolutely different in the *modus operandi* of habit, and that of instinct; and the comparison of the weaver and spider is much more exact than has been imagined. The two orders of phenomena might be so far confounded, that an instinct might appear to be positively formed from habit. A person who was exercised from infancy in collecting and hiding the remnant of his meals, would end by doing so as mechanically and as uselessly as the domestic dog.

These notions are by no means in opposition to the principles of mental philosophy which have been generally received. A distinction has always been admitted between the intellectual and active powers: two separate classes of phenomena belong to each, and, consequently, two different assemblages of organs, as the seats of these phenomena.

It appears extremely probable, that it is in the organs of activity that the instinctive faculties reside. The phenomena of habit seem to present, not only an excellent illustration, but something very like a direct proof of this. The frequent operations of the intellectual power, or of any other cause on the active organs, must be immensely strengthened by the repeated influence of the one, and the exercise of the others, and must end by becoming a necessary mode of action. In the same manner instinctive actions are the result of a necessary mode, but of a mode which, instead of being acquired, is primitive and essential to the nature of the beings which exhibit these actions. Thus the instinctive action and the habitual action may be considered, in the actual *modus operandi*, precisely the same ; but the difference is, that the instinctive action was from the beginning, what the other, from repeated practice, becomes in the end.

A single additional example will suffice to render this theory completely clear and intelligible. When a man, after having perfectly understood, and fixed in his memory the principle of riding, mounts a horse for the first time, none of his motions or attitudes, notwithstanding all his science, are what they ought to be. His body swings forward or backward, to the right or left, instead of remaining in a vertical position. His legs are moving about, instead of being fixed ; the motions of his hands and feet do not accord, and in a word, there is no sort of harmony between him and his horse. At first it is only by a very great mental effort that he is enabled to make any one of the motions requisite in a given case, then another in accord with the first, and finally, all that are prescribed by the art. By degrees the same effort of mind becomes less and less necessary ; the motions which were the most slow and difficult, are now performed with facility and promptitude, the moment the mind judges them to be necessary. At last, after a practice of a certain duration, intelligence takes no further share in this exercise. Everything that is requisite is performed, as it were, of itself. If the horse makes a contrary motion, this instantly produces a corresponding motion on the part of the rider, and that with the quickness with which the eyelid closes to protect the eye, or the head turns to avoid a blow. From this time, all the intellectual principles with which the exercise commenced, are transformed into simple associations of motion, in fact, into the merest mechanism.

We have seen in the text that the family or group of ordinary Pachydermata with cloven feet are assimilated to the Ruminantia, particularly in their ostelogical characters.

characters. We proceed to the genera and species of this group.

To the generic characters of the genus HIPPOPOTAMUS, including but a single existing species, it does not seem necessary to add anything here to the notice of them in the text; we shall proceed, therefore, briefly to describe the species from M. F. Cuvier and Mr. Burchell.

The locomotion of the *Hippopotamus* is as inelegant as its make. It inhabits principally the muddy banks of rivers, which it quits only by night in search of pasture, and at the least noise or slightest indication of danger, dives to the bottom of the water, and from time to time brings the nostrils only to the surface to breathe; hence it is extremely difficult to kill it. It is herbivorous, but lives also on the roots and bark of water-trees and plants.

The Hippopotami are gregarious in their habits, and are found probably throughout Africa, except the most northern part. Anciently they were common in Egypt.

" The Hippopotamus," says Buckhardt, " is very common in Dongola. It is a dreadful plague there on account of its voracity, and the want of means in the inhabitants to destroy it. It often descends the Nile as far as Sukkot. In 1812 several of them passed the Bahr el Hadjar, and made their appearance at Wady Halfa and Den, an occurrence unknown to the oldest inhabitants. One was killed by an Arab by a shot over its right eye; the peasants ate the flesh, and the skin and teeth were sold to a merchant of Sioutt. Another continued its course northward, and was seen beyond the cataract at Assouan at Derau, one day's march north of that place."

The Hippopotamus is not common at Shendy, though it occasionally makes its appearance there. During Mr. Buckhardt's stay there, one was in the river in the vicinity of Boeydha, which made great ravages in the fields. It never rose above water in the day-time, but came on shore in the night, and destroyed as much by the treading of its enormous feet as it did by its voracity; the people have no means of killing them. At Sennaar, where Hippopotami are numerous, they are caught by trenches, slightly covered with reeds, in which they fall during their nightly excursions. The whips called korbadj, which are formed of their skins, are made at Sennaar, and on the Nile, above that place. Immediately after being taken off, the skin is cut into narrow strips about five or six feet in length, gradually tapering to a point: each strip is then rolled up so that the edges unite and form a pipe, in which state it is tied fast and left to dry in the sun. In order to render these whips pliable, they must be rubbed with butter or grease. At Shendy they are sold at the rate of twelve or sixteen for a Spanish dollar; in Egypt, where they are in general use, and the dread of every servant and peasant, they are worth from half a dollar to a dollar each. In colder climates, even in Syria, they become brittle, crack, and lose their elasticity.

We have but little information on the sensitive organs of these animals; in conformity, however, with their brutal character these seem but little developed. The eyes are small, as are also the external ears; the nostrils are bulging, and surrounded to all appearance with cartilage and muscle, to enable the animal to shut or close them according to their position in or out of the water. The whole body is nearly denuded except at the end of the tail and the ears, and there are some rough hairs on the lips. It is believed that these animals bring but one young at a time, which the mother carries about on her back. They attain ten or eleven feet in length, and stand four or five in height. Mr. Burchell found about six bushels of chewed grass in the stomach. The food passes in a very undigested state, and even then has more the appearance of mingled grass and straw. The largest intestine, when inflated, measures about eight inches in diameter. A good figure from a living specimen is a desideratum in Zoology. We shall add Mr. Burchell's observations upon them in Southern Africa.

The monstrous size and almost shapeless mass of even a small Hippopotamus, when lying on the ground, appears enormous. The animal is entirely of one uniform colour, which may be correctly imitated by a light tint of China ink. The hide, above an inch in thickness, is hardly flexible; the ribs are covered with a thick layer of fat, known to the colonists as a rarity by the name of *zeekoe-spek* or sea-cow pork. This can only be preserved by salting, as in attempting to dry it in the sun in the same manner as the other parts of the animal are usually treated, it melts away : the rest of the flesh consists entirely of lean.

It is very seldom that the Hippopotamus is wounded in any other part than the head, but this does not happen from the impenetrable nature of the rest of the hide, a reason which has often been assigned, and originally invented, like many other such tales, for the purpose of exciting wonder. The truth is, that as the Hippopotamus hardly ever quits the river but at night, and by day seldom ventures more than its head above the surface of the water, there is no place left for the marksman ; for no bullet, owing to its great rapidity, can penetrate that element in a direct line when fired obliquely, but first rebounds as it were from the surface of it.

The Hippopotamus, when rendered wary by the suspicion of approaching danger, raises out of the water only his nostrils, eyes, and ears, which being all placed in the same horizontal plane towards the upper part of the head, it may with probability be concluded that nature assigned them this position with a view to ensuring the safety of the animal by enabling him to breathe, see, and hear, without exposing himself much, on which account they are not so easily shot as many other animals. Their great size is nothing in favour of the marksman; and, unless he aim with as much precision as if it were but a hare, he fires in vain.

When no more than the upper part of the head is seen above the water, it appears very much like the head of a horse, and sufficiently justifies (says Mr. Burchell) the name of Hippopotamus, given to it by the Ancients, who, as this circumstance seems to prove, could rarely have had a sight of the entire animal, otherwise they would have discovered, that, of all quadrupeds, this bears, in form and general appearance, the least resemblance to a horse. Nor can anything be more inapplicable than the colonial name of *zeekoe,* sea-cow, to which animal it has not the slightest resemblance. M. F. Cuvier, however, conjectures that its ancient name has reference to its voice, which Adanson informs us is like neighing. The name Cheropotamus, or River-Hog, has been suggested instead of Hippopotamus, or River-Horse. A more perfect analogy gives a preference to the former, but both are objectionable in strictness, by conveying a notion of generic identity, when a generic analogy or partial similitude only exists.

It would be interesting to be better acquainted with the biography of this mammiferous tenant of the African rivers. Its habits and location must bring it very much into contact with the Crocodile and its consimilars, on whose natural domicile the Hippopotamus seems to intrude. Were it not superior in strength to these formidable and ferocious reptiles, it would soon be driven from its watery retreat, but we hear of no struggles between these animals. As in all others, their relative capabilities seem known instinctively to each other; and while the strongest has no inclination to attack, the weakest is afraid of hostility. It may be remarked, that whatever superiority of intellectual powers is to be found relatively among the lower animals, such superiority is not in any case sufficient in degree to get the better of physical force; while in man, whose bodily powers, however modified and adapted to his station, are weak, intellect bids defiance to strength, and triumphs over all physical opposition.

The Hippopotamus, however related to the omnivorous swine, appears to live solely on the vegetable kingdom. It might be suspected, from its watery habitat, that it was partially piscivorous; but this is not the fact, and if it be driven to the river at all in search of food, it appears to be food only of a vegetable description which affects moist situations. In common with the Rhinoceros, the Swine, and even the Elephant, this animal is fond of wallowing in the mud, and may resort to this habit as well for a protection against parasitical insects and flies, as from an instinctive impulse.

The males are said to fight desperately for the possession of the females: the latter bring but one at a birth, after a gestation, as it is said, of about nine months. The young take to the water almost immediately after birth. On the whole, however, it appears we have yet much to learn on the natural history of this formidable beast.

The name of SWINE, or SUS, has been extended by naturalists, from the animal commonly so called, to all those which have any generical relations with it. Still these relations are not so close as to prevent the different species from forming very distinct and characteristic groups, whose organic modifications are sufficiently important.

Independently of the Swine, properly so called, this genus contains the *Babyroussa*, the *Phaco-chœres*, and the *Pecaries*. These last, notwithstanding the arrangement of the Baron, we shall venture, on the authority of his brother, to include among the Swine; for the Pecaries, notwithstanding their peculiar traits, have the constitution and the principal characters of the Hog.

To add anything to the generic description in the text, as far as conformation is concerned, would be equally superfluous and tedious. The senses of these animals, that of smell excepted, are obtuse enough. The imperfection of that of feeling is much increased by the thick coat of fat which usually covers the body.

The Domestic Hog may serve to give a pretty exact idea of the other species of this genus. Their proportions and gait are equally clumsy. Heaviness and length of head, short neck, limbs short and thin in proportion to the body, are their principal external traits. Their usual pace is a trot ; they go with the head downwards, and the eyes directed forward. They delight in humid, marshy, and muddy places, where they dig for roots and worms

The intelligence of these animals is limited, and they are not in general very susceptible of education. Yet they are easily tamed, and become attached to those who treat them well. They feed almost indiscriminately on animal or vegetable substances; but it may be considered that roots and grains form their principal nutriment. They are found in very large herds, and usually in unfrequented places. The voice of all the species resembles more or less that of our Domestic Hog. They are found in all parts of the globe, with the exception of New Holland.

The *Wild Boar. (Sus Scrofa var. aper.)* The Wild Boar is of a brownish-black over the whole body. It retires into the thickest part of the forest, where it chooses a retreat, from which, when attacked, it never sallies forth but in the last extremity. December or January is the rutting time of these animals; gestation continues about one hundred and twenty days or more, and there are six or eight young ones at a birth. These are striped irregularly with longitudinal bands of different depths of brown

C.Hamilton Smith Esq.ʳ del.ᵗ

THE WILD BOAR.

SUS APER.

London Published by G.B.Whittaker.Sep 1825.

on a ground in which white and fawn colour are mingled. This disappears with the second change of hairs.

The old boar lives usually alone; but the females unite together, and form with their little ones very numerous troops for mutual defence. This takes place especially during the nonage of their offspring. The mothers then become furious; the strongest oppose the danger, pressing one against the other, and placing the little ones behind.

The chase of the Wild Boar is extremely dangerous. The great strength of the animal, and its powerful tusks, render it very formidable to the dogs and hunters. When forced to quit its retreat, it retires at first but slowly, and destruction is the lot of the dogs which press it too closely. The instant it is wounded it stops, and rushes on him from whom it appears to have received the blow. It is on such occasions that the Boar is truly terrible : he tramples under foot and tears in pieces everything in his way. It is at about four years old that the boars are most difficult to be hunted : they will run for a very long time, and their tusks being at that age straighter and more trenchant, inflict much deeper wounds. The old boars do not run to the same distance, and cannot wound so severely with their curved tusks. The chase is usually performed by large dogs of the mastiff breed.

The Wild Boar seeks his food in the evening; it commits desperate devastation in the cultivated fields which adjoin his native forests, and, when pressed by hunger, will not hesitate to attack even living animals.

This species is found in the temperate regions of Europe and Asia, and we are also assured that it exists in Syria, India, and the northern parts of Africa. There can be no doubt that this species is the root of our Domestic Hog. In the South Sea Islands, however, it may be observed that there is a small black variety with short legs, and the Wild Boar is not known in those islands. The mode indeed in which these remote islands first received their animal population is very mysterious; but whatever that mode might be, analogy seems to lead to the conclusion that it is to the species in question to which the black variety of the South Sea Islands owes its origin, as well as all the rest. The Wild Boar also, however fierce in disposition—however different in person, evinces a capability of being tamed, which further intimates its relationship to the domesticated races, with all of which it will propagate a perfect posterity.

The specimen we have figured was a very fine one, killed in the Island of Peacocks, near Potsdam. We have just said that the Wild Boar appears to form the original stock whence all our domestic swine have sprung: for these animals produce together a race that reproduce permanently. The different modifications produced in these animals by domestication have not been much studied, and accordingly naturalists admit but four or five principal varieties in the species.

The intelligence of the Hog seems to have gained but little from the superintendence of man. They return of their own accord from the fields to their sty, recognise their keepers and follow their call. We are told that in some parts of Scotland they are harnessed in company with the Horse and Ass. Gestation continues nine months, and twelve, or sometimes even fifteen, constitute a litter.

In the variety of the Common Hog are many races which it would be interesting and important to examine more closely than has hitherto been done. Their most usual colour is a dirty white, but some are altogether black, and some are pied. The principal races are:

(a.) *The English*, which acquire an extraordinary bulk, and sometimes arrive at the weight of twelve hundred pounds. They are whitish, and the body much elongated. There are two other races in this country. The one of small size, produced by crossing the Chinese Pig with the Wild Pig of North America. The other larger, from the crossing of the English Pig with the Chinese.

(*b.*) *The Jutland race* is distinguished by an elongated body and pendant ears, but especially by the curve of the back and length of the limbs. It is a considerable object of commerce.

(*c.*) *The race of Zealand* is smaller than the preceding. Its ears are rather raised, and the back well furnished with silky hair. The individuals of this race, when fat, yield from one hundred and sixty to two hundred pounds of lard.

(*d.*) *The races of Poland and Russia* are generally very small, and of a reddish colour.

(*e.*) *The black race with short limbs.* This race is distinguished by the shortened proportions of the head, the folds of skin above the eyes, thick jaws, small neck, broad back, few hairs, except bristles, long body, and small straight ears. It is proper to the south of Europe, and the race of Bergamo is confounded with it. From these Pigs are, or ought to be made, the Bologna sausages.

(*f.*) *The races of France* are principally that of the valley of Auge in Normandy; the head of which is small and pointed; the ears narrow, the body long and thick, the hair white and scanty, and the bones in general small; weight about six hundred pounds: that of Poitou, with a strong head and projecting forehead, large and pendant ears, elongated body, rough hair, legs large, and yet the weight not within a hundred pounds of the last: that of Perigord, with black and rough hair, short and thick neck, thick but compact body. This last race, mingled with the preceding, produces an intermediate race, which is pied, and very much in request.

(*g.*) *The race with a single toe*, or rather with three united, is doubtless one of the most important to the naturalist. This singular race, which was known to Ari-

stotle, and whose existence has been admitted by all naturalists, has hitherto been but imperfectly described.

The toes of the Common Pig, like all perfect toes, are
formed of three phalanges.   Two of these toes much
shorter than the others, and which, in walking, are not
placed on the ground, are situated on each side of the two
middle toes, a little backwards.   The two greater toes
touch, and exceed the others in length by the two last
phalanges.   Now the small lateral toes have suffered no
change in the Solipede Pig; it is in the structure of the
middle toes that the characters of this race consist.   Two
phalanges are singularly developed between the second and
third; the extremity of one toe, being extended by a hoof,
which serves as an intermedium to unite the two others.
This union, however, is but imperfect, and seems produced
only by the compression which the supernumerary hoof
occasions, for amidst all these irregularities, the traces of
the three hoofs are clearly distinguishable.

(h.) *The Turkish Pig*. This may, moreover, be considered
as a variety of the Common Hog, by reason of its peculiar
traits.   It is found pretty extensively in Hungary, and
Turkey in Europe.   The individuals of this race have a
short and narrow head, ears erect and pointed, legs slender
and short, body very short, and hairs frizzled, of an iron
gray, and sometimes black or brown.   It appears that this
Pig is more easily fattened than ours.

(i.) *The Siamese Pig* is small, long bodied, very low on the
limbs, tail pendant, ears erect and very small, few silky
hairs, colour generally black, sometimes white, rarely
spotted.   The flesh is delicate and well tasted   This
variety is found in all the South Sea Islands very fruitful,
but not profitable, on account of its small size.

(k.) *The Pig of Guinea*. This variety, not much known,
seems to be of the Common Hog, though some authors have
made it the type of a different species.   It is the *Sus*

Guineensis of Brisson and Klein, and the *Sus Porcus* of Linnæus. It is known only by the account and figure given by Marcgrave. Its size is small, like the Siamese, but it is particularly distinguished from all other races by its elongated and pointed ears, and its tail descending almost to the ground. Its coat is frizzled, but soft in comparison with other pigs. It is of a reddish colour. The head seems rather slender. It is said to be frequently exported from Guinea to America.

The Masked Boar. (*Sus Larvatus.*) This species is distinguished by a fleshy prominence on the fore part of the head, entirely enveloping the upper half like a mask. It is a native of the Cape, nearly the size of the European Boar, and has all its proportions. The only distinction is in the fleshy protuberances. From the head to the eyes it is of the usual figure, but from under the eyes commences this protuberance, which gradually diminishes towards the snout. Thus there appear to be two heads, the half of the one being as it were enclosed in the other. This Boar also appears, according to the figure of Daniel, to have on each side of the face, under the eyes, two other very large excrescences, the surface of which is very irregular and wrinkled. The peculiar characters of the skull correspond with this facial mask. They principally consist in the great developement of the external edge of the alveolus of the upper canine. This edge, in the Common Boar, does not exceed an inch in elevation ; whereas in the species of which we are treating, it is developed into a long apophysis, and terminates in a broad tuberculated swelling, corresponding to the similar tubercles of the middle part of the bones of the nose.

The head of this species is, moreover, distinguished by a large arch formed by the cheek-bones, and by the long surface to which the muscles of the trunk are attached.

The habits of this animal are little known, but it appears to be very wild, dangerous, and intractable.

There is a specimen in the British Museum under the name of *Sus Larvatus,* which, when compared with the figure in Daniel's African Scenery, to which Cuvier refers for the type of Sus Larvatus, is considerably different, so much so, as to render the identity of the two doubtful; but the dried specimen in the Museum may, in its fresh state, have had the immense protuberance observable in Daniel's figure.

Had not the Baron himself adverted to the osteological characters which distinguish his division, Phaco-chœres, from the Common Swine, and placed the Sus Larvatus among the latter, apparently from actual observation, we should have strongly inclined to the opinion that the animal of Daniel's was referrible to the S. Æthiopicus of Gmelin.

But assuming the distinctness of the Asiatic species, or Sus Larvatus, may it not then be the *Leucrocuta* of Pliny, lib. viii. cap. xxi. Leucrocutam pernicissimam ferocem asini fere magnitudine cruribus cervinis collo cauda pectore leonis capite melium biscula ungula ore ad aures usque recesso dentium loco osse perpetuo. Hancce ferocem humanus voce tradunt imitari.

Babyroussa. (Sus Babyrussa.) Though this animal was not unknown to the ancients, and has been frequently noticed by the moderns, it has never been brought into Europe, and its head alone is to be found in the cabinets of Natural History; its character and habits have not been described.

Pliny evidently mentions it in his eighth book, chap. lii. when he says that Wild Boars are found in India, which have two horns on the forehead, similar to those of a heifer, and tusks like those of the common Wild Boars.

Ælian mentions one under the name of τετραχερως (four-horned); and Cosmes, the solitary, who lived in the commencement of the sixth century, in his description of the

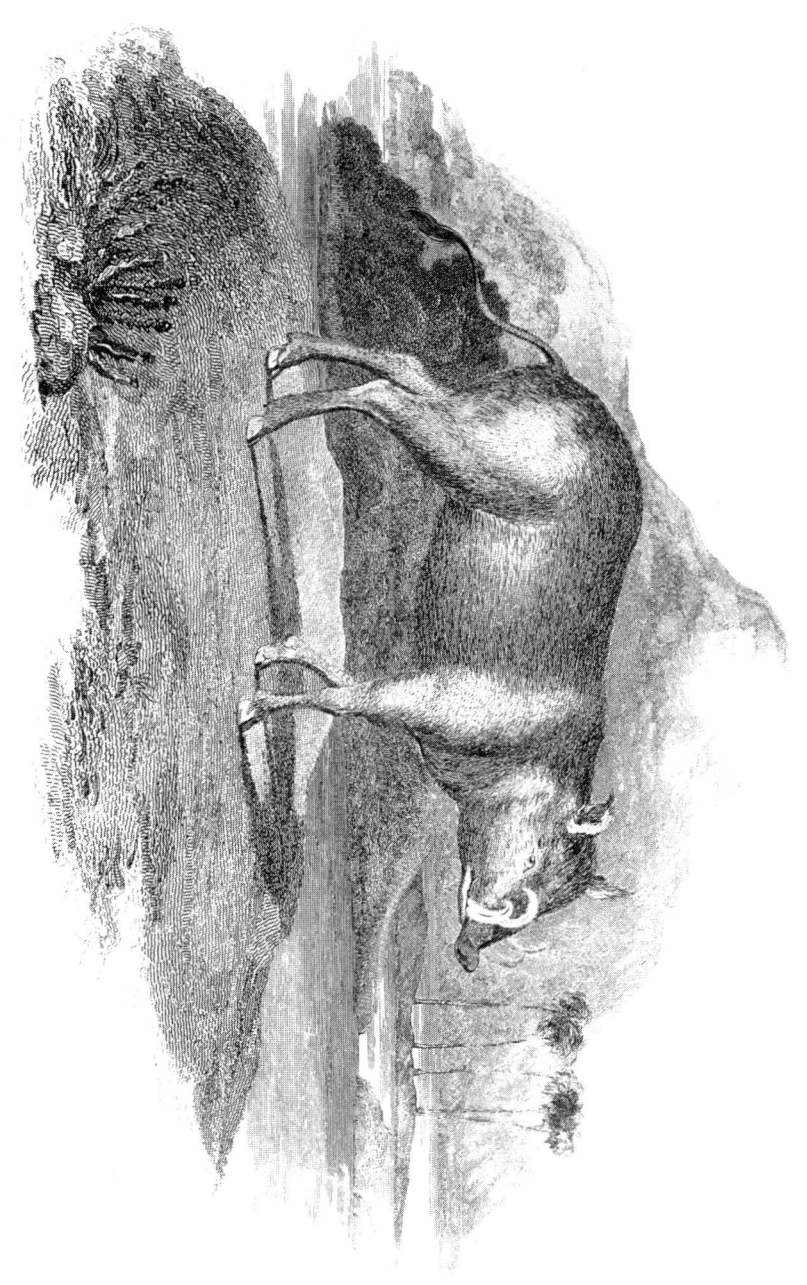

THE BABIROUSSA.

SUS BABIRUSSA.

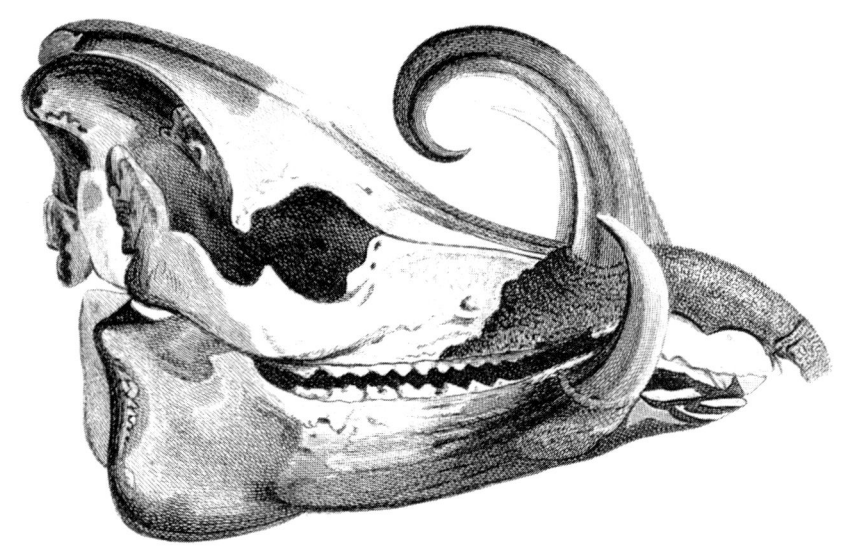

SKULL OF THE BABIROUSSA.

Griffith fc

British Mus.

London Published by B.G. Whittaker May 1826.

animals of India, treating of the χοῖρελαφος or Swine-deer, states that he had both seen and eaten of that animal. Valentin alone has entered into some details concerning this curious species.

The forms of the Babyroussa are not so clumsy as those of the other species of this genus ; but it seems perfect exaggeration to compare it to the stag in lightness. There is an entire affinity between it and the other Hogs. It has their heaviness and peculiarity of gait ; like them it has the thick neck, the head terminated by a snout, and small eyes. Its limbs are a little more elevated, and the neck somewhat longer. The head is likewise more narrow, and the long tusks of the male give it a very peculiar physiognomy.

The general colour of this animal is a reddish-ashen. The hairs are short and woolly, but a few long silken ones are observed to escape from among the others. Its skin is thin, and not covered underneath with a coat of lard. Its scent is very fine, and the flesh has a very agreeable taste. The voice exactly resembles the grunting of a Pig. It feeds on herbs and leaves, and Valentin asserts that it does not root, which, however, appears doubtful.

The Babyroussas, when hunted, throw themselves into the sea, being excellent swimmers. They thus pass in the Indian Archipelago from one island to another. They are easily tamed, and on this subject we may remark, that the extent of brain in a Babyroussa is nearly double that of the brain of the Wild Boar ; but the head of the latter is greater by the extent of the frontal and occipital sinuses. We have annexed a figure of this animal from Major Smith's collection, and another of the skull from a specimen in the British Museum, which has some of the skin remaining upon it.

We have already ventured on an observation relative to the possible identity between the S. Æthiopicus of

Gmelin, and the S. Larvatus or Masked Boar of Daniel, which our author refers to two different subgenera, the latter to the Common Hogs, and the former to Phaco-chœres or Warted Hogs, on which we shall now say only a word or two. We have engraved a figure of the Æthiopian Boar from a specimen in the Museum of the Missionary Society, which accords so much in its dried state with the specimen in the Museum before mentioned, and which is referred to Daniel's species, as to have given rise to the hesitation we have stated on the subject. It may be, indeed, that the specimen in the British Museum should be referred to the African species of Gmelin.

The other species of Phaco-chœres, or the Cape Verd Boar, the Sus Africanus of the text, is thus described in Gmelin: Sus Africanus, dentibus primoribus duobus, corpus setis longissimis tenuibus tectum, caput elongatum, nasus gracilis, dentes canini lati, eboris duritiæ, superiores crassi, obliqui, truncati, molaris in utroque maxilla utrinque vj. Anteriores maximi, maxilla superiores inferiori multo longior; auriculæ angustæ, erectæ, acuminatæ, apice telis longissimis barbatæ, cauda gracilis, floccosa, primum crurium articulum attingens. There is a fragment only of this species in Paris.

We have, not without hesitation, engraved from a drawing by Howitt of certainly a specimen of this species, though a very old and ill-stuffed one, which was lately in Riddell's Museum, and came from the Leverian. The principal ground of our hesitation has been from the indistinct delineation of the teeth, or rather from the appearance of two canine teeth in the lower jaw. The molares anteriores maximi of Gmelin's description may, indeed, especially when the shrinking consequent on long keeping a stuffed specimen is considered, account for the appearance the animal exhibited, which the artist has copied without thorough investigation. As, however, there

ÆTHIOPIAN BOAR.

London Published by G. & W.B. Whittaker, Feby 1824.

THE CAPE VERD BOAR.

SUS AFRICANUS. Gm.

is no other figure, and the bearded ears and lengthened hair of the animal are very peculiar in its kind, the figure may perhaps not be uninteresting.

The subgeneric characters of the dorsal gland or opening, the teeth, and the toes, which separate the PECCARIES from the common swine, are noticed elsewhere. We have only here to give some further detail of the species. Until the time of D'Azara it was presumed that only one species of the Hog with a dorsal gland existed in America: that naturalist, however, has distinguished the Common Peccary from the white-lipped species or Tajassu.

The Peccary and the Common Hog resemble each other in general form and manner of living, and in their taste and readiness for an omnivorous sort of diet; they also dig with the nose, and eat and drink in the same manner; their bristles, moreover, become erect in anger—they all respire strongly, and grunt when irritated: but the Peccaries are observed to be more readily tamed, and to submit to the authority of man with more celerity than the Wild Boar.

" It is said," says D'Azara, " that their flesh is good, but that it is necessary to deprive them of the dorsal pipe or gland immediately they are killed, otherwise the meat will taste of the secretion produced there, but the Indians it appears eat them without doing so."

The Peccaries differ from the Wild Boar and Domestic Hog in having the head shorter and thicker, the angle at the buttock closer, and the body, neck, ears, and legs, shorter. They have no visible tail, but D'Azara states that a very short one may be found on close inspection, which is flat; their bristles are nearly stiff enough to penetrate a considerable resistance.

Both the species inhabit the large and thick forests of America; the Common Peccaries in numerous herds conducted by a male leader, and the white-lipped species in pairs, or in small numbers only—the two species keeping always distinct from each other, and not even resorting to one place at the same time.

It is said of the common species, that when one of the herd is alarmed, he makes a signal with his feet, which is repeated by all the rest; they then are on their guard, and if attacked, unite and surround the Jaguar, the Puma, or even the man, who may so assail them, and will speedily destroy him, unless he escape up a tree, or have the good fortune to kill the leader, when the rest take to flight, which they will not do, though many of the common herd be killed. This report, however, though probably not altogether baseless, seems exaggerated, as D'Azara's personal experience evinced. It is said moreover, that the Jaguar, the great predatory of the American forest, will follow these herds in silence, and seize the opportunity of an individual being in the rear to seize and kill it in an instant, when it immediately takes to a tree for refuge till the herd have passed and left their dead companion behind. If this trait of character were sufficiently established, it would give rise to many curious reflections.

The white-lipped species, on the contrary, not availing itself of the benefit of combined efforts, flies at the first attack, and defends itself only in the last resort. It is also less powerful than the other species. The other personal characters which distinguish these two species will appear by a description of each.

The first species is sometimes called the *Collared Peccary.* *(Dicotyles Torquatus.)* This animal, colour excepted, has all the externals of a young Wild Boar. Its magnitude, however, does not exceed that of a middle-sized Dog. The hairs are thick and bristly, and their large rings, alternately black and white, give to the skin of the animal an uniform division of these colours. We must, however, except a white and narrow band which surrounds the neck, and takes an oblique direction from the elevation of the

THE TAGNICATI OR WHITE-LIPPED PECARY.

DICOTYLES LABIATUS. Cuv.

shoulders to the front of the limbs: the dorsal line is also blacker than the rest of the body. There is a gland on the back which produces an odorous matter smelling like musk, according to M. D'Azara; but according to M. F. Cuvier, the odour thus emitted is very fetid, like that of the perspiration from the arm-pits. It is emitted in greater abundance when the animal is angry, because the muscles of the skin are then contracted for the purpose of elevating the dorsal bristles. The male and female exactly resemble; they live together in the forests. Once a year the female brings forth two young ones, of an uniform reddish tint. These animals are very easily tamed, and rendered as domestic as the Hog itself.

Two of these animals in the French Menagerie lived on the best possible terms with the dogs and other animals in the yard; they returned of themselves to their stall, came when called, and appeared fond of being caressed. But they loved liberty, always endeavouring to escape when forced to return, and sometimes attempting to bite; they wounded a young boar which had been placed along with them; they delighted in heat, and suffered much and grew very thin from the effects of cold. Bread and fruits constituted their usual nutriment; but, like the Domestic Hog, they would eat almost any thing. They were habitually silent; but, when frightened, would utter a sharp cry, while they expressed their satisfaction by a slight grunting.

The *Tajassu (Dicotyles Labiatus.)* The colour of this species is, in general, black; sometimes under the flanks and belly, and between the eye and ear, silken hairs are seen, which have in the middle a whitish ring, giving to these parts a grayish tint, and the lower jaw is entirely white; the male and female are similar: the young ones are born about the month of April. Their tint, in the upper parts, is of a greyish red; the lower jaw is often white, as well as the under part of the body. The young Tajassu does not assume the colours of the adult, until a year has elapsed; until then it bears some resemblance to the Peccary. The matter produced by the dorsal gland of this species is inodorous.

It seems extremely probable that all the species of the Boar are not yet known to naturalists. Dampier, for instance, speaks of the Wild Boars of Mindanao, of a hideous figure, with large tufts or protuberances on the eyes. Dapper relates, that in the kingdom of Quoja, a particular species of the Wild Boar is to be found, called by the negroes *Cauja Quinta*, and the majority of travellers to the Guinea coast, speak of Wild Pigs not easily referrible to the species which are known.

The name of Hog, or Pig, has been sometimes given to animals which differ essentially from those which we have just described. Thus the Europeans have called the Cavy the Water Hog. The animal improperly called a Guinea Pig among ourselves, and *Cochon d'Inde* by our neighbours, is known to be one of the glires or rodentia. The Tatous are called Hogs in Armour by the Spaniards. The Hollanders of the Cape call the Porcupine the Iron Hog. The Porpoise has been called the Sea Hog, and the same name has been given by Molina to a species of the Phoca. The Hog Ape of Aristotle is a species of the Quadrumana not easily recognised at present, but is probably a Baboon, as may be judged from its figure on the Mosaic of Palestrina.

Having thus entered into all the details concerning this genus which it appeared important to notice, we shall indulge in a few brief general reflections on the animals which compose it.

Man, in a social state, was not satisfied with having reduced to subjection animals which appeared untameable, with having made the proud and impetuous steed the companion of his labours, his journeys, and his combats,—the massy and powerful Ox his drudge in agriculture,—with searching through rocks and precipices for the Ram and the

Goat to form colonies around him for an abundant supply of nutriment and clothing,—nor with having modified, subdued, and softened, the carnivorous nature of the Dog, and transformed him into a guardian, a guide, a companion and a friend, an active and intelligent agent of his commands, whose fidelity no bribery can corrupt, and whose attachment can neither be shaken by chastisement nor ingratitude.

Repeated success only inflamed the ambition of man, and spurred him forward in pursuit of new accessions. Having subjugated the most useful species, and satisfied his most pressing wants, still desirous of beholding the most perfect abundance around him, he proceeded to lay other tribes of the animal kingdom under contribution. He drew the formidable Boar from his native forests, and by care and by attention to the quantity and quality of its nutriment, has rendered its flesh one of the most common and most savoury articles of diet.

Let us pause a moment to contemplate these conquests over the brute creation.—Unlike the conquests over our fellows, achieved by violence and cemented by blood, conquests which argue nothing but perverseness and degeneracy, these peaceable victories most evidently demonstrate the natural superiority of man, the strength of his judgment, and the fertility of his imagination. If in his physical conformation he approaches to the animal tribes, how is he not raised above them by his mental power! By the power which at pleasure can change their nature, can tame the wild and subdue the obstinate, and make his will their law! The strongest and most intelligent species of the brute could never yet achieve the subservience of another; force is nothing. The majority of species thus appropriated by us possess the greatest physical strength, but they must yield to the power of mind ; to that untired activity, and sleepless intelligence, which emanate from the Divinity himself, and the perfection of which constitutes the exclusive prerogative of man.

The astonishing fecundity of the animals now under consideration, is one of their most obvious and remarkable characters. They live and multiply in every climate of the world, with the exception of the Polar regions ; accordingly we find that, though their natural life would, if permitted, extend to fifteen or twenty years, yet they are capable of reproduction from nine months or a year old. Their lubricity is extreme, and even furious. The rut is almost perpetual, and the female, even in a state of pregnancy, will seek the male. It is even said that she will occasionally admit the advances of a male of a different species. The production of fifteen or even twenty in a litter is not unfrequent, and instances have been known even of thirty-seven. The celebrated Vauban has made a calculation of the probable production of an ordinary sow, during the space of ten years. He has not comprehended the male pigs in his estimate, though they may reasonably be supposed as numerous as the females in each litter. Moreover, six young ones only, male and female, have been allowed to each, though generally they are more numerous. The result is, that the product of a single sow in eleven years, which are equivalent to ten generations, will be six million four hundred and thirty-four thousand eight hundred and thirty-eight pigs. Taking it however in round numbers, and allowing for accident, disease, and the ravages of wolves, four hundred and thirty-four thousand eight hundred and thirty-eight, there will remain six million of pigs, which is about the number existing in France. " Were we to extend our calculations," says Vauban, "to the twelfth generation, we should find as great a number to result as all Europe would be capable of supporting ; and were they to be continued to the sixteenth, as great a number would result as would be adequate to the abundant peopling of the

globe." A remarkable instance of the fecundity of these ani
mals occurred in this country about twenty-eight years
ago. A sow belonging to Mr. Thomas Richdale, Kegworth,
Leicestershire, had produced, in the year 1797, three hundred and fifty-five young ones in twenty litters; four years
before, it brought forth two hundred and five in twelve
litters, and afterwards it had eight litters more. The
number produced in these last added to the first, made the
three hundred and fifty-five.

This remarkable fecundity, united with the deficiency of
all other useful qualities in this animal, and the excellency
of its flesh, points it out as a most obvious source of nutriment. Let the consumption be ever so great, there will
always be an ample supply for the demand. The ease, too,
with which these animals are brought up and fed, renders
them a most advantageous property to the poorer classes of
society. In the country there are few families that cannot
rear a single pig every year, and thus obtain a cheap and
nutritious diet, not to mention the profit arising from the
lard, fat, *&c.*, of the animal. In some countries the principal source of existence, to the poor peasant, is his Pig.
In Ireland these animals are brought up and fattened to a
large size, and then brought to market by the owner, and
sold at a tolerable price; with part of this a younger,
leaner, and worse-conditioned Pig is purchased, fattened
in the same way, and sold at a profit. Happy for the poor
peasant, if this only property be not seized by some inex-
orable landlord, or some tithe-farmer, or *middle-man,* a
species of vermin, for the extirpation of which Ireland
might well exchange her boasted exemption from less pernicious reptiles.

Among the ancients, the Hog was in very high esteem.
It was the peculiar sacrifice to Ceres, the goddess of harvest.
In the Island of Crete, Hogs were regarded as sacred.
In ancient Rome, very particular attention was bestowed

upon them, and the art of rearing and fattening them was much studied, an art which the Latin writers on rural economy have termed *Porculatio*. Under the Emperors, gluttony and epicurism were carried to an excess equally cruel and disgusting. Among the rich there were two very famous methods of dressing this animal. The one consisted in serving up a Hog entire, with one side roasted, and the other boiled. The other mode was called the *Trojan*, in allusion to the Trojan Horse, whose interior was filled with combatants. The inside of the Hog, from which the viscera had been withdrawn, was stuffed with victims of all kinds, such as thrushes, larks, beccaficoes, oysters, &c., the whole being bathed in the best wine and the most exquisite gravy. So great was the expense of this dish, that it became the subject of a sumptuary law, while the barbarous modes of torturing this poor animal to death, for the purpose of imparting a higher flavour to its flesh, passed unpunished and unregarded. It is not possible to read the anecdotes found in history of such infernal gluttony, without horror; but we shall forbear any mention of atrocities which make us blush at belonging to the species capable of committing them.

In hot climates the flesh of swine is not good. M.Sonnini remarks, that in Egypt, Syria, and even the southern parts of Greece, this meat, though very white and delicate, is so far from firm, and so surcharged with fat, that it disagrees with the strongest stomachs. It is therefore considered unwholesome, and this will account for its proscription by the legislators and priests of the east. Such an abstinence was doubtless indispensable to health under the burning suns of Egypt and Arabia. The Egyptians were permitted to eat pork only once a year, on the feast day of the Moon, and then they sacrificed a number of these animals to that planet. At other times, if any one even touched a Hog, he was obliged immediately to plunge into the Nile with his clothes on, by way of purification. The swine-herds formed an isolated class, the outcasts of society. They were interdicted from entering the temples, or intermarrying with any other families. This aversion for swine has been transmitted to the modern Egyptians. The Copts rear no Pigs, no more than the followers of Mahomet.—The Jews, who borrowed from the Egyptians their horror of Pigs, as well as many other peculiarities, continue their abstinence from them in colder climates, where they form one of the most useful articles of subsistence.

We have seen that the species of the genus RHINOCEROS vary in regard to the leading character of dentition, both in the number and form of the teeth. The incisors are either altogether wanting, or are four in each jaw, two very strong, and two small and weak, and the pairs varying in their relative situation. They have, however, no canine teeth: the cheek-teeth are seven in each jaw, on each side: the upper teeth are square, with several convex lines; the lower have their crowns furnished with transverse prominences.

The teeth therefore afford no very good generic character. Their distinct toes, while they separate these animals from the swine, connect them as a group or family with Hyrax, or Tapir, or the Common Pachydermata, without a proboscis; but the character which very properly and very strikingly marks them as a genus, is that from which they are named, Rhinoceros, (ριν κερας,) or nasal horn. The species differ in this respect, however, some having one, and others two horns, but they agree in the very singular position of this organ, or weapon, which is on the nose.

The remaining characters we would here shortly refer to as generic, are their heavy body, thick legs, head short, with the occiput elevated, and the cerebral cavity small, eyes small, ears moderate, tail short, round at the base, and compressed laterally toward the extremity; two inguinal mammæ.

The horns of this genus present a singular character: they do not envelop a bony axis, like the horns of the ruminating animals, nor do they partake of the osseous nature of the horns of stags, but they appear to be formed of horny fibres, like thick hairs closely agglutinated together. There is much room for observation on the structure of these horns, but instead of attempting any remarks of our own upon the subject, we shall illustrate it much better by transcribing what has been already so ably observed by Mr. Burchell.

" The horn of the Rhinoceros (says he) differing in structure from that of every other animal, and placed in a situation, of which it is the only example, had long appeared to me to be an anomaly very deserving of examination; and therefore on the present occasion (the first in which he had inspected one recently killed) it was the first object of my curiosity and attention. The view which I now began to take of its structure and nature, was afterwards, in the course of my journey, further confirmed by the following mode of reasoning, which, to render it less complicated, I shall confine to the class of Mammalia, or, as it is more commonly called quadrupeds. Dispersed over the skin of all animals are pores, which I have supposed to secrete a peculiar fluid, which may be designated by the name of corneous matter. This secretion, or fluid, is designed by nature for the forming of various most useful and important *additamenta*, all of which continue growing during the whole life; have an insertion not deeper than the thickness of the skin, and are further distinguished by the absence of all sensibility and vascular organization, being purely exuvial parts, like the perfected feathers of birds. In all these parts, the growth takes place by the addition of new matter at their base. When these pores are separate, they produce hairs. When they are confluent and in a line, they produce the nails, the claws, and the hoofs, the fibrous appearance of which naturally leads to the supposition of their being confluent hairs; and the same may be said of the scales of the Manis.

" The quills of the Porcupine, Hedgehog, and other animals, may be regarded as hairs of extraordinary size. When the pores are confluent and in a ring, they furnish the corneous case of the horns of animals of the ruminating class ; and when confluent on a circular area, they supply matter for the formation of a solid horn, such as we see on the Rhinoceros. An examination of the structure and appearance of this latter will be found to support my explanation of its nature; as about its base, it is in many instances evidently rough and fibrous like a worn-out brush. It grows from the skin only, in the same manner as the hair, a circumstance which entirely divests of improbability the assertion of its being sometimes seen loose, although by no means so loose as some writers have supposed. Nor is it at all extraordinary that the Rhinoceros should possess the power of moving it, to a certain degree, since the Hog, to which, in a natural arrangement it so closely approaches, has a much greater power of moving its bristles, which, if concreted, would form a horn of the same nature. With respect to the idea which I had entertained, of a single horn being an anomaly, it arose from the consideration that all the osseous parts of animals, excepting the spine, were in pairs ; those which appear single being in fact divided longitudinally by a suture : so that any bony process, such as that which supports the corneous case of horned animals, must, to be single or in the central line of the face or head, stand over a suture ; a case which no anatomist has hitherto discovered in nature. The single horn of the Rhinoceros is therefore no anomaly, because having no connexion with, or not deriving its origin from the bones, and being, as I have endeavoured to show, only concreted hair, nature might, if its mode of life required, have given it other horns of the same kind on any part of the body, without at all disturbing that system and those laws which she has followed in the structure of every quadruped.

" It is this rule of nature, and consequent reasoning, which will not allow me to believe that the Unicorn, such as we see it represented, exists anywhere but in those representations, or in imagination; and many circumstances concur to render it highly probable, that the name was at first intended for nothing more than a species of Rhinoceros."

The common Two-horned Rhinoceros, the *R. Bicornis* of Linnæus, when named, was supposed to be the only species distinguished by two horns ; modern discoveries, however, have refuted this notion, and our author substituted the epithet *Africanus* for that of *Bicornis:* this, however, appears to be an insufficient distinction, as still more recently Mr. Burchell has described a second species, with two horns, proper to South Africa.

This species is destitute of incisive teeth, and even of an intermaxillary bone; the skin is excessively thick, but not so much so as the Asiatic species. Mr. Burchell found that musket-balls, of a mixture of lead and tin, penetrated this skin easily, though they were flatted by striking against the bones; but he thinks that balls of lead alone, or if fired with a weak charge of powder, might possibly be turned by the thickness of the hide: it is perfectly smooth, and destitute of those extraordinary folds which mark its Asiatic congener. On the central line down the face, and above the nostrils, is placed the first and largest horn, the lower edge of which is nearly on the same horizontal plane as the eye; from this lower edge is continued the flexible upper lip, which the animal makes considerable use of as an organ of touch, and for seizing its food; the lower horn represents a long sharp-pointed cone, crescented or inclining gently backward from the base ; a very short space above this horn, is situated the second or upper horn, not half the length of the former, but nearly as big at the base, conical in shape, and perfectly straight, and placed nearly above the eyes, in a concavity described from the upper edge of the base of the lower horn to the top of the head. The horns being placed upon, but not emanating from the bone of the nose, are partially moveable, more so than in the Asiatic species, arising probably from a greater smoothness of the bone in question in this.   No hair appears upon the animal except at the edge of the ears, and extremity of the tail.

The first view of this beast suggests the idea of an enormous hog, to which besides, in its general form, it bears some outward resemblance, in the shape of the skull, the smallness of the eyes, and the proportionate size of its ears; but in its shapeless clumsy legs and feet, it more resembles the Hippopotamus and Elephant.   It is in fact in many less obvious particulars closely allied to all these.

The dimensions of one measured by Mr. Burchell, were in length over the forehead and along the back from the extremity of the nose to the insertion of the tail, eleven feet and two inches, but in a direct line not more than nine feet three inches; the tail, which at its extremity was complanated or flattened vertically, measured twenty inches, and the circumference of the largest part of the body eight feet four inches.

The senses, and a partial account of the character of this species, were described and given to Mr. Burchell by a native South African of experience, than whom no one can be better enabled to give us information on the subject : we shall therefore insert his account.   The outline of this account accords with Bruce's interesting description of this animal :—

" Their smell is so keen and nice that they know, even at a great distance, whether any man be coming towards them; and on the first suspicion of this take to flight. Therefore it is only by approaching them against the wind, or from the leeward, that the hunter can ever expect to get within musket-shot. Yet, in doing this, he must move silently and cautiously, so as not to make the least noise in the bushes as he passes through them; otherwise their hearing is so exceedingly quick, that they would instantly take alarm, and move far away to some more undisturbed spot. But the dangerous part of the business is, that when they are thus disturbed, they sometimes become furious and take it into their head to pursue their enemy; and then, if they once get sight of the hunter, it is impossible for him to escape, unless he possess a degree of coolness and presence of mind, which, in such a case, is not always to be found. Yet, if he will quietly wait till the enraged animal make a run at him, and will then spring suddenly on one side to let it pass, he may gain time enough for re-loading his gun before the Rhinoceros get sight of him again; which fortunately, it does slowly, and with difficulty. The knowledge of this imperfection of sight, which is occasioned perhaps by the excessive smallness of the aperture of the eye (its greatest length being only one inch) in proportion to the bulk of the animal, encourages the hunter to advance without taking much pains to conceal himself; and, by attending to the usual precautions just mentioned, he may safely approach within musket-shot. This creature seems to take as much pleasure in wallowing in the mud as the Hog."

The best accounts and figures, till lately, published of the *Indian Rhinoceros*, are by Parsons, (*Philosophical Transactions*,) by Edwards, (*Gleanings*, vol. i.) and by Thomas, (*Philosophical Transactions*, 1800.) Since the times, however, of these publications in 1815, a specimen was exhibited alive in this country, from which our engraving from a drawing by Mr. Landseer is taken. This individual was afterwards taken to Paris, and M. F. Cuvier in his great

THE INDIAN RHINOCEROS.

work on Mammalia has given us every particular of its natural history and description. There is no reason to suppose that the Asiatic species differs in its natural habits from its African congener; but as our notice of the latter is principally from Mr. Burchell, whose observations were made on the animal in its wild and natural state, we prefer adopting M. Cuvier's account of his specimen in a captive condition, to repeating the older notices of Parsons, Edwards, and Thomas.

This Rhinoceros was young, and was habitually very gentle, obedient to its keeper, and sensible of his attentions and caresses. He was nevertheless occasionally seized with paroxysms of violence, during which it seemed quite necessary to keep out of his reach: the cause of this occasional violence could not be discovered, unless it were to be traced to a blind impulse or desire of that liberty which he had never enjoyed, and which excited in him an effort to break his chains, and to burst from the unnatural captivity in which he was detained. Bread or fruit, however, soon calmed him. He knew those who were most liberal of food to him; as soon as he saw them, he stretched out toward them his long upper lip, opened his mouth and put out his tongue. The cage in which he was kept being very small, he had little opportunity of displaying the extent of his mental faculties, and his keeper took no other pains than to induce him to forget or misconceive his own strength and to obey; but judging by the attention he bestowed on everything passing about him, and by the distinction he was able to make of persons especially, it might be fairly presumed that his intelligence would have acquired a much greater developement, under more favourable circumstances; but his immense power, and the apprehensions constantly entertained, that in one of his paroxysms he might break his prison, procured him at all times a very gentle treatment. Nothing was required of him without reward; and the slight movement which was allowed him, was a further cause for requiring but little of him, as for example, to open his mouth, to turn his head to the right, or to the left, to lift up his leg, &c. &c.

He was thicker and still more unwieldy in his proportions than the Elephant, although he was less in general size. His height was about five feet six inches, and his length nearly eight feet. The whole body was covered with a very thick tuberculous skin nearly naked, and disposed in irregular folds. His colour when dry, was gray with a violet tint, but it was found necessary to keep his skin constantly lubricated, to prevent its chapping; under the folds the animal was flesh-coloured. On the outer side of the limbs, on the knees and head, the tubercles of the skin were greatly elongated, and resembled horny threads attached parallelly to each other; the few hairs to be found principally on the tail and the ears, were stiff, thick, and straight; there were a few others, however, on other parts of the body, which were curled, and, although thick and hard, had a woolly appearance.

The senses of this animal, except that of touch, appeared tolerably delicate. He frequently made use of that of smell, and preferred saccharine fruits, and even sugar itself, to all other food. He collected and held everything intended for the mouth with his moveable upper lip; and when he ate hay, he formed it first into little bunches, which he placed between his teeth by means of his tongue.

The power of this species is frequently displayed to a surprising degree when hunting it. A few years ago, a party of Europeans, with their native attendants and elephants, when out on the dangerous sport of hunting these animals, met with a herd of seven of them, led, as it appeared, by one larger and stronger than the rest. When the large rhinoceros charged the hunters, the leading elephants, instead of using their tusks or weapons, which in ordinary cases they are ready enough to do, wheeled round, and received the blow of the rhinoceros's horn on the posteriors; the blow brought them immediately to the ground with their riders, and as soon as they had risen, the brute was again ready, and again brought them down, and in this manner did the contest continue until four of the seven were killed, when the rest made good their retreat.

By comparing the tenour of these short observations of them in their wild condition, and in a state of confinement, we may gather sufficient data on which to form a tolerable estimate of the character of these animals. Endowed with amazing powers of body—powers which can repel, if not overcome, the active ferocity of the Lion and the ponderous strength of the Elephant, but at the same time seeking their sustenance not by the destruction of animal life, but in the profuse banquet of the vegetable kingdom, they might naturally be expected to avail themselves of their physical power principally in self-defence. Accordingly we find, that against the first aggressor the Rhinoceros is a terrible enemy ; but if left to the ordinary bent of his own inclination, if unmolested in short, he does not wantonly seek occasion to exercise his strength to the injury of other creatures.

A new species of Rhinoceros was discovered by Mr. Burchell a description of which has not yet been given by him, as it is intended to accompany a part of his very interesting work which is not yet published ; but he has communicated to M. de B. Blainville some account of it, which was printed in the *Journal de Physique* for August, 1817.

This species was first seen in 24° south latitude, inhabiting the immense plains of the country, which are perfectly dry during the greater part of the year. The animal frequents the spring daily, not merely to drink, but also to roll in the mud, which, sticking to its naked skin, serves to defend it from the burning sun of this clime. In size it is nearly double that of the common *R. Bicornis*. These two species are known by the Negroes and Hottentots as distinct, under different names. The principal specific character, next perhaps to size, is in the truncated form of the lips and nose, whence Mr. Burchell names the species *R. Sinusis.* The natives state that it feeds on tender grass, while the other species eats branch and bushes — a statement which seems verified by the difference in the form of the mouth of the two. The head, separated from the first vertebra, was so heavy, that four men could not raise it from the ground, and eight were necessary to get it up into a waggon. In the character of the double horns, and in the absence of the remarkable folds which distinguish the Asiatic species, this agrees with the common African species.

Some others are inserted in the Table, whose specific pretensions are not altogether certain, and of whose general history little is as yet known.

If human passions could be imputed to the Author of nature, we might well picture to ourselves the mixture of ridicule and contempt with which he must frequently have occasion to regard the pigmy efforts of us *soi-disant* lords of the creation when we would subject organized creation to artificial systems and arrangements. Any attempt to divide living beings by distinct lines of demarcation, as we mark out the minor divisions of a great kingdom on a map, are hopeless, unless we submit to endless exceptions, to frequent inconveniences, and occasional absurdities.

Anatomy is certainly the surest guide in the arrangement of the animal kingdom ; but, even if the osteology of every animal were perfectly known, and the whole accurately compared, there can be no doubt that in their osteological as well as in their superficial characters, they would be found to approach each other in some particulars, and to recede from each other in others; to be in fact so irregularly blended together on the one hand and separated on the other, that real divisions between them could no where be found ; that a certain number of analogies alone must be permitted to constitute a generic separation, and that such separation, when so constituted, to be received correctly, must be looked on with the eye of a liberal latitudinarian, and not with that of a systematic bigot.

Thus when we find the Daman, an animal not much bigger than a Hare, placed with the Elephant and the Hippopotamus, the Horse and the Rhinoceros, we almost involuntarily start ; and if told that this is on account of anatomical analogies, we perceive at once that consistency in one particular is sacrificed to the same principle in another, and that disparity of size and physical capability must be tolerated if we would divide animals by their osteological analogies.

Our indefatigable author has now ascertained that the Damans of North and South Africa, which, whether one or two species constitute the genus HYRAX in their anatomical characters, are remarkably assimilated to the Rhinoceros and the Tapir, and, consequently, to all the genera, more or less, of the order *Pachydermata.* It is beside our purpose here to enter into all the points of similarity in the anatomy of these animals compared with each other, which Cuvier has investigated in his *Fossil Osteology;* suffice it to say, that these points seem amply sufficient, in spite of the existing and wide differences in dimensions, to warrant the removal of the genus *Hyrax* from the *Rodentia,* in which it had been previously placed, and the transfer of it to the present order.

The Dutch Colonists at the Cape call the South African or *Cape Hyrax, Klip-daassie,* or the Rock Badger. Kolbe, the first author who mentions it, calls it a Marmot, which was adopted by Vosmaer and Buffon. Blumenbach, in his *Manuel*, left it with the Rodentia, as did also Pallas, though he adverts to the differences of its internal character from the rest of that order. Herman was the first to separate it generically under the name Hyrax ('Υραξ, Mouse of the Etolians,) a far-fetched etymology, which was adopted by Schreber and Gmelin, though they kept the new genus in the order Rodentia.

It is obvious, therefore, that the Daman is one of those animals which is intermediate between two or more genera, the very plague of naturalists, though they merit from their singularity particular attention, with reference both to the animals themselves and the class in general. Independently of disparity in size from others of the present order, it differs from them in the nails, which, instead of covering the bottom of the toes, scarcely cover the whole of their upper side; its motions are plantigrade, while other Pachydermata move on the toes, and have the carpus and the tarsus adapted only to pronation: its fur is thick and soft, and it has labial whiskers, in which particulars it differs also from others of the order, as it does also from most of them in great agility and activity. If, therefore, its osteological characters are to prevail in its classification, other dissimilarities must be endured.

The result of a comparison of this singular animal with others, to which it is more or less assimilated, must suffice here in the stead of a detailed account; for a note of the particulars of its generic characters, we must refer to the text and table, and proceed to a notice of the species.

The *South African*, or *Cape Hyrax*, is the best known. It is not much larger than a hare; the make is clumsy, rather long, low on the legs, with a short neck, and a thick head terminated by an obtuse muzzle; the fur is uniformly grayish-brown, with the inside of the ears

C. Hamilton Smith Esqr del.^t

J^t Basire sc.

THE SYRIAN HYRAX.

HYRAX SYRIACUS. Brua

London. Publish'd by G & W. B. Whittaker Feby 1824.

white, and a blackish band is sometimes found on the back. It inhabits the clefts of rocks, and is frequently devoured by animals of prey. Such as have been brought to Europe have been easily tamed, and have become much attached to their keeper; they were active and cleanly, and fed exclusively on vegetable substances.

The *Hyrax* found in *Abyssinia,* which Bruce treats as distinct from the South African species, is considered by the Baron Cuvier as specifically the same. Although, however, the Baron has had the examination of five perfect skeletons, and of ten heads, of these animals, it does not appear in his Fossil Osteology that he had perfect specimens of both, which he could compare with each other; and the points of difference, if any, would be principally in the number of the front toes, which are said to be three in the Syrian Hyrax or Ashkokoo of Bruce, and four in the other.

There is a specimen of the Syrian Hyrax, in the Museum at Frankfort, which we have engraved, from a drawing by Major Smith, which drawing distinguishes only three anterior toes.

To verify this drawing and Bruce's description, we have inspected a specimen of the Syrian Hyrax lately deposited in the British Museum, which specimen has clearly only three toes on the anterior feet, without any exterior tubercle or representative of a third: the acuminated nail on the first posterior toe is not so sharp as that represented in the drawing, and the colour is more like that of a rabbit than the specimen of the South African species in the same establishment, which is uniform in colour and much darker than the other.

We are indebted to M. D'Azàra for a good account of the *American Tapir,* called *Anta* by the Brazilian Portuguese, which account we shall in substance subjoin, referring to the text and table for the generic characters.

2 G 2

The Tapir, called Mborébi in Paraguay, is not numerous in that province, and is generally found alone, or sometimes in company with one other. It sleeps during day concealed in the most sequestered and umbrageous places, and goes forth at night in search of water-melons, gourds, and pasture. If taken young, it may be almost immediately tamed, it goes about the house, and does not seek, even when adult, a greater degree of liberty; it will suffer any one to touch and caress it, without however shewing any preferable affection for any particular person. It never bites, and if annoyed, merely utters a sharp hissing noise, very disproportioned to its size. In this domestic state, it eats flesh, both raw and cooked, and is indeed omnivorous, not excepting even rags of silk or worsted, and will gnaw a stick or cask, whence it appears to be still more gluttonous than the hog, and less capable of choice of its aliment. It is known to swallow a nitrous kind of earth called *barrero*, and D'Azara found a large quantity of this earth in the stomach of one. Hence, says the Spanish naturalist, it may easily be concluded, that no one will bring up an animal so completely noxious and unprofitable, which has nothing attractive, and whose only good quality is of a negative description—it requires neither attention nor care.

The female brings one young in the month of November, and has the entire care of it, without any assistance from the male; even its maternal energies, however, seem not very strong in defence of the young, though it is said that, when hard pushed, it will kick and will also seize the dogs by the spine, lift them from the ground, and by means of a violent shake, will tear the skin.

Its natural defence against the large American Felinæ appears to be by rushing into the thickest part of the wood, through which it thrusts its body with comparative facility; while the pursuer, whose skin and body is less prepared for this kind of obstruction and resistance, soon becomes wea-

THE TAPIR.

London, Publish'd by G & W. B. Whittaker Feb.? 1824.

ried, if not injured. According to D'Azara, it does not seek open paths, but breaks, thrusts aside, and tears everything it meets with the head, which it carries very low : by these means, and by an acute sight, especially in twilight, and ready hearing, it avoids its natural enemies.

The Tapir is hunted by dogs, and sometimes taken by the hunter lying in ambush among the water-melons on which it feeds during the night. When shot, it is observed never to fall immediately, and D'Azara saw one which had received two balls through the heart, run afterwards upwards of two hundred paces. The Indians, whose appetite is by no means delicate, eat its flesh. The Tapir is very strong, and, as it is found in woods, both where the soil is dry and where it is the contrary, there seems reason to conclude that its habitation is indifferent, provided at least that there be plenty of cover, in which it delights. It is an able swimmer, but is not observed to dive: when wounded, it generally takes to the water.

The head is compressed laterally, especially on the upper side, the cheeks being convex. Between the shoulders commences a prominence, which passes gradually, increasing along the neck, over the top of the head, and terminates level with the eyes: this prominence is composed of a very hard thick skin from the shoulders to the occiput, but the rest is bony; a very short stiff mane passes along its whole length.

The *Malay Tapir*, says Sir Stamford Raffles, resembles in form the American, and has a similar flexible proboscis six or eight inches in length. Its general appearance is heavy and massive, somewhat resembling the Hog. The eyes are small, the ears roundish, and bordered with white. The skin is thick and firm, thinly covered with short hair: there is no mane on the back, as in the American species: the tail is short, and almost destitute of hair; the legs are short and stout.

It appears, that until the age of four months it is black, beautifully marked with spots and stripes of a fawn colour above, and white below; and at six months old it assumes its permanent colours of black, with a simple large white patch, covering the back and sides, but not meeting under the belly as represented in the figure taken from the specimen in Paris.

The living specimen described by Sir Stamford, was sent when young from Bencoolen to Bengal, and became very tractable. It was allowed to roam occasionally in the park at Barrackpore, and the man who had charge of it stated that it frequently entered the ponds, and appeared to walk along the bottom under water, without making any attempt to swim. The flesh is eaten by the natives of Sumatra. It is known by different names in different parts of the country: by the people of Liniun, it is called *Saladang;* by those of the interior of Maima, *Giudol;* in the interior of Bencoolen, *Babi Alu;* and at Malacca, *Teunu.*

The first time this species appears to have been noticed was in 1772, but it was then, and long afterwards, treated as an Hippopotamus, till about 1816, when it first attracted attention. As it reaches eight feet in length and six in circumference, it is the more remarkable that it should so long have escaped the notice of zoologists, and might indeed support an argument in favour of the position, that the number of supposed extinct species, whose osseous remains are to be found, are not in fact extinct, but exist now, though unknown to scientific men. This is not the place to discuss that question, and we merely allude to it for the sake of observing, that, whoever will take the pains to investigate the science of animal fossils, will soon find that the recent discovery of a Tapir in America will by no means invalidate the doctrine of extinct species.

Respecting the genus of the Horse, which is sufficiently

characterized in the text, we shall enter into no minute details. A few general remarks upon it, a brief description of the different species so important to mankind, and such observations on the races of that one which has given a name to the genus as may prove interesting, will be amply sufficient for our present purposes.

Horses compose a very natural, but a very isolated division among the Mammalia. They cannot be divided into partial groups; they constitute a single genus, so distinct, and so important in its characters, that it cannot easily be linked, by any connecting traits, to any other group or genus. The different positions which it has hitherto occupied in the systems of naturalists, may serve to prove this fact. Linnæus unites the Horse with the Hippopotamus, to form a genus in his order *Belluæ*. Erxleben places it between the Elephant and the Dromedary. Storr forms it into a distinct order, coming after the Ruminantia, under the name of *Solipedes*. This last arrangement was at first adopted by the Baron, but, subsequently, he placed this genus, as we have seen, among the Pachydermata.

Though totally herbivorous, the Horses have not more than one stomach, and they do not ruminate. Their four feet are monodactylous; there are, however, vestiges of two other toes under the skin. These two last characters approximate them more to certain of the Pachydermata than to any other mammifera. The molars have flat coronals, six on each side in both jaws. The complicated character of these teeth renders a written description of them unintelligible, and their notoriety renders it unnecessary. There are eight incisors in each jaw, and two canines in the males, and sometimes in the female of the domesticated species.

Their eyes are large, and sight excellent, and although they are not nocturnal animals, they can distinguish objects very clearly by night.

Their ears being large, and having great mobility in the external conch, the sense of hearing is consequently extremely fine. It is probably that one which they possess in the greatest perfection, a fact observable in all animals naturally timid*. At the slightest motion, the least appearance of an object that is new to them, they stop and listen with the utmost attention.

Their sense of smell is very delicate, and they frequently exercise it, especially on objects which excite in them any suspicion Their sense of feeling seems also very fine.

In their forms, proportions, and movements, are indicated an equal degree of strength and agility. The Horse is the only animal in which a thick body is not incompatible with gracefulness. The rounded crupper, the firm humerus, the broad sternum, the powerful femur, and the long and sinewy leg, all evince vigour and elasticity. The head is not comparatively so light and elegant as that of some other animals, but it is capable of the varied yet united expression of docility and pride, of courage and of caution.

Our domestic Horses of the middle size, and common race, may serve to give us a correct idea of the forms and characteristic traits of the different species of this genus, though not of the physiognomy. Among them all the differences consist alone in colour, or in the proportions of some external parts of the organs of sense and motion, or, finally, in their intellectual dispositions. The Saddle-Horses, of such fine forms and elegant proportions, whose motion is so light, and whose docility is so great, and the thick

* I have made the same observation on individuals of this character among the human species. Timid females have usually a very fine sense of hearing. People born blind, or long blind, are usually in the same predicament. This sense is also more nearly connected with mental quickness than is usually supposed. I never knew a very stupid person whose sense of hearing was not naturally dull.—E. P.

and clumsy Horses which we employ in the draught, are equally the result of domestication. Their characters are preserved by the attention of man : abandoned to themselves and to nature, they would soon resume the primitive forms of their species, and lose all those valuable qualities which are owing to breeding and education.

Horses live in numerous troops, and inhabit the open champaign countries. These troops are each conducted by a male chief, continually at their head in travel or in fight. These chiefs are indebted for their elevation to their strength and courage, and when their natural force is abated by the advance of age, their authority passes to him who in his turn exhibits those qualities in the highest degree. There is rarely any dispute concerning this right of succession. The individual possessed of the requisite endowments rises naturally from an inferior to a higher rank, and is, finally, placed at the head of his fellows, by the mere force of circumstances, without foresight or volition having in any shape furthered or retarded his advancement. The authority of these chiefs is very considerable, although naturally restricted by the interests of the troop. They are constantly and everywhere followed. If there is a necessity of seeking fresher pastures, or a milder climate, the chiefs take the lead : and in the hour of combat they are the first to expose themselves to danger. A secret instinct teaches these animals that their strength consists in their union. Accordingly, whenever they are menaced by a ferocious beast, or any other of their enemies, they instantly combine in close order, and if any succumb, it is generally the weakest, he who has not sufficient strength to fly, or who is too slow in his movements, if there be a necessity of grouping for mutual defence.

Their principal enemies are the larger Felinæ, the Lion, the Tiger, the Panther, or the Leopard, which they can generally escape with facility or resist with success. Their fleetness soon leaves their pursuers at a distance, and they can strike with the hinder feet with immense force, and bite with great violence and effect.

All the species of this genus are originally natives of Asia and Africa. Three of them are aborigines of the vast and elevated plains of central Asia ; the two last are peculiar to the most southern regions of Africa. None existed in America or New Holland previously to the discovery of these countries by the Europeans, for we can no longer regard the *Huemul* of Chili, described by Molina as an *Equus Bisulcus*, or cloven-footed Horse. It is evidently a ruminant animal of the genus of the Lama, if not the Lama itself.

The genus of the Horse and that of the Camel are the only ones which have each of them furnished two domestic species. These species will couple and produce. But notwithstanding this circumstance, and notwithstanding all the advantages afforded by domestication for the development of certain parts, and the formation of varieties, these species have never yet been converted into each other. The individuals produced from this connexion remain always the same, and never reproduce. This fact is sufficient to overturn the system of some naturalists, that the diversity of species is owing to accidental causes, and that nature did not originally establish a separate type for each. It proves that this theory rests upon nothing but vague conjecture, and has no well-authenticated phenomenon on which to rest as a foundation. Conditions more favourable to the production of such a conversion of species are not to be found in the whole Animal Kingdom, than are afforded by the domestication of the Horse and Ass, and the circumstance of their coupling. The physical difference between these two animals is trifling, and consists merely in the proportions of a small number of their organs. In their intellectual qualities they are more

removed from each other: but among the separate races
of the Horse the differences are more decidedly remarkable. Compare the Sardinian Horse, small, compact, and
nervous, with the large and soft Horses of Holland, or the
fleet and elegant Spanish Courser with one of our own dray
Horses. In the midst, however, of all these differences
which have been so often reproduced, we never find a race
arisen with the long ears of the Ass, or any other of the
peculiar qualities of this species. The same remark is
equally applicable to the Ass; there are many varieties in
this species also, but no horse has ever been found among
them. It may be replied, perhaps, that varieties have
ceased to be formed; but, independently of this assertion
being unfounded, the *onus* of indicating the period when
the existing varieties were formed, still lies with the ad-
vocates of the system we are opposing. In fact, all the
examples that can be gathered in this way are totally unfavourable to this hypothesis. The skeletons of the animals preserved by the ancient Egyptians, which existed
three or four thousand years ago, have the same characters
as those of existing species. We can go no further back
for proofs. The fossil remains found in the ancient strata
of the earth belong to species which exist no longer. In
the genus *Equus* we shall begin, as first in importance, with
the Horse properly so called, *Equus Cabalus*. Lin.

This species appears to have been aboriginal in Great
Tartary; but the troops of Wild Horses found there, at
the present day, are supposed, with good reason, to have
sprung from individuals that had escaped from the trammels of domestication. This opinion is founded on the
difference of colour among these animals, and the facility
with which they can be reduced to servitude. This being
the case, it is not possible for us to become acquainted
with the species of the Horse in its original purity, entirely
exempted from the influence of man, and as it came from

Nature. Still all our continents, New Holland excepted, possess Horses which have been in an independent state for many generations, and might, consequently, serve to give us a tolerable idea of what the Horse originally was ; but the misfortune is, that our information concerning them is so very imperfect, that no very precise notions are attainable on the subject. The observations of travellers on this point are so various, that they appear to be speaking of different species or varieties, and it is next to impossible to form any concordance between them. Nor are they sufficiently ample to establish another fact, which, however, appears by no means improbable : this is, that Horses in the wild state have not universally the same characters, but vary according to climate and other local circumstances to whose influence they may be exposed. These are points which it would be extremely interesting to investigate, and which would tend to illustrate, not only the history of this particular species, but also that of animals in general. There is nothing so much wanting to Natural History, as researches into the influence of external causes on organization.

Pallas has described a wild Mare caught in the country situated between the Jaïk and the Volga, which was very young, and afterwards proved extremely docile. The Wild Horses inhabiting these regions are fawn-coloured, reddish, or light-bay; in summer they proceed as far as possible to the northward, to escape the heat and the flies, and procure better pastures. The Colt described by Pallas was of an Isabella or light-bay colour, and the mane and tail were black. Compared with a Domestic Colt of the Calmuck breed, and of the same age, its stature was greater, its limbs stronger, the head larger, and the ears longer; these last it habitually carried in a couchant posture, like a Horse that was about to bite. The forehead was convex, a thick mane descended as far as the withers, and the tail was in nothing different from that of a Tame Horse; its hoofs were smaller and more pointed, and the hairs were generally frizzled, especially towards the crupper and tail.

Leo Africanus and Marmol make mention of Wild Horses in Africa; but all they say is, that these Horses are smaller than the domestic races, of an ashen or white colour, with short and bristling hair. This is a very insufficient description, not to mention that these authors make use of the very same expressions in speaking of the Wild Ass.

The Wild Horses of America are rather better known. Many travellers have entered into details concerning them, especially M. D'Azara, who has done so with his usual precision. From the first period of the arrival of Europeans in the New World, many Horses were left to themselves, and propagated very rapidly. They were formerly very common at St. Domingo, and even then differed in some traits of character from the Spanish race, to which they owe their birth. The head was thicker, and the ears and neck longer; but where these animals have more particularly multiplied, is in the southern continent of America, and to the south of La Plata. There they may be sometimes met in troops to the number of ten thousand each. They also proceed from some Spanish race, but have lost much of the elegance, beauty, lightness, and grace of their primitive stock. They are not so tall, their heads are thicker, the limbs more clumsy, the ears longer, and the coat much rougher. Their usual colour is chestnut-bay, and sometimes, but very rarely, black. These numerous troops of Wild Horses are found in the immense and thinly-inhabited plains which extend from the shores of La Plata, to the country of the Patagonians. Each inhabits a canton or district peculiar to itself, which it defends from all foreign intrusion as its own especial property, nor will it ever abandon it, except when compelled by hunger or some enemy of very superior strength. They march in serried columns, and when disturbed by any object, they approach it within a certain distance, having the strongest individuals at their head, examine it attentively, describing one or many circles around it. If it does not appear dangerous, they approach with precaution ; but if the chiefs recognise any danger, and give an example of flight, they are instantly followed by the entire troop.

The instinct which induces Horses thus continually to unite in families, renders it very dangerous for travellers to fall in with these wild troops, for it exposes them to the liability of losing their own Horses for ever. The moment these hordes perceive any domestic Horses, they call to them with the utmost eagerness, approaching as near them as prudence will permit. If the others are not guarded with the utmost care, they will take to their heels, and it is utterly vain to attempt to catch them again.

These Wild Horses can be tamed and brought back to a domestic state with great facility, even though they are adult when caught. The South Americans are extremely dexterous in taking them with long cords, or as they are called, *lassos*, which they throw with wonderful address and precision, and thus entwine the animals which they are desirous of possessing. Those of the wild studs are watched by men in the districts they inhabit, appointed for this express purpose, and who have no other occupation. They are mounted on some of those Horses which have been already tamed, and they reconduct the troop to the lands of the proprietor whenever they happen to wander. Those men are also employed to catch them when there is a necessity. They mount on horseback, summon the troop to a quarter from which it cannot escape, mingle among them, provided with the instrument above mentioned. They fling it on the neck of the animal, which, finding itself caught, fastens the knot still tighter by its endeavours to break loose. He falls at last when respiration fails, the men throw themselves upon him, bind him, and put a strong halter round his neck.

In each of these wild troops the chief possesses peculiar privileges. He is the grand sultan, and his harem is very extensive. Should any other have the temerity to invade his rights in this way, or annoy him in his amours, he would soon pay the forfeit of his audacity. In a case of this sort the rage of the chief knows no bounds ; he immediately attacks his unhappy rival, obliges him to fly, and not unfrequently deprives him of life.

Sometimes, like a proud conqueror, he deigns to admit him in his train, as if to witness his enjoyments. He would not, in all probability, prove so generous, if he could reflect and foresee that the conquered enemy of to-day may become in his turn the conqueror of to-morrow, and take an ample vengeance for the affronts he has received.

It appears from the details we have entered into, that the Horse, in a state of nature, would be about the middle size ; have a larger head and ears, stouter limbs, and a coarser skin than the domestic Horse. Its intelligence, however, does not seem to be much affected by such a state; those Horses are easily reduced under the yoke of servitude, but infinite pains are necessary to restore to them their lofty stature and elegant proportions.

The rut takes place in the spring, and gestations continue for twelve months. The colt is born covered with hair, with the eyes open, and sufficient strength to enable it to support itself and walk. Some days after the birth, the two middle incisives appear in each jaw ; two others, at the right and left of the first, come in three or four months after, and the two last in about six months. These are the milk teeth, which are reproduced in the same order between two or three years' time, at intervals of six months. The colt is suckled almost twelve months, and its growth is fully developed towards the fifth year. The Wild Horse may live from thirty to forty years. In youth, the age is known by the incisors: these teeth have a hollow in their upper part, which is gradually effaced by use, and the degrees are so regular as to correspond with a determinate space of time.

The milk-incisors are whiter and narrower than those which follow. At about fifteen months old, those which first appeared begin to lose their cavity by the effect of wear. Those which come after are not marked until the twentieth month. Finally, after two years, the cavity of these last is in its turn effaced; the adult teeth lose their cavities in a similar order. Any differences in these various changes are characteristic of peculiar races, and sometimes of individuals which arrive sooner at the adult state. After the twelfth year there are no certain rules to judge of the age of Horses.

The wild troops have no fixed place of repose; they usually select dry and sheltered situations for that purpose, at the foot of a rock for example, or the edge of a wood, where they may be protected from the winds. They have the same dread of storms that most other animals have. At the approach of the tempest they are agitated and restless; they seek the wildest and most sequestered spots to conceal themselves. If the storm bursts forth before their retreat, if a violent clap of thunder should be heard, the terrified troop betakes itself to instantaneous and rapid flight; the wind and the lightning cannot outstrip them. Wretched indeed would be the living object that should intercept them in their route.

The mothers which have young ones never quit the troop; these last are courageously defended by the rest, and rarely fall by the tooth of any carnivorous animal. If the enemy is formidable, and they cannot escape, they unite in a close and circular band, approach their heads together, and presenting their heels to the adversary, deal out kicks with equal force and abundance. If he be not dangerous, they amuse themselves as before mentioned, by forming a circle around him, and, finally, trample him underfoot. They generally look out for dry pastures and firm soils, productive of short but fine herbage; they will feed on the buds and bark of trees, and in winter on the dead leaves, mosses, and even the young branches and wild fruits of various kinds.

The senses of the Horse, as we have remarked in our observations on the genus, are acute and delicate; their voice assumes different tones, from various causes: the females neigh less frequently and with less force than the males, and castration produces a similar modification in the latter.

The intellectual character of horses, consists in clearness of perception, and excellence of memory; their education is entirely founded on the association of those impressions which they have received.

When we consider the varieties produced by domestication in this species, there is cause for great astonishment; we see them sometimes reduced to the stature of the Deer, and sometimes increased to the bulk of the Dromedary, exhibiting the elegance and lightness of the Stag, or the weight and corpulence of the Ox: some races have the head small and slender, the eyes lively, the ears fine and directed forward, and the nostrils wide and mobile; others, on the contrary, have the head heavy, the eyes dull, the ears large and inclined backwards, the nostrils narrow, and closed: some have the frontal ridge arched, others straight; sometimes the hair is smooth and scanty, sometimes abundant and frizzled, sometimes long and silky; and every variety of colour resulting from fawn, black, and white, is found among them in an infinity of shades and proportions.

In the gait of horses there is also a very considerable

variety; besides the walk, the trot, and gallop, movements too well known to need description, some horses raise both legs at one side together, and bring them to the ground together, the two legs on the opposite side executing the same motion; this is called ambling: some horses gallop with the fore-legs, and trot with the hind; this is a very defective gait, and argues a weakness in the hind quarter: other horses step with each leg separately in succession; the left fore-leg for example comes first, then the left hind-leg, then the right fore, and then the right hinder: this is a very secure gait in rough roads. There are many other artificial movements which are the result of education; but it would be foreign to our purpose to notice them, or many other particulars of the manage, which have but a remote connexion with the natural history of the Horse.

The Horse is susceptible of very great attachment for the human species when properly treated. A good deal, however, that is repeated on this head in authors, borders somewhat on the marvellous: as for instance the stories of Bucephalus; the anecdote of a horse belonging to a Scythian Prince, which trampled to death the murderer of his master; and that of the horse of Nicomedes, which through grief for his death suffered itself to perish of hunger; the following anecdote, however, on this subject is well authenticated:—

The Tyrolese in one of their insurrections, in 1809, took fifteen Bavarian Horses, they mounted them with as many of their men; but in a rencontre with a squadron of the regiment of Bubenhoven, when these horses heard the trumpet and recognised the uniform of the regiment, they set off at full gallop and carried their riders, in spite of all their efforts, into the Bavarian ranks, where they were made prisoners.

Numerous anecdotes of this description have been related by many writers and are very generally known: one in particular, related by Southey in his history of the late Peninsular war, is very interesting.

Education develops the powers of horses to a very considerable extent; a good Arab Horse will make a journey of fifty leagues in twenty-four hours, and the Tartar Horses frequently make journeys of many days without stopping, except to eat of a few mouthfuls of barley : some of our English racers have been known to run eighty feet in a second, a speed almost surpassing the swiftness of the wind.

The moral qualities of horses are not less diversified than their physical : some possess the most undaunted intrepidity ; others are so timid that everything affrights them; some are distinguished for memory, for prudence, and extreme susceptibility of education; others on the contrary are giddy, obstinate, and dull : these differences should certainly be deemed characteristic of races, for they are permanent and hereditary ; but such points have, unfortunately, not been sufficiently studied.

Every country possesses certain races of horses, whose characters depend on the localities that surround them, and the wants of the inhabitants of each respective climate. The Arabs have peculiarly devoted themselves to the development and preservation of such qualities as fit the Horse for the saddle, as lightness, vigour, and docility. In agricultural countries, those races which are proper for draught and labour have been especially attended to ; the northern countries, where vegetation is abundant, have produced horses of the largest size, the southern have proved less favourable to the development of the body, and more so to that of energy and vigour.

We shall conclude this article on the species, by a brief view of the different races.

The *Arab Horse* is, without contradiction, the first in the world; it may not, however, be considered handsome, according to our general notions of the beauty of Horses: the

2 H 2

head is square, the frontal ridge is hollowed, rather than prominent, the chest narrow, and sometimes resembling that of a stag. This conformation which has been regarded as a defect, is common to all animals intended for swift and frequent motion ; the knowledge of the first laws of animal physiology is sufficient to vindicate its necessity. This horse is distinguished by a fine skin, smooth hair, and the sanguiferous vessels are very apparent : the apophyses to which the muscles are attached are strongly defined, the muscles themselves are also clearly designated under the skin ; the articulations are large and strong, and exempt from those defects so often observed among our common horses. The limbs are fine and have no more hair than the rest of the body, and the foot is excellent and sure ; the common stature is about four feet six or seven inches.

The Arab Horses receive a moderate portion of nutriment—generally about five or six pounds of barley in the evening, and sometimes in the tent a little minced barley-straw. They generally travel from eighteen to twenty leagues in the day, sometimes more ; they do not sweat much, and are for a long period fit for service. Their wind is exceedingly good—we might say almost inexhaustible.

A horse of this description will in running carry his head and chest so as completely to cover his rider. His tail turns up in the air, and curls in a manner which we have vainly attempted to imitate by an operation equally useless and barbarous. This horse, in short, possesses in the most eminent degree the qualities of endurance, vigour, and admirable temper. This union of characters, which are applicable to every service, and which always prove hereditary, places the Arabian Horse without a rival at the head of his species.

The Arabs divide their horses into two races. One, in no great repute and appropriated to servile uses, they name *Kadischi*, which means *horses of an unknown race.* The second kind they call *Kochlani, Kohejle,* or *Kailhan,* which means *Horses whose genealogy is known for two thousand years.* This race, say the Arabs, originated from the studs of Solomon. The individuals composing it are sometimes sold at such enormous prices as appears almost incredible. They boast that these horses are capable of performing the most wonderful journeys, of sustaining the greatest fatigues, and passing entire days without nutriment; and of their impetuosity in attacking the enemy, and fidelity and attachment to their masters, many marvellous tales are related.

In the breeding of the *Kochlani* Horses, the Arabs use the utmost precaution to avoid being deceived on the point of genealogy. The Mares are covered in the presence of a witness, who remains near them twenty days to make sure that they are not dishonoured by any vulgar Stallion. The same witness must also be present at the accouchement ; and a certificate of the legitimate birth of the colt is made out within the seven first days subsequent to that event. Whenever all the prescribed formalities have not been rigorously performed, the colt is considered to be *Kadischi ;* and whatever advantages he may possess, he is yet a serious loser in consequence of the non-authentication of his birth. These precautions prove how amazingly jealous the Arabs are of preserving their better race of horses in the most untainted purity of descent.

This race is principally cultivated by the Bedouin Arabs between Bassora, Merdin, and Syria. They sell the Stallions without any difficulty, but as we have already hinted, at a most enormous price; but they will by no means consent to sell the Mares, and these last are never obtained by strangers, except fraudulently or by dint of excessive bribery. These Mares enjoy the exclusive privilege of transmitting the purity of the race to their descendants, and the genealogies are always reckoned from the mothers.

It is to be observed that the Mares of the Kochlani race are never covered by Kadischi Stallions ; should that, however, take place accidentally, the colt is considered of the race of the father. It is frequent enough, however, to have the Kadischi Mares covered by Kochlani Stallions, and in this case the colt is considered as of the race of the mother. All this proves what a high idea the Arabs entertain of the Kochlani race, and how utterly they exclude from it mixtures of all kinds.

Sometimes it is possible to procure horses of this race very cheap. This proceeds from the extreme superstition of the Arabs; they regard certain marks upon their horses as signs of good or evil fortune, and will often sell for a trifle a most excellent horse who happens to have a mark of the latter kind, while a very defective one will be kept if he exhibit a lucky mark.

The wandering Arab of the desert places his highest felicity in the possession of good horses, and is much more attached to them than the inhabitant of towns. They are his companions rather than his servants, and it is with great difficulty and regret that he can be induced to part with any of them.

The intermixture of the Arabian improves every other race, even those which are considerably larger, and altogether of a different make and constitution. The manner in which this improvement takes place is worth the attention of a philosopher. It is not always clearly observable in the first generation. An Arab Stallion, for instance, coupling with a mare of any of our common race, will not give birth to a handsome colt; but this colt, who generally possesses the good qualities of the sire, will, as if he had within him undeveloped the full germ of the better race, originate in his turn an offspring handsomer than himself, and possessing in the same degree all his moral and physical good qualities.

Tartar Horses. Under the name of Tartars (though not strictly correct) we may comprehend all the wandering tribes who inhabit the immense plains of Central Asia.

Men who have no fixed habitation, who live in tents, are shepherds, and who, when they have consumed all the pastures of one district, remove to another in search of new,—of the horses of these people we know nothing, except from the relations of such travellers as have visited them. Their accounts, it must be confessed, are all more or less imperfect; still we have sufficient data to conclude that of all the horses in the world, those of the Tartars approach most nearly to the savage state. They seem to be small and but indifferently made, but sober and indefatigable. According to some writers, they are the fittest of all horses to support the longest and most violent journeys without eating or drinking. Brought up among all the other animals of the horde, exposed from their infancy to the inclemency of the weather, accustomed to little food, and to follow their mothers in the most rapid and extensive excursions, they acquire the most astonishing power of sustaining fatigue. Besides, it must be remarked, that these people, who esteem their horses for no qualities but what are truly serviceable, and who live in a great measure on their flesh, preserve only the most vigorous. The others, not being able to endure the trials to which they are exposed, are soon killed and eaten, to prevent them from consuming the provision of more valuable animals. The most vigorous horses, those which undergo the given ordeal, being alone preserved for service and reproduction, their offspring must necessarily partake of their hardy qualities, and the race thus continue one of the best perhaps on earth for the endurance of privation and fatigue. Their education is not much attended to: left alone, as it were, until the moment in which it is absolutely necessary to take and break them in, they are indocile and restive. But the Tartar horseman cares little for this; all he asks of them is to run fast and long; he gives himself no trouble about anything else.

The *Persian Horses* are next to the Arabian, from which

they descend, in the highest estimation in the East. For a certain distance they can run as fast, and even faster; but they are far from possessing the endurance of the Arab.

The Persian Horse has a finer head, and the crupper better turned than the Arab. Towards the north of Persia there is a strong race of horses, not very unlike those of Normandy in France, which are suffered to pasture undisturbed from eight to nine months in the year in the fertile meadows of Chirvan and Mazendaran. The horses of this race are considered excellent for cavalry.

The Persians take the same pains in the preservation of their races as the Arabs.

The Persian Horse was brought into England during the reign of Elizabeth, and produced by its intermixture a most excellent breed. But the Arab Horse has been preferred since we have had the means of procuring it, and have discovered its peculiar advantages.

The *Barbary Horses* have the chest better made than the Arab, or, more properly speaking, it is rounder; they are less calculated for running, and more esteemed for the the manage than any other exercise. The forehead, instead of being hollowed, is rather prominent, and their head is finer than the Arab. Their shoulders are flat, their crupper rather long, and they are rather frequently long-jointed. The figure of the Barbary Horse is more showy and imposing than that of the Arab; it is nearly of the same stature, few of them being above four feet nine. It is not very forward and sprightly, but requires to be excited and put in train by degrees. Then it will exhibit the same vigour, swiftness, and lightness, which it inherits from the Arab Horse, from which it appears to descend. The best Barbary Horses are found at the present day in the kingdoms of Morocco and Fez, but the Moors do not take near so much care of their horses as the Arabians.

The *Turkish Horses* approximate to the Arab, from

which they likewise descend. Like them they have the straight and rather slender forehand, their body is longer, and the reins more elevated, but their qualities are exactly the same.

When we consider the situation of the inhabitants of Europe, we shall find that they have need of much greater varieties in the characters of their horses than the nations of which we have just been speaking. More civilized than the Asiatics, and at least equally numerous, though shut up within a smaller space of territory, having no deserts to cross, and plenty of provender for their horses, they had no necessity to look for the qualities of temperance and lightness in those animals, which would render them unfit for the most ordinary labours : possessed of no other beasts of burden, such as Camels and Dromedaries, they were necessitated to look for horses which by their loftier stature and greater bulk could draw or carry the heaviest loads. They reserved the finest, the lightest, and handsomest of such horses for the service of the saddle ; but, losing sight of the origin of these animals, neglecting the first sources from which they came ; no longer considering what regions were most favourable for preserving the forms and constitutions which they had received from the hand of nature ; they paid no attention to that mode of education which was best adapted to counteract the influence of a colder or a moister climate, nor to the effect of a more abundant, but less stimulating nutriment, which, while it increased corporeal bulk, added nothing to energy and spirit. Thus by degrees the original type disappeared, and horses fit only for the purposes of draught alone remained ; or, at least any others became extremely rare. This degeneration has affected all the horses of Europe more or less, according to the measures adopted by governments to remedy it, or the spirit and disposition of the people.

The *Spanish Horses* enjoyed for a long period the

highest reputation of any in Europe; But the little attention which has been lately paid to their education has considerably diminished the number of good horses in Spain. The head of the Spanish Horse is a little large and strong, the ears lie somewhat low, and are generally rather long. The forehand is strong, somewhat too fleshy, and the crupper turned rather like that of the Mules. They have a little belly, and their stature is ordinarily about four feet six or eight. Their motions are exceedingly supple, and they have much fire, docility, grace, and action. Their vices are probably more to be attributed to defective education, and a bad system of farriery than to nature. They are equally well adapted for the manage and for cavalry.

The Horses of Spain have been in esteem from the time of the Romans, but it seems very probable that there is an intermixture of Barbary blood in the modern race of that country. The provinces of Andalusia, Grenada, and Estremadura, have always, exclusively of the rest of Spain furnished the most distinguished horses. The district of Xeres in particular has always produced the most esteemed. Two races are found there perfectly distinct, one remarkable for its elegance and fine proportions, which does not retain its full development until six or seven years old, is preserved in all its purity at the Chartreuse of Xeres, and by a small number of proprietors. It is rather long in the limbs, which, while it detracts from the solidity of its figure, adds to the gracefulness of its motions, and is considered as a perfection by the Spaniards. The other race is large, less elegant, formed more for strength, and more multiplied.

The *Horses of Germany* have never been in any great request. The breeds, however, have been, to a certain extent ameliorated within the last hundred years. Most of the princes of Germany have very excellent horses in their studs, and most of their stallions are chosen among the Arabs, the Barbs, Turks, and Spaniards. From such sources the productions must of course be good. Prussia in particular, in consequence of the very great expense she has gone to in procuring these stallions, has now a most excellent and serviceable breed of horses. The chief objection made in general to the German Horses is that they are somewhat short-winded.

The *Horses of Italy* were formerly in much greater repute than at present, and were considered equally good for the manage and for the carriage. Those of the kingdom of Naples were especially esteemed ; but they have much degenerated since it became customary to renew the races, not with Arab stallions, but by crossing them with horses from other countries of Europe.

Switzerland possesses a good race of Draught-horses: some are even very fit for the carriage. These Horses are generally compact, well limbed, vigorous, and sober ; but the jaws, feet, limbs, &c., are in general too much loaded with hair. The German and the ancient Italian Stallions were the origin of this race.

The *Dutch Horses* are good for draught, but nothing else. The best are in Friezland, Berg, and the country of Juliers. Their feet are generally large, they eat much, have little endurance, and are said to be subject to many diseases. The Scandinavian Horses have much the same character ; they are a little finer in the limbs, and trot well, which adapts them for the use of the carriage.

In *France* there are horses of every description, and all of them extremely serviceable animals. But we must be excused from subscribing to the very exaggerated pretensions of our neighbours on this subject. In beauty, swiftness, elegance and grace, the French horses are totally surpassed by many others. Their hardiness, however, and the absence of too much delicacy in their education, fit them very eminently for severe service, and they seem to be better adapted to endure the hardships of a campaign than ours. The best horses in France are furnished by Limousin and Normandy. The first produces very fine saddle-horses, and Normandy both excellent saddle-horses and capital carriage-horses. The Norman saddle-horses are not so good for hunting as the Limousin, but they are better for military uses, and much stronger. The Norman Horses have, it must be observed, been very much intermixed, especially with ours. Franche-Comté and the Boulonnois produce very good draught-horses. Auvergne, Poitou, and Burgundy, very excellent poney-hacks. Rousillon, Bugey, Foret, the countries of Auch, Navarre, Bretagne, &c., also produce good saddle-horses, but in less estimation than those of Limousin and Normandy.

England is without contradiction the country of the whole world, in which the art of breeding and educating horses is carried to the utmost pitch of nicety and refinement. This refinement however, has been pushed to an excess among us, the beneficial result of which, as to the species of horses in general, may well be questioned. We have devoted ourselves, in a peculiar manner, above all other nations to the production of one quality in our horses, and that is speed. In this particular we have completely succeeded, and the English Racer, never has, and probably never can be equalled. It is observable, that the same quality is the principal desideratum in almost all our other breeds, the heavy cart-horses excepted. The necessity of rapid intercommunication among us, and our ardent ambition to facilitate this purpose as much as possible, have been among other causes, greatly productive of the cultivation of this quality in our horses. The rapidity with which our mails and public coaches fly from one end of this island to the other, surpasses all that the world has ever seen in this kind, and equals everything that fable or romance has feigned. But, alas ! the poor animals are no great gainers ; and, it may safely be asserted that more horses are consumed in England in every ten years, than in any other country in the world, in ten times that period, excepting those which perish in the glorious barbarities of war.

If the Tartars take too little care of their horses, we take too much. There are few horses, perhaps in the world, so liable to take cold as ours, of the better breeds. This delicacy of constitution in our horses, the result of too tender an education, was thoroughly exemplified during the late war. The French Horses, though, in other points much inferior, evinced the most decided superiority in enduring the fatigues and privations of campaigning.

To enter into the details connected with the history of the English Horses, would far exceed our limits, and in many respects be foreign to the object of our work. We must indeed, content ourselves with the briefest possible sketch of the subject, and, in truth, it has been so often amply treated of, and is so generally known, that our readers may easily console themselves for our brevity.

Our horses were originally altogether unfit for the saddle. By the importation of Arabs, and other Asiatic horses, a wonderful amelioration gradually took place, and the result of their crossing, and the crossing of their offspring with our indigenous breed, has been the production of four principal classes of horses, the characters of which are exceedingly well defined.

The first is the Racer, immediately proceeding from an Arabian or Barbary stallion, with an English mare, already crossed with a Barb or Arab, in the first degree, or the result of two crossings in the same degree. This we term first blood, or the nearest possible, to the foreign stock.

The next is the Hunter, the result of the crossing of a stallion of the first blood, with a mare of a degree less near the original source. The conformation and admirable

qualities of these horses are too well known to need description.

The third is the result of the crossing of the Hunter with
more common mares, more strongly limbed, and more approaching our indigenous race. This forms our chaise and
carriage horse. We export more of these two last classes
than any other, especially into France.

The fourth is our Dray-Horse, the result of the crossing
of the last, with the strongest mares of our country. The
gigantic proportions and immense powers of these horses
are only equalled by their intelligence and docility. It may
safely be said, that this breed of horses is not to be paral-
leled on the face of the globe.

In the mixture of all these classes, even among the most
inferior individuals, the influence of the Arab blood is still
to be observed. It shews itself either in the conformation
of some peculiar parts, or the preservation of some peculiar
qualities.

The regeneration of our native horses is undoubtedly
now carried to its highest possible pitch. We have now,
little or no occasion to import any more Asiatic horses.
Our best blood stallions are found to answer all purposes
of amelioration equally well.

The peculiar races of horses are preserved by the formation of studs. The mode of conducting such establishments, and of educating these valuable animals, come not
within our province to describe; but the rules on which
they are founded, must be derived from the natural constitution of the animals themselves. In general, the
qualities peculiar to each race, are propagated by generation, from which it is obvious, that the various races must
not be indiscriminately mingled. The development of
qualities, whether physical or moral, is always produced
with the greatest certainty by gentle and almost insensible
gradations. The most entire liberty must be left to na-

ture, when she has any tendency to produce the end we aim at. Such principles should form the basis of every rule respecting the direction of studs ; but the majority of men who are engaged in the education and breeding of horses, seem utterly to misunderstand or overlook them. By kindness, gentleness, and patience alone, can we succeed in subduing and instructing these valuable animals, when they are not naturally vicious. Force may doubtless constrain them to obey, but they will at the same time lose their most valuable qualities, their ardour, courage, docility and intelligence. Let us observe the difference between an animal which has been habitually influenced by the whip, and him who has always been guided by the skilful hand of his rider. The one will love his master, and delight in obedience to his commands ; the other will cease to obey when he has ceased to tremble, and when he has once discovered the secret of his strength, he will take ample vengeance on his oppressor. The operation, indeed, of properly breaking horses, is one of extreme difficulty, and one which has seldom been well performed, because it has been mostly intrusted to the hands of empirics. Its success depends upon principles which are drawn from the physical and moral qualities of the animal which such empirics have never studied. The most important point to attain, is that all the perceptions of the Horse should be clear and precise, without this, memory is useless, produces only vague associations, and can do nothing but misguide. It is on this account that gentleness and patience are so essential to the education of horses. Nothing is so calculated to confuse and falsify the impressions we attempts to make, than to accompany them continually with chastisement and terror.

The second species of this genus which we shall notice, is the *Dziggtai, Equus Hemionus* of Pallas. The tail has hairs only on its extremity, and there is a black dorsal line which enlarges on the crupper. In winter the hair is very long, but of a smooth and shining appearance in summer. The colour of the body is an uniform light bay, but in winter it partakes more of the red.

Messerschmit was the first who indicated this animal, but it is to Pallas that we are indebted for its exact description. The name, in the language of the Mongoles, signifies *large ear*, and has been given by that people to this animal, whose ears are longer than those of the Horse, but straighter and better formed than the ears of the mule. Its stature is that of a horse of the middle size, and its comformation seems like that of the mule, to partake of the Horse and Ass. It is by no means improbable that this is the Wild Mule of the ancients. The head is strong and a little heavy, the forehead flatted and narrow, the chest large, the back long and curved, and the crupper slightly attenuated. The limbs are light, the shoulders narrow, and the hoofs resemble those of the Ass. The mane is short and thick, the tail almost two feet long, and very like that of a cow, both are black. All the proportions of this animal exhibit much elegance and lightness, and its limbs possess the most astonishing suppleness and capacity of speed. It runs literally with the rapidity of lightning, carrying its head erect, and snuffing up the wind. It easily escapes the hunters, for the fleetest courser that ever scoured the desert, would in vain attempt to overtake it Its air is wild and fiery, expressive of its unbounded energy and tameless character.

Their character is peaceable and social. Their troops are generally twenty or thirty in number, sometimes one hundred. Each has its chief, who watches over its safety, conducts its progress, and in danger gives the signal of flight. This signal consists in leaping three times in a circle round the object which inspires fear. If the chief is slain, which will sometimes happen, as he approaches very near the hunters, the troop will disperse, and thus afford greater facility of destruction. The Mongols, the Tungooses, and other nations who border on the Great Desert, hunt these animals for the purpose of eating their flesh, which they consider no small delicacy.

These nations, however, have not yet succeeded in taming the Dziggtai, not even when they have taken it very young. These animals would assuredly make the best ponies in the world, were it possible to subjugate and domesticate them. Their character seems absolutely untameable, and those individuals, with whom the experiment has been tried, have killed themselves in their shackles, rather than endure them. Pallas, however, is of opinion that the thing might be done, if proper means were adopted, and we perfectly agree with him.

There is no animal we believe, but what may be tamed to a certain point. All that is wanting is the proper method. We cannot therefore conclude with Sonnini, that this species will be utterly annihilated before any of its individuals shall be reduced to the service of man. But the art of taming animals is one that, generally speaking, is very unscientifically exercised.

The Ass. (Equus Asinus.) Until latter times the Ass was known to the moderns only in a domestic state. The ancients indeed speak of Wild Asses under the name *Onager*, but, as usual, give us no description of them, contenting themselves with relating a few particulars, from which, nothing can be gathered. Modern travellers also speak of Wild Asses, but trouble themselves with details just as little as the ancients. Dapper mentions some seen in the islands of the Archipelago, and Leo Africanus and Marmolle, mention others in Africa. The existence of such animals in Asia, is indubitably confirmed by Olearius, Pietro della Valle, and others; but these writers favour us with no description. Pallas, in his travels in the southern parts of Russia, has given us a tolerably exact account of the Wild

Ass of those regions, but he admits that the *Koulan* is no more than the genuine Ass, abandoned to nature, and exempted from all traces of domestication.

This animal is of the magnitude of a middle-sized horse. The head is heavy, but the ears are not quite so long as those of the Common Ass. The colour is gray or brownish-yellow, with a brown dorsal stripe and one or two bands across the shoulders. It passes the winter in the warm parts of Persia and India, and advances in summer to the North of the Oural, where it finds fresh and abundant pastures. These animals live in numerous troops.

The domestic races of the Ass appear to have been subjected to man from time immemorial. Many varieties are produced in this animal by domestication, though fewer than in the Horse, which, being better able to support the vicissitudes of climate and of season, is, consequently, more liable to modification from such causes. The countries most suitable to the Ass, are those of the South. Accordingly it is in Persia, Egypt, and Arabia, that the strongest and finest varieties of this species are to be found. Some very different from the small and feeble natives of our climates, almost equal the Horse in magnitude and stature. Spain also possesses some fine races of the Ass, which are also occasionally to be found in the southern provinces of France, but in proportion, as we advance northward, this animal diminishes in size, and becomes more and more difficult of preservation.

To describe this animal as known to us, would be superfluous. Its senses are in general excellent, its apprehension is slow, but the perceptions which it is capable of are clear and precise. To this fact, may be attributed the sureness of its footsteps, and the general good sense that it exhibits. Its timidity causes the caution which we find it to possess, and also the resistance which it sometimes manifests, and which we, without any reason. confound with obstinacy.

These animals have generally a robust constitution, are subject to few maladies, and their temperance is extreme. These qualities proceed from the utter absence of all delicacy in its education. No domestic animal is more neglected or exposed to such ill-treatment. The nutriment rejected by other animals, is considered as too good for it, and it is overwhelmed with hardships and chastisement. It renders us indeed no other kind of service than the Horse, and can only do that in proportion to its strength, which is not great. It is doubtless the facility with which it is nourished, and the strength of its temperament, which induces us to preserve it at all. Accordingly we generally find it the companion of the poor, of whose misery and fatigue it is the constant sharer. The Orientals, however, much esteem this despised animal, and treat it with attention. It is employed by them in all the services of the Horse and is most generally in use for the saddle. To a certain class of men, it is the only animal allowed for riding, and especially to Europeans.

The Ass is very susceptible of education; it has all the requisite qualities, delicate senses, and an excellent memory; it remembers all the roads it has travelled, and its timidity prevents it from taking others willingly. It fears water, to which, however, it may gradually become accustomed. If its eyes are covered, it refuses to proceed, and if overloaded, it will accelerate its pace, and go on until it falls.

But for the Horse, it would certainly have been the first of our domestic animals; our attention would have developed in it new qualities, and improved the old. The Wild Horse and Wild Ass are very nearly of the same size; their strength is equal, and their disposition not much different. It may be even said that the Ass has more solid good qualities than the Horse. But the intelligence of the latter is greater, and has gained the superiority which it ought. Brute strength should have no proportion in esteem, in comparison of intellectual power.

Of all that relates to reproduction, &c. it is unnecessary to speak. The discordant bray of the Ass is too well known: that of the male is louder and more horrible than that of the female.

This species was not known in more Northern Europe, in the time of Aristotle. In proportion as our marshes were dried, our forests cleared, and our climate became milder it became established among us. The Greeks possessed fine races, which passed from them into Italy. Some are found even in Sweden at the present day; a proof how much the constitution of the animal may be modified by gentle and insensible gradations. The Ass has been transported into America, where it has met with the same scurvy usage as in Europe.

The *Couagga*. (*Equus Quagga*. Gm.) For essential characters, see text.

This species reminds us of the forms and proportions of the Horse, by the lightness of its figure, and the smallness of its head and ears, but it has the tail of the Ass. Its height to the withers is about four feet. The colour on the head and neck is a deep blackish-brown; the rest is of a clear brown, growing paler below, and underneath a pretty white. The head and neck are striped with a grayish white; they are longitudinal on the forehead and temples, transversal on the cheeks; between the eyes and mouth they form triangles; there are ten bands on the neck, the mane is blackish; there is a black line along the spine descending to the tail.

The Couaggas go in troops in the more solitary regions of Southern Africa, often more than one hundred in number; they never mix with their congeners the Zebras. Their cry resembles very much the barking of a dog; they are easily tamed, and rendered subservient to domestic uses: and in confirmation of this assertion, it may be observed that among the equipages occasionally exhibited in the gay season in Hyde Park, and other fashionable places of resort, may be seen a curricle drawn by two Couaggas, which seem as subservient to the curb and whip as any well-trained horses.

When we consider that this species is thus capable of highly beneficial services in a domesticated condition ; that its natural courage is evinced in its wild state, by the manner in which, according to the report of travellers, it repels the Hyæna and the Wolf,—an endowment which would be of great value to the animal, if completely subjected to man; that this species is an inhabitant of the hottest parts of the earth, and is therefore likely to be of service where the Horse loses his capabilities by climate; we may naturally be surprised that the Couagga has been suffered by us to retain its liberty so long. Naturalists now, however, have discovered ths pliability of its disposition, in conjunction with its physical powers, and practical men will probably in time, take advantage of the discovery, by adding the Couagga to the number of species subdued to the general profit, convenience or pleasure of mankind.

The *Zebra*. (*Equus Zebra.*) For the description of the Zebra, we cannot do better than translate Buffon, a writer whose tact could at once seize all that was interesting and important, and the felicity of whose style could adorn the dryest details.

" The Zebra is perhaps of all quadrupeds the best made, and the most beautifully clad by the hand of nature. To the figure and graces of the Horse, it adds the light elegance of the Stag, and the black and white bands with which its body is ornamented, are arranged with such wonderful symmetry, that we might almost be disposed to imagine, that rule and compass had been employed in their formation. These alternate bands are narrow, parallel, and exactly separated; they extend not only over the body, but the head, thighs and legs, and even over the ears and tail. They follow so exactly the contours of the different parts, enlarging more or less according to the development of the muscles, and the roundness of the different forms, that they exhibit the entire figure in the most advantageous point of view. In the female these bands are alternately black and white, in the male they are black and yellow, but always of a lively and brilliant tint. They also rest upon a ground of short, fine, and copious hairs, whose lustre considerably augments the general beauty of the colours."

The Zebra is an inhabitant of Africa, and appears to exist from Abyssinia to the Cape of Good Hope, but is found most abundant in this great Peninsula. It appears to have been known to the Romans under the name *Hippo-tigris* but as Pliny makes no mention of it, it is not probable that it was often seen by that nation.

We are very little informed concerning the natural habits of this species, a result arising from its wild and untameable character. Unlike the Couagga, it spurns the yoke of servitude, and proves one of the most rebellious subjects of the vain lord of the creation. Many attempts have been made at the Cape, to discipline this animal, but to no purpose. Sparmann relates an instance of a rich citizen, who had brought up, and to a certain extent tamed some zebras, with the view of appropriating them to the harness or the saddle. He one day had them yoked to his chariot, although they had been utterly unaccustomed to anything of the sort. But he had nearly well paid for his folly, for the Zebras rushed back to their stalls, with such terrible fury as to leave him, when he escaped with life, very little inclination to repeat the experiment.

Mr. Barrow, however, seems to think that the Zebra might be tamed, notwithstanding its vicious and obstinate character, if proper means were resorted to. More skill, perseverance,and patience than the Dutch peasant possesses, are necessary to subdue an animal naturally haughty and courageous, or to tame one that is naturally timid. It is not by whips, nor goads, nor spurs, that a vicious animal taken in a state of nature is to be conquered. Wounds and ill-treatment only increase its resistance and obstinacy.

In corroboration of this, M. F. Cuvier cites an instance of a female which was perfectly tame and gentle, and suffered herself to be mounted without difficulty.

Three instances have occurred in Europe of female zebras producing mules. The first took place in this country. Lord Clive, on his return from India, brought with him a female zebra from the Cape. The experiment was first tried with an Arab horse, but failed. Asses were then tried, but with no better effect. At last, by painting one of these asses like a zebra, the plan succeeded. The result was a foal which resembled both father and mother. It had the form of the first, and the colour of the second, excepting that the tints were not so strongly marked. After his lordship's death, this mule was lost sight of, and its fate is unknown.

The second instance took place at Turin, between an ass and a female zebra, but the offspring did not survive.

The third instance took place in the Menagerie of Paris. From a female zebra, and a Spanish ass of the largest size, proceeded a very well-formed female mule. This animal proved a little larger than the mother, but as it grew up, had much of the form of the father. It was excessively indocile.

The experiment was then repeated with a horse. Conception took place, but in the eighth month of gestation the Zebra died. On opening the body, a male foetus was found, without hair, but having the head marked with black and white stripes.

The *Dauw.* (*Equus Montanus.*) " This beautiful animal," says Mr. Burchell, " has hitherto been confounded by naturalists with the Zebra. When these were first described by modern writers, the Couagga was considered to be the female zebra, while both that and the true zebra, bore in common among the colonists, the name of Quakka. The *Wilde Paarde*, named *Dauw* by the Hottentots, and a much scarcer animal than the other two, was never suspected to be a different species, though it be far more distinct from the Quakka and Zebra, than these are from each other.

" The hoofs of animals," says Mr. Burchell, " destined by nature to inhabit rocky mountains, are, as far as I have observed, of a form very different from those intended for sandy plains ; and this form is in itself sufficient to point out the *Dauw* as a separate species. The stripes of the skin will answer that purpose equally well, and show at the same time the great affinity and specific distinction of the *Ass*, which may be characterized by a single stripe across the shoulders. The Quakka has many similar marks on the head and fore part of the body : the Zebra is covered with stripes over the head and the whole of the body, but the legs are white ; and the Wilde Paarde is striped over every part even down to the feet. The Zebra and Wilde Paarde may be further distinguished from each other, by the stripes of the former being brown and white, and the brown stripe being double, that is, having a paler stripe within it ; while the latter, which may be named *Equus Montanus*, is most regularly and beautifully covered with single black and white stripes : added to this, the former is never to be found on the mountains, nor the latter on the plains."

END OF THE THIRD VOLUME.

Printed by W. CLOWES, Northumberland-court, Strand.

LIST OF PLATES IN THE THIRD VOLUME.

LIST OF PLATES IN THE THIRD VOLUME.